Karsten Levsen

Fundamental Aspects of Organic Mass Spectrometry

Progress in Mass Spectrometry
Fortschritte der Massenspektrometrie

Edited by/Herausgegeben von
Herbert Budzikiewicz

Band 1
M. Hesse: Indolalkaloide
Teil 1 (Text), Teil 2 (Spektren)

Volume 2
S. E. Drewes: Chroman and Related Compounds

Band 3
M. Hesse u. H. O. Bernhard: Alkaloide
(außer Indol-, Triterpen- und Steroidalkaloide)

Volume 4
K. Levsen: Fundamental Aspects of Organic Mass Spectrometry

Vol. 4

Karsten Levsen

Fundamental Aspects of Organic Mass Spectrometry

Verlag Chemie
Weinheim · New York · 1978

Priv. Doz. Dr. Karsten Levsen
Institut für Physikalische Chemie
der Universität
Wegelerstr. 12
D-5300 Bonn 1

This book contains 87 figures, 25 tables, and 63 schemes.

CIP-Kurztitelaufnahme der Deutschen Bibliothek

Levsen, Karsten
Fundamental aspects of organic mass spectrometry.
– 1. Aufl. – Weinheim, New York: Verlag Chemie,
1978.
(Progress in mass spectrometry; Vol. 4)
ISBN 3–527–25789–6 (Weinheim)
ISBN 0–89573–009–X (New York)

© Verlag Chemie, GmbH, D-6940 Weinheim, 1978

All rights reserved (including those of translation into foreign languages). No part of this book may be reproduced in any form – by photoprint, microfilm, or any other means – nor transmitted or translated into a machine language without written permission from the publishers.

Alle Rechte, insbesondere die der Übersetzung in fremde Sprachen, vorbehalten. Kein Teil dieses Buches darf ohne schriftliche Genehmigung des Verlages in irgendeiner Form – durch Photokopie, Mikrofilm oder irgendein anderes Verfahren – reproduziert oder in eine von Maschinen, insbesondere von Datenverarbeitungsmaschinen, verwendbare Sprache übertragen oder übersetzt werden.

Registered names, trademarks, etc. used in this book, even without specific indication thereof, are not to be considered unprotected by law.

Printer: Zechnersche Buchdruckerei, D-6720 Speyer
Bookbinder: J. Schäffer OHG, D-6718 Grünstadt
Printed in West Germany

To
my wife and my parents

Preface and Introduction

It is only little more than two decades ago that the usefulness of mass spectrometry for the structure elucidation of organic compounds was fully recognized. Already during these first mass spectrometric studies of organic compounds it became evident that besides its unquestioned analytical importance this technique opened up a new and fascinating field of research, the study of the chemistry and physics of organic ions in the gas phase. The steadily increasing number of mass spectrometers available in chemical laboratories throughout the world led to a rapid expansion of this new discipline.

This volume intends to give a detailed, but easily understood introduction into the theoretical aspects and methodical concepts used in organic mass spectrometry today. It is addressed to chemists who are starting research in this area and already have some basic knowledge of organic mass spectrometry and the instrumentation used in this field. Hence experimental methods are only described in some detail where they deviate from standard techniques.

If a subject of this broadness is to be treated in a single monograph of manageable size some restriction in the selection of topics is inevitable. Thus the book is confined to the field of positive ions. Ion-molecule reactions, although a particularly important area of organic mass spectrometry, are only discussed where they are of relevance for ion structure determination. Whereas established fields in organic mass spectrometry are treated only briefly newer theoretical and methodical approaches are reviewed in detail.

After a concise discussion of the ionization process (Chapter I) a major part of this book (Chapters II and III) is devoted to the Quasi-Equilibrium Theory which allows the fragmentation of gaseous organic ions to be rationalized and to be predicted quantitatively. After a presentation of the basic assumptions and mathematical approaches used in this theory, experiments, which allow its validity to be tested, are discribed in detail. Furthermore, it will be shown how the theory can be used to explain a variety of phenomena encountered in organic mass spectrometry such as the dependence of fragment abundances on both the internal energy and ion lifetime, isotope effects, substituent effects, and metastable ions.

In the last two chapters (Chapters IV and V) both the basic aspects and methods used for the elucidation of decomposition mechanisms and structures of gaseous ions are treated. Originally the emphasis in organic mass spectrometry was laid on the elucidation of fragmentation mechanisms whilst the structures of the ions involved in these fragmentations were only inferred indirectly if at all. With the advent of new, more reliable techniques the determination of the structure of gaseous ions, and, more recently, even their detailed potential energy surfaces is one of the central topics in organic mass spectrometry today and is hence treated in particular detail.

When discussing the fundamental aspects of organic mass spectrometry it is impossible to avoid some overlapping with previous books. Thus several subjects treated in Chapter III of this book have been discussed earlier in detail in excellent monographs (Ref. 2 and 104 in Chapter III). I have attempted to update these previous presentations.

My original intention of making the list of references as complete as possible had to be abandoned in view of the large number of relevant publications. Rather I have attempted to have at least several new areas of research refereed exhaustively. Relevant literature until February 1978 has been covered.

Various sections or even larger parts of this manuscript have been thoroughly read and critically reviewed by many colleagues (Prof. Beckey, Dr. Bowen, Prof. Cooks, Dr. Howe, Dr. Nibbering, Dr. Rosenstock, Dr. Schwarz, Dr. Van de Sande and Dr. Williams). Their comments have made an invaluable contribution to improving the book. I am extremely grateful for their cooperation. My particular thanks are due to Prof. Beckey for his continued support and interest over many years. Various authors provided preprints and reprints of their current work, thus helping to update the content. I am especially indebted to the editor, Prof. Budzikiewicz, both for stimulating comments and continuous encouragment during the preparation of the manuscript.

When asked by the publisher I agreed only reluctantly to write this book in English. Although an English version will certainly reach a larger number of readers there was the danger that this could only be achieved at the cost of precision in the formulation. Hence I am grateful to Dr. Taylor who invested considerable efforts and time to improve the English. Finally I would like to thank Miss Chambers and my coworkers for numerous hours of proof reading.

Bonn, March 1978   K. Levsen

Contents

Chapter I. The Ionization Process   1

1	The Ionization Process, Franck-Condon Factors	1
2	Adiabatic and Vertical Ionization Potential	3
3	Ionization Efficiency Curves	4
3.1	Threshold Laws	4
3.2	Hot Bands	6
3.3	Autoionization	7
4	Determination of Ionization Potentials	8
4.1	Optical Spectroscopy	9
4.2	Threshold Experiments	9
4.2.1	Photoionization Thresholds	9
4.2.2	Electron Impact Thresholds	9
4.3	Electron Spectroscopy	10
5	The Internal Energy Distribution of the Molecular Ion	12
5.1	The Energy Deposition Function	12
5.2	The Thermal Energy Distribution	17
6	Substituent Effects on Ionization Potentials	19
6.1	The Inductive Effect in Aliphatic Compounds	19
6.2	Mesomeric Effects in Aromatic Compounds	20
7	References	22

Chapter II. The Quasi-Equilibrium Theory   25

1	The Theory	25
1.1	Introduction	25
1.2	Basic Assumptions	26
1.3	The Rate Expression	27
1.4	Enumeration of States	29
1.5	Calculation of Rate Constants	32
1.5.1	Configuration of the Molecular Ion and the Transition State	33
1.5.2	The Vibrational and Rotational Frequencies in the Molecular Ion and the Transition State	35
1.5.3	The Symmetry Factor	36
1.5.4	The Activation Energy	37
1.6	Modified QET-Phase Space Theory	37
1.7	Calculation of Breakdown Graphs and Mass Spectra	39
2.	Experimental Tests of the QET	41
2.1	Experimental Verifications of the Basic Assumptions of the QET	41
2.1.1	Radiationless Transitions and Energy Randomization	42
2.1.2	Isolated Electronic States	45

2.1.3 Ion Lifetimes 47
2.1.3.1 The Principle of Ion Lifetime Measurements 48
2.1.3.2 Long Ion Lifetimes ($t > 10^{-7}$ s) 50
2.1.3.3 Short Ion Lifetimes ($t < 10^{-7}$ s) 53
2.2 Quantitative Tests of the QET 57
2.2.1 Mass Spectra 58
2.2.2 Breakdown Curves 64
2.2.2.1 Electron Impact 64
2.2.2.2 Photoionization 66
2.2.2.3 Charge Exchange 67
2.2.2.4 Photoelectron-Photoion-Coincidence (PEPICO) 69
2.2.3 Determination of the Rate Constant, K(E) 71
2.2.4 Kinetic Energy Release Distribution 74

3 Appendix. The Rate Equation 79

4 References 81

Chapter III. Application of the Quasi-Equilibrium Theory to Organic Mass Spectrometry 89

1 K Versus E Curves 89

2 Internal Energy and Ion Lifetime Dependence of Competing Rearrangement Reactions and Direkt Bond Cleavages 92
2.1 Energy Dependence 92
2.2 Ion Lifetime Dependence 96
2.3 Atom Scrambling 98

3 Distribution of Internal Energy between Fragments 102
3.1 Internal Energy of Product Ions 102
3.2 Degree of Freedom Effect 105

4 Determination of Appearance Potentials 108
4.1 The Kinetic Shift 108
4.2 Experimental Determination of the Kinetic Shift 110
4.3 The Competitive Shift 112
4.4 The Thermal Shift 113

5 Kinetic Isotope Effects 113

6 Substituent Effects on Fragmentation Reactions 118
6.1 Substituent Effects on Appearance Potentials 118
6.2 Substituent Effects on Fragment Abundances 119
6.3 Competing Fragmentation of Disubstituted Aromatic and Aliphatic Compounds 125
6.4 Substituent Effects on the Kinetic Energy Release 127

7 Metastable Ions 128
7.1 The Origin of Metastable Ions 128
7.2 Kinetic Energy Release 129

7.2.1	The Peak Shape of Metastable Ions	129
7.2.2	Determination of Kinetic Energy Release	130
7.2.3	Kinetic Energy Release Distribution	132
7.3	Energy Partitioning	134
7.3.1	Sources for Kinetic Energy Release	134
7.3.2	Energy Partitioning	135
8	Collision Processes	138
8.1	Collision Induced Dissociation	138
8.2	Charge Transfer Reactions	141
8.2.1	Charge Exchange of Doubly Charged Ions	142
8.2.2	Charge Stripping and Charge Inversion	143
9	References	144

Chapter IV. Reaction Mechanisms 152

1	The Mechanistic Approach	152
1.1	Product Stabilities and Bond Strengths	152
1.2	Stevenson's Rule	156
1.3	Proton Affinities	159
1.4	Charge Localization – Radical Site Localization	160
2	Methods for the Elucidation of Ionic Reaction Mechanisms	163
2.1	Isotopic Labelling	163
2.2	Steric Blocking	164
2.3	Metastable Ions	165
2.3.1	Fragmentation Pathways	165
2.3.2	Kinetic Energy Release and Energy Partitioning Data	166
2.3.3	Kinetic Energy Release as Potential Energy Surface Probe	170
2.3.4	Kinetic Energy Release and Molecular Orbital Symmetry Considerations	170
2.4	Isotope Effects	172
2.5	The Detection of Functional Group Interaction in Apparent Direct Bond Cleavages	177
2.5.1	Energy Dependence	177
2.5.2	Abundant Metastable Ions	179
2.5.3	Relative Fragment Abundances	180
2.5.4	Appearance Potentials	182
2.5.5	Symmetry Arguments	183
2.6	Field Ionization Kinetics	186
2.6.1	Mechanisms for Hydrogen Randomization	187
2.6.2	Suppression of Atom Randomization	190
2.6.3	Differentiation between Hydrogen Randomization and Non-specific Hydrogen Rearrangements	192
2.6.4	Other Systems	194
2.7	Molecular Orbital Calculations	196
2.7.1	Qualitative Prediction of Fragmentation Pathways	196
2.7.2	Hydrogen Scrambling and Skeletal Isomerization	198

2.7.3 Substituent Effects   199
2.8   Reaction Mechanisms and Ion Structures   199
2.9   Structure of Neutral Fragments   202

3   References   202

Chapter V. The Structure of Gaseous Ions   209

1   Introduction   209

2   Fundamental Considerations   210
2.1   Definitions   210
2.2   Isomerization of Gaseous Ions   212

3   Methods for Ion Structure Elucidation   215
3.1   Internal Energy and Ion Lifetime   215
3.2   The Mechanism leading to the Formation of an Ion   219
3.3   Thermochemical Properties (Heats of Formation)   219
3.3.1   The Principle   219
3.3.2   Possible Sources of Error   220
3.3.3   Examples   221
3.4   Degradation Reactions   225
3.4.1   Decomposition Mechanisms   225
3.4.2   Comparison of Mass Spectra   226
3.4.3   Metastable Ion Characteristics   228
3.4.3.1 The Principle   228
3.4.3.2 Critical Evaluation of the Method   229
3.4.3.3 Examples   230
3.4.3.4 Rate-Determining Isomerization prior to Decomposition   232
3.4.3.5 Time-Dependent Metastable Ion Characteristics   235
3.4.4   Kinetic Energy Release   237
3.4.4.1 The Principle   237
3.4.4.2 Critical Evaluation of the Method   238
3.4.4.3 Examples   239
3.4.4.4 Kinetic Energy Release and Molecular Orbital Symmetry Considerations   242
3.4.5   Collisional Activation   243
3.4.5.1 The Principle . 243
3.4.5.2 The Technique   244
3.4.5.3 Critical Evaluation of the Method   244
3.4.5.4 Examples   245
3.5   Ion-Molecule Reactions   249
3.5.1   Ion-Molecule Reactions Studied by Ion Cyclotron Resonance Spectrometry   249
3.5.1.1 The Principle   249
3.5.1.2 The Technique   249
3.5.1.3 Critical Evaluation of the Method   251
3.5.1.4 Examples   252
3.5.2   The Ion Cyclotron Resonance Photodissociation Technique   256
3.6   Molecular Orbital Calculations   258

3.6.1	Molecular Orbital Calculations and Ion Structure	258
3.6.2	The Method	260
3.6.3	Examples	261
3.7	Other Techniques	264
3.7.1	Isotope Effects	264
3.7.2	Charge Stripping	264
3.7.3	Experimental Determination of Rate Constants	265
4	Experimental Results	265
4.1	Hydrocarbon Ions	266
4.1.1	Aliphatic and Alicyclic Hydrocarbon Ions	266
4.1.2	The Influence of a Heteroatom on the Isomerization of a Hydrocarbon Chain	272
4.1.3	The Time Scale for Isomerization of Hydrocarbon Ions	273
4.1.4	Aromatic Hydrocarbon Ions	278
4.2	Heteroatom-Containing Ions	279
4.2.1	Stability of Heteroatom-Containing Ions	279
4.2.2	Potential Energy Surfaces	280
4.3	Examples for Ion Structure Elucidation	284
5	References	293

Subject Index 306

Chapter I. The Ionization Process

1 The Ionization Process, Franck-Condon Factors

Loss of an electron from an atom or molecule A, for instance by photo-ionization or electron impact leads to the generation of an ion $A^{+\cdot}$. If the electron is removed from the highest occupied orbital of the molecule the minimum energy necessary for this process is termed the first ionization potential, IP. Removal of an electron from lower lying molecular orbitals of the neutral molecule requires the supply of an energy, corresponding to the second IP, third IP, etc[*].

The heat of formation of the ion $A^{+\cdot}$ (at 0 °K) is given by[**]

$$\Delta H_f(A^{+\cdot}) = IP(A^{+\cdot}) + \Delta H_f(A) \qquad (I-1)$$

The heats of formation of the neutral species are generally known from thermochemical measurements and have been tabulated [1,2]. The removal of an electron occurs within about 10^{-16} s[***] and is hence about two orders of magnitude faster than a vibration within a molecule (10^{-13} to 10^{-14} s). Thus the ionization takes place at constant internuclear distance (Franck-Condon Principle).

The ionization process can be most conveniently explained for a diatomic molecule. Fig. I-1 shows schematically the potential energy curves for the ground electronic states of the ion and the molecule. In Fig. I-1a the potential energy curve of the ion has a minimum at an internuclear distance which is only slightly larger than that of the molecule (which holds approximately if a non-bonding electron is removed by the ionization process), whilst in Fig. I-1b the minimum of the first ionized state is shifted considerably to larger internuclear distances (as expected if a bonding electron is removed). In principle, electronic transitions are possible from the ground vibrational level (v = 0) of the molecule to the various vibrational levels (v' = 0, 1, 2, 3....) of, for instance, the first ionized electronic state, and the

---

[*] The second IP should not be confused with the ionization potential of doubly charged ions.

[**] This definition implies that the heat capacities of the ion and the neutral are identical, which is often very nearly so.

[***] At 50 eV an electron has a velocity of 4.2×10^8 cm s^{-1} and will traverse a molecule with a diameter of a few Å in about 10^{-16} s [3].

relative transition probabilities may be represented to a good approximation by the squares of the normalized vibrational overlap integrals, also called *Franck-Condon factors*. These relative transition probabilities can be readily calculated for smaller molecules [4]. It follows from the Franck-Condon Principle that the relative transition probabilities (Franck-Condon factors) depend strongly on changes in the internuclear distance after ionization, as also illustrated in Fig. I-1 schematical-

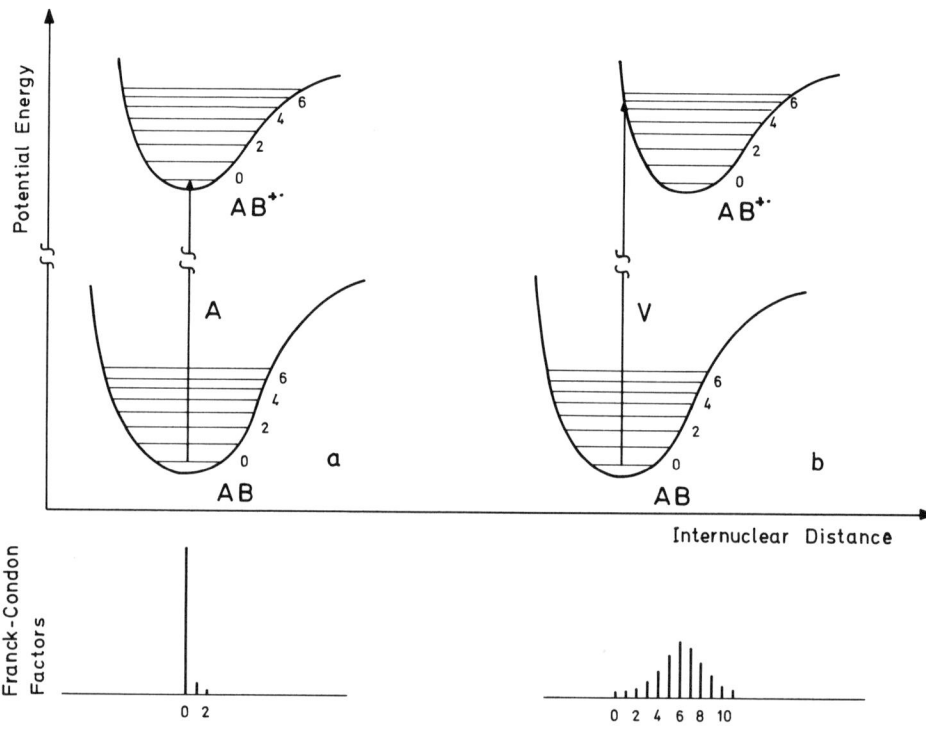

Fig. I-1. Schematic potential energy curves for a molecule and its ion in their electronic ground states assuming (a) small and (b) large bond length changes after ionization (A = adiabatic ionization potential, V = vertical ionization potential). The distribution of Franck-Condon factors for various transitions is shown schematically in the lower part of the figure.

ly where the Franck-Condon factors for the transition to the various vibrational levels of the ionized state are represented as bars. Such a graph of the distribution of Franck-Condon factors is known as *Franck-Condon envelope*. In Fig. I-1a the largest Franck-Condon factor is observed for the 0-0 transition, in Fig. I-1b however for the 0-6 transition. In reality the Franck-Condon progressions are much more sensitive to internuclear distance changes than shown in the schematic Fig. 1.

2 Adiabatic and Vertical Ionization Potential

The energy difference between the ground vibrational levels ($v' = v = 0$) of the lowest electronic states of a molecular ion and the corresponding molecule is defined as the *adiabatic ionization potential* (A in Fig. I-1). In contrast, the *vertical ionization potential* of a diatomic molecule corresponds to the transition from the ground state of the molecule to the state of the ion for which the Franck-Condon factor is largest [5] (V in Fig. I-1b). In some instances the vertical and the adiabatic ionization potential may be identical (see Fig. I-1a). However, the simple potential energy diagram represented in Fig. I-1 for a diatomic molecule no longer applies to polyatomic molecules. Thus it has been pointed out by Vestal [6] that for such polyatomic molecules the vertical ionization potential in this form may be less well defined. According to molecular quantum mechanics the vertical ionization potential of a polyatomic molecule is the energy required to remove an electron while holding the nuclei fixed in their position. This vertical ionization potential is closely approximated by the maximum of the Franck-Condon envelope.

For thermochemical calculations, e.g. for the determination of the heat of formation of an ion according to equation I-1, knowledge of the adiabatic ionization potential is desirable. However, it is obvious from Fig. I-1b that the determination of the adiabatic ionization potential may be difficult with photoionization and especially with electron impact techniques if a bonding electron (e.g. in alkanes) is removed during the ionization process, as the Franck-Condon factor for the adiabatic transition is then apparently considerably smaller than when the electron is non-bonding. It has, however, been pointed out by Vestal [6] that the adiabatic transition must not necessarily be zero, but may be as large as other transitions. However, because of the very rapid increase in the density of states with energy the probability for an adiabatic transition may be several orders of magnitude smaller than the maximum transition probability, which makes it difficult or impossible to determine the adiabatic ionization potential in some cases when using photoionization or electron impact methods. In these instances (such as in methane) the reason for not determining the adiabatic ionization potential is therefore not that the adiabatic transition is inaccessible but rather that it is very improbable.

If however a non-bonding electron is lost during the ionization (e.g. from a heteroatom) the Franck-Condon factor for the adiabatic transition is apparently very large compared to other transitions to neighboring vibrational levels.

3 Ionization Efficiency Curves

3.1 Threshold Laws

The ionization efficiency curves represent the relation between the intensity of an ion signal (e.g. the molecular ion) and the energy of the impacting particle. There are considerable differences between the curves for photoionization and those for electron impact in the threshold region. A simple relationship between the ionization cross section, σ, and the energy of the impacting particle for a transition to a given electronic, vibrational, and rotational state has been given by Wigner [7] and Geltman [8].

$$\sigma(E) \sim A(E - E_o)^{n-1} \qquad (I-2)$$

(E_o = threshold energy, n = number of electrons ejected from the collision complex, A is a constant).

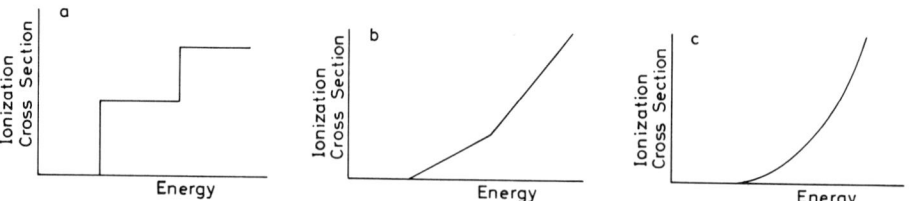

Fig. I-2. Theoretical ionization cross sections of an atom (no vibrational states) for photoionization (a) as well as single (b) and double ionization (c) by electron impact assuming the validity of Geltman's law [8].

Fig. I-2 represents for an atom (no vibrational and rotational excitation) the theoretical threshold behavior for photoionization (n = 1) as well as single (n = 2) and double ionization (n = 3) by electron impact. A simple step function is expected for photoionization. If at higher internal energies transitions to the first excited state are possible a second step function is superimposed onto the first one (Fig. I-2a). A linear threshold law is found for single ionization by electron impact (Fig. I-2b). The break in the curve indicates transitions to the first excited electronic state. Finally a quadratic law is expected for double ionization by electron impact (Fig. I-2c).

Although the interpretation of much experimental data is based on these threshold laws one should keep in mind that these laws describe the ionization behavior quantitatively only near the threshold. Unfortunately little is known how far above the threshold eq. I-2 is valid. Moreover, the threshold laws are not universally valid. Thus good agreement is found for argon, krypton, and xenon [9], poorer agreement for helium and neon [10].

In contrast with the threshold behavior of atoms the *photoionization threshold of molecules* can no longer be described by a simple step function (even if only the transition to a single electronic state is considered) because when starting from the vibrational ground state of the molecule various transitions are possible not only to excited vibrational states (Fig. I-1) but also to numerous rotational states of the ion. Thus the total transition probability in the threshold region is the sum of many individual step functions of closely similar energy, leading to a staircase like photoionization efficiency curve. This is

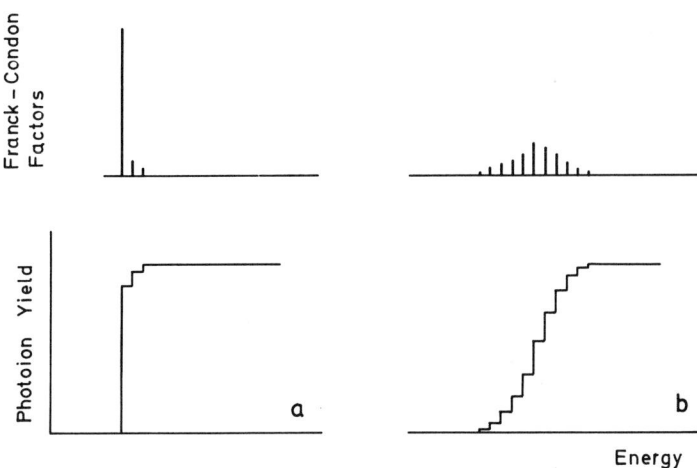

Fig. I-3. Franck-Condon factors and hypothetical photoionization yield curves expected for (a) small and (b) large bond length changes in a vertical ionization.

illustrated schematically in Fig. I-3 for the case of a small (a) and a large (b) change in the internuclear distance after ionization considering only vibrational excitation. The height of each individual step is proportional to the Franck-Condon factor of the corresponding transition shown in the upper part of this figure. Finally, the effect of rotation on the shape of the photoionization efficiency curve is to produce some additional tailing of the onset and rounding of the steps [11]. As far as experimental evidence is available today the photoionization threshold law for molecules does seem to be a step function for each transition, in contrast to the atomic case [11]*⁾.

In the case of *electron impact ionization* where a linear threshold law was assumed for a single ionization it is expected that for molecules a series of straight line segments will be observed in the ionization efficiency curve with the onset of each new segment representing transitions to a new vibrational level. The slope increases successively in accord with the increasing Franck-Condon factors (Fig. I-4). Although

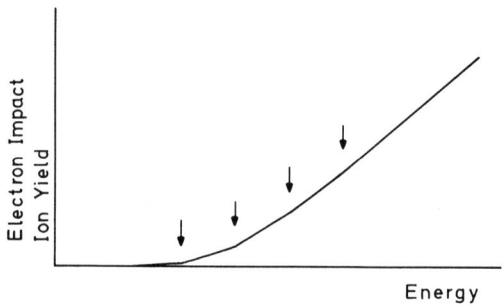

Fig. I-4. Hypothetical electron impact ionization yield curves for a molecule.

the observation of straight line segments has been claimed in some instances, their existence has not been confirmed unambiguously [11].

3.2 Hot Bands

The step function character of the photoionization efficiency curves makes it possible in most cases to detect the 0-0 transition (transition from the vibrational ground state of the molecule to the vibrational ground state of the ion in its lowest electronic state) and thus to determine the adiabatic ionization potential. In the above discussion it has however always been assumed that the molecule is in its vibrational ground state prior to ionization although this is not true for larger molecules even at room temperature. If, for instance, part of the molecules is already in the first vibrational state prior to ionization, then 1-0 transitions may also occur, i.e. photoions may be produced below their adiabatic threshold leading to a small step or tailing in the ionization efficiency curve below the step for the 0-0 transition

---

*
  Some exceptions including a number of ketones, are known however [85].

and this is termed a *vibrational hot band*. Changing the temperature can be used to study and correct for hot bands.

3.3 Autoionization

The shape of the ionization cross section curve may be further complicated by the occurrence of *autoionization*. If a molecule is excited by the impinging photon or electron to a superexcited neutral state*[)] lying above the lowest ionic state (and possibly above several excited ionic

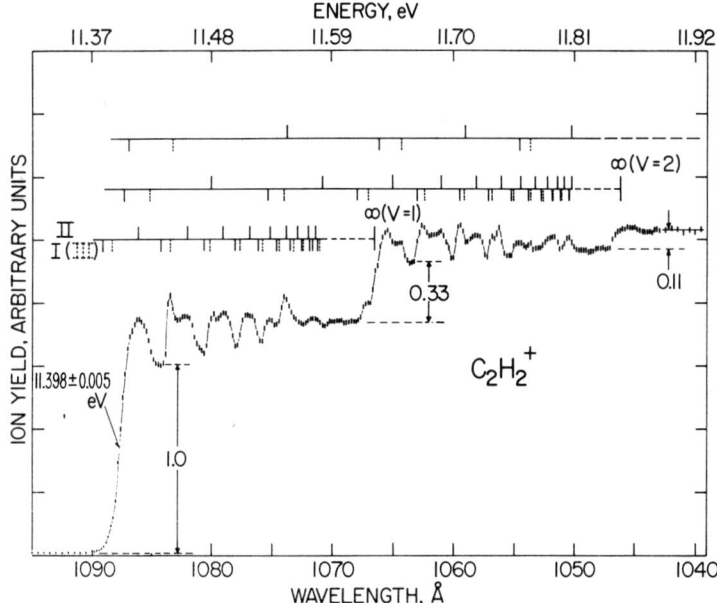

Fig. I-5. Photoionization yield curve for acetylene in the threshold region showing superposed autoionization structure [86]. (By courtesy of Elsevier Publishing Company)

states) this excited state of the molecule may suffer a radiationless decay either into an ion and an electron of the same total energy by autoionization**[)] or by ion pair formation or by predissociation into neu-

---

*
 The excited state of the molecule may be either due to electronic or vibrational excitation [12].

**
 Autoionization normally occurs by Rydberg levels of the molecule [12] converging to excited states of the ion, which may make it difficult to detect the steps of an excited state due to direct ionization.

tral fragments. Autoionization is a resonance process leading in the case of photoionization to peaks which are superimposed onto the normal staircase structure of the ionization cross section curve. This is illustrated in Fig. I-5 for acetylene [86]. The cross section curve not only shows three steps corresponding to transitions to the ground vibrational level and to two excited vibrational levels of the ion (in its electronic ground state), but also the superimposed autoionization peaks. Very little is known about the cross section behavior of autoionizing levels which may differ from system to system.

It is apparent that the occurrence of autoionization may complicate the interpretation both of photoionization and electron impact ionization cross section measurements. Therefore it is noteworthy that photoelectron spectroscopy experiments suffer far less from interference from autoionization phenomena than photoionization work where the wavelength or energy of the photon is varied, thus sweeping over many accessible excited neutral states. A constant photo wavelength is used in photoelectron spectroscopy and autoionization will only occur if an excited neutral state happens to lie at the ionizing photon energy. Similarly no autoionization is observed in threshold photoelectron spectroscopy (a technique described in Section I-4.3). Thus it has been suggested that the extent of autoionization can be determined by comparing threshold photoelectron spectra with photoionization yield measurements [13].

If predissociation of an excited neutral state (into neutral fragments) is faster than autoionization the latter will not be observed. Predissociation may also have an important effect on the ionization cross section*).

4 Determination of Ionization Potentials

The problem of accurate ionization potential measurements has been thoroughly discussed in several recent articles [11,14,15]. Three principal groups of methods are in current use for determining ionization potentials of organic molecules [11].
(1) Optical spectroscopy.
(2) Threshold experiments.
(3) Electron spectroscopy.

---

*
  If the Rydberg states are depopulated by predissociation the steps in the photoionization cross section curve otherwise smeared off by autoionization can be better detected.

4.1 Optical Spectroscopy

The most accurate data are obtained using spectroscopic techniques which have been applied to ionization potential measurements of smaller molecules. The adiabatic ionization potential is determined by finding the convergence limit of a Rydberg series. Although an energy resolution of 10^{-4} nm can be achieved most ionization potentials are reported with a resolution of 10^{-2} nm [11].

4.2 Threshold Experiments

The threshold laws for photoionization and electron impact measurements have been discussed in Section 3.1.

4.2.1 Photoionization Thresholds

Even for large organic molecules the photoionization threshold curve has a reasonably sharp onset, which most probably corresponds to the adiabatic threshold, for substances where a non-bonding electron is removed. If, however, a bonding electron is lost during the ionization (e.g. in alkanes) the photoionization threshold curve has a very gradual onset and it is not apparent whether the onset point includes the 0-0 transition corresponding to the adiabatic IP or not [11]. Another uncertainty in the determination of the adiabatic ionization potential arises from the possible presence of hot bands which may be responsible for the gradual onset of the photoionization curves of the higher alkanes. For aliphatic hydrocarbons, including methane, the adiabatic ionization potential is not yet firmly established. However, in many other cases adiabatic ionization potentials can be determined which agree with spectroscopic data to within 5-10 meV.

4.2.2 Electron Impact Thresholds

There is a double disadvantage in the use of electron impact techniques: (1) the linear threshold law makes it more difficult to determine the true onset of the ionization cross section curve; (2) conventional electron beams have a large energy spread due to the Boltzmann distribution of the electrons coming from the hot filament, the filament potential drop, the field penetration and surface adsorption. The second shortcoming (2) can be overcome by using an electron monochromator [17, 18] (double hemispherical sector) and a calibration gas. Despite the unfavorable threshold behavior and the tacit assumption that the threshold behavior of the calibration gas is similar to that of the molecule under study, ionization potentials which agree with experimental photoionization thresholds to within 30 meV have been reported using this technique. Quasimonoenergetic ions can be obtained with the RPD (retarding potential difference) [19,20] and the EDD (energy distribution difference) [21] technique in which a small portion of the broad electron energy distribution is selected by measuring the difference in ion currents at nominal electron energies E and $E + \Delta E$.

As the equipment necessary to produce monoenergetic or quasimonoenergetic electrons is not available in most laboratories, polyenergetic electron impact techniques are commonly used to determine ionization potentials. In these cases both a calibration gas and a variety of extrapolation methods are used to correct for the broad electron energy distribution (semilog plot [22], extrapolated voltage difference [23], energy compensation [24], critical slope [25], second derivative [26]). These methods have been described in detail by Kiser [15]. The reliability of the various extrapolation methods has been tested by Occolowitz et al. [27] both theoretically and experimentally. Using model ionization efficiency curves the authors showed that the critical slope method [25] should give the most reliable values. This conclusion is supported by their experimental results which show that the ionization potentials obtained by the critical slope method are close to published photoionization data. In general the accuracy of the various extrapolation methods ranges from 0.1 to 0.5 eV.

4.3 Electron Spectroscopy

A group of very accurate new methods for ionization potential measurements has been added to the above described techniques with the advance of electron spectroscopy, the most important being the photoelectron spectroscopy [28-31] which is now used in two variations. In the original and still mostly applied version ionization is achieved by photons of constant energy. (In most cases the He I resonance line corresponding to an energy of 21.22 eV is used. More recently also the He II line at 40.81 eV has been used as photon source). The energy distribution of the ejected electrons reflects the distribution of accessible energy levels of the neutral target molecule according to equation

$$E_{ion} = h\nu - E_{electron} \qquad (I-3)$$

A wide variety of analyzers has been employed for the energy analysis of the ejected electrons the most efficient being the 127° electrostatic sector analyzer. Because of contact potential contribution calibration of the energy scale with gases of known ionization potential is necessary. A schematic drawing of a photoelectron spectrometer is shown in Fig. I-6.

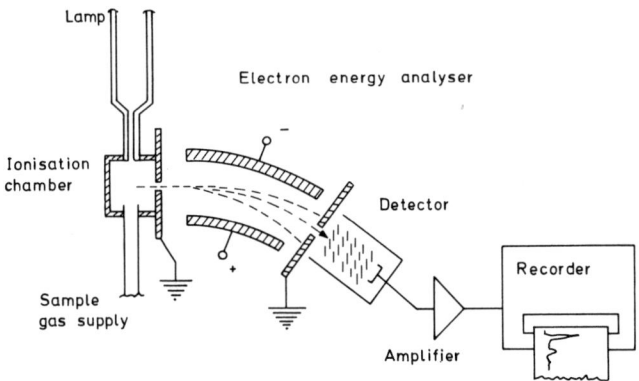

Fig. I-6. Schematic layout of a photoelectron spectrometer [31]. (By courtesy of Butterworths.)

In the resonance photoionization technique (also called threshold or zero kinetic energy photoelectron spectroscopy) [32-36] the photon energy is varied and only those photoelectrons with zero kinetic energy are detected. Thus the photon energy corresponds directly to the energy of the transition to an accessible ionic state (without interference from autoionization). The zero energy electrons can be detected by two different methods. In the first, all electrons are accelerated through a known potential drop and those whose energy corresponds to the potential drop itself are selected by a 127° electrostatic analyzer [32-34]. In the second method [35,36] (Fig. I-7) the electrons are accelerated

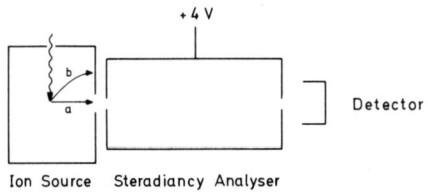

Fig. I-7. Schematic arrangement for the detection of zero kinetic energy electrons in photoelectron spectroscopy experiments [35].

by a small uniform electric field toward a set of collimating holes. Most hot electrons are emitted with a significant velocity component perpendicular to the applied field and will follow trajectories such as b in Fig. I-7 and consequently not be detected. On the other hand, the zero kinetic energy electrons have no perpendicular initial velocity component (trajectory a) and will thus reach the detector.

In a photoelectron spectrum the ejected electrons arise from allowed electronic transitions to all vibrational levels of the ion which are Franck-Condon accessible. Assuming that the photoionization threshold law is a step function which is flat from the ionization threshold energy up to the photon source line energy (which is not exactly the case as discussed below) the energy of the electrons ejected from a given electronic state will have a distribution determined by the Franck-Condon factors. Thus with sufficient energy resolution one obtains in principle a "bar graph" of experimental Franck-Condon factors [11] which simplifies the detection of the 0-0 transition and thus the adiabatic ionization potential considerably. However, hot bands may also be present in photoelectron spectra and can be detected by varying the temperature. The accuracy of ionization potential measurements by photoelectron spectroscopy is approaching 2 meV in favorable cases, but is typically of the order of one or several hundreths of an eV.

5 The Internal Energy Distribution of the Molecular Ion

5.1 The Energy Deposition Function

In addition to the thermal energy present in the molecule before ionization, excitation energy will be transferred to the molecular ion during the ionization process if the energy of the impinging particle (electron or photon) is higher than the ionization potential. The probability of depositing a given energy E onto the molecular ion as function of the energy V of the colliding particle is expressed by the energy deposition function, $P(E,V)$.

The energy deposition function depends both on the probability distribution $Y(E)$ for electronic transitions from the ground state of the molecule to the different vibrational levels of the various electronic states of the ion (described by the Franck-Condon factors) and the threshold behavior, i.e. the variation in the probability of transferring an energy $E + I$ (I = adiabatic ionization potential) to the molecular ion as function of the energy V of the impinging particle, expressed as function of the energy difference, $f(V - E - I)$ [6], i.e.

$$P(E,V) = f(V - E - I) Y(E) \tag{I-4}$$

A knowledge of the energy deposition function is of vital importance for understanding the fragmentation of an ion as described by the Quasi-Equilibrium Theory (QET) discussed in Chapter II of this volume.

Unfortunately a reliable experimental determination of the energy deposition function is difficult. It has been shown originally by Morrison [37] and later by Chupka et al. [38-41] that the energy distribution within the molecular ions is given by the *first derivative* of the to-

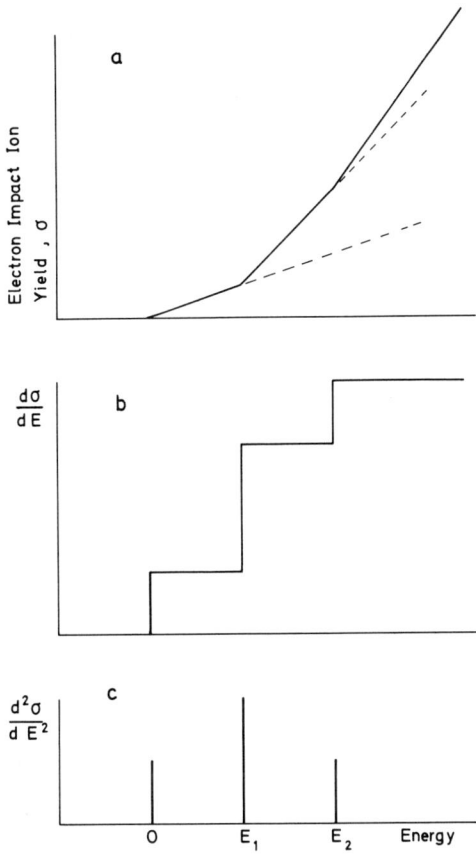

Fig. I-8. Hypothetical ion yield curve obtained with monoenergetic electrons (a) assuming the validity of Geltman's law [8]. Transitions to the ground or one of two excited electronic states are assumed to be possible with relative probabilities of 1:2:1. The second derivative of this curve (c) represents the internal energy distribution of the ions.

tal ionization efficiency for photoionization and the *second derivative* of the total ionization efficiency for electron impact ionization, provided that the Geltman threshold laws are valid. The justification of this procedure will be made plausible using Fig. I-8. Neglecting vibrational and rotational excitation we assume in this figure that upon ionization transition to the ground electronic state or one of two excited electronic states is possible with relative probabilities of 1:2:1 and that the transition probability is the same for electron impact and photoionization. Thus three groups of ions with internal energy zero, E_1 and E_2 are formed leading to three straight line segments in the total ionization efficiency curve for electron impact (Fig. I-8a) where the successive slope increase corresponds to the relative population of the three states. Similarly the total ionization efficiency curve for photoionization (Fig. I-8b) shows three steps the heights of which again represent the relative transition probabilities to the three electronic states. It is now apparent that the first derivative of the electron impact cross section curve is the curve of Fig. 8b (i.e. the photoionization cross section curve), the second derivative of Fig. 8a (or first derivative of Fig. 8b) finally is the curve of Fig. 8c which shows three peaks at the energies I, $I + E_1$ and $I + E_2$ thus representing the energy distribution of the molecular ions. Fig. 8c is identical with the photoelectron spectrum of the molecule under study if the Geltman threshold law is valid over the whole energy range.

In reality the derivative methods give only a rough approximation of the energy distribution because the threshold laws are obeyed only approximately, considerable deviations from the theoretical behavior being expected, especially at higher energies, and substantial autoionization may also occur especially under electron impact conditions[*]. The energy distribution derived by this procedure is of course a poorer approximation for electron impact than for photoionization especially if polyenergetic electrons are used.

Even if the energy distributions obtained by the second derivative method constitute only a rough approximation of the real situation, they at least give some indication of the range of energies transferred to the molecular ion upon electron impact ionization. Fig. I-9a shows the second derivative of the total ionization cross section for EI of propane as function of the energy of the ions [38]. The figure suggests that most propane molecular ions have internal energies in the range

---

[*] Thus any structure in the second derivative curve of the total ionization efficiency for electron impact may be due either to an excited state or to autoionization.

I-5 Internal Energy Distribution

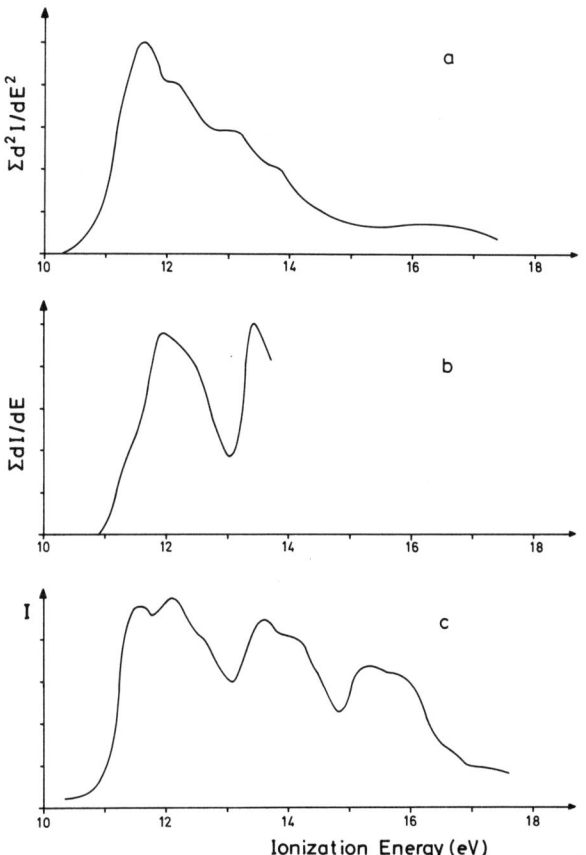

Fig. I-9. Internal energy distribution curve for propane obtained by electron impact [38], photoionization [40] and photoelectron spectroscopy [44].

of 0-6 eV. However, the energy distribution curve shows little structure. The first derivative of the photoionization cross section of propane from 11 to 14 eV (corresponding to 0-3 eV internal energy) is shown in Fig. I-9b [40]. In contrast to the electron impact results there is a pronounced minimum at 13 eV ($\sim$ 2 eV internal energy).

The most direct information on the energy deposition function at least for photoionization comes from photoelectron spectroscopy. Provided that the threshold law for photoionization is a step function up to the photon source energy (21.22 eV) the photoelectron spectrum represents the energy distribution directly. (However in order to estimate the energy deposition function from the photoelectron spectrum one has to correct for the energy-dependent transmission function of the elec-

16   I *The Ionization Process*

tron energy analyzer). Direct equation of the photoelectron spectrum with the energy deposition function is, however, not possible in general as the relative intensities of transitions to various electronic states (i.e. the height of the peaks in the photoelectron spectrum) depend markedly on the photon source energy [43]. Thus photoelectron spectra give a reliable estimate of the photoionization deposition function only for the particular photon energy used (e.g. 21.22 eV). As mentioned above it is advantageous that the photoelectron spectroscopy technique suffers far less from autoionization than photoionization and electron impact yield measurements.

Fig. I-9c represents the photoelectron spectrum of propane [44] showing (in contrast to Fig. I-9a) two distinct minima. Within the comparable energy range there is a good correspondence between the energy distribution derived from the photoionization efficiency curve (Fig. I-9b) and the photoelectron spectrum (Fig. I-9c).

For organic mass spectrometrists the electron impact energy deposition function is of greater interest than that for photoionization. As mentioned above the second derivative method does not give reliable information on this function. Therefore the suggestion that the photoelectron spectrum may be used as crude approximation of the energy distribution after electron impact is of interest [45]. Thus approximate energy distributions of substituted 1,2-diphenylethane molecular ions produced by electron impact have been derived from modified photoelectron spectra using the EI threshold law, thermal energy convolution and

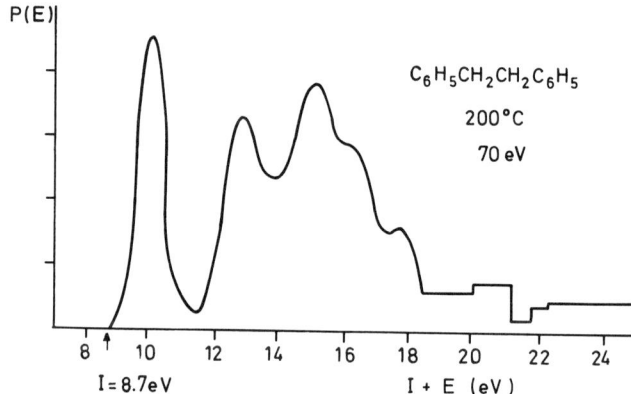

Fig. I-10. Internal energy distribution of 1,2-diphenylethane molecular ions formed by using electron energies of 70 eV. The curve has been obtained from the corresponding photoelectron spectrum which has been convoluted with the thermal energy distribution and adjusted to reflect the ion abundances in the 70 eV mass spectrum [45].

empirical observation of ion abundances [45]. Fig. I-10 represents the
energy distribution function derived by this method for 1,2-diphenyl-
ethane showing a pronounced minimum at 11 eV. The existence of such a
minimum in the internal energy distribution of 1,2-diphenylethane is
corroborated by a strong temperature dependence of the metastable ion
for benzyl ion formation, the appearance potential of which happens almost to coincide with the location of the minimum in P(E) [45]. However
at present it is not completely clear whether and to what extent the
energy distribution in a molecular ion after EI can be equated with the
suitably modified photoelectron spectrum of that compound. Arguments
against [42,46,47] and in favor [6,45,48] of this assumption have been
raised. Thus although the same types of ionization processes occur in
electron and photon impact optically forbidden transitions may occur
under electron impact conditions leading to the population of additional ionic states [47]. Moreover, it has already been pointed out that
the relative transition probabilities (i.e. the peak height) of the
photoelectron spectrum are energy dependent and will be different for
20 and 70 eV.

Meisels et al. [42] suggested that more reliable energy deposition
functions may be obtained by folding the energy loss function into
photoelectron spectra, where the energy loss functions are calculated
from the Bethe-Born theory of collisions using experimentally determined photoionization cross sections. The authors report energy deposition
functions for methane, ethane and ethylene.

Summarizing, reliable electron impact deposition functions are not
yet available. However, the present results and comparison with photoelectron spectroscopy clearly show that the majority of ions formed after 70 eV electron impact have a range of internal energies of roughly
0-10 eV while only a small fraction of molecular ions will be formed
with internal energies in excess of 10 eV which explains why the pattern of EI spectra changes little at energies above 20 eV. There is also some strong indication that this energy distribution shows pronounced minima with many compounds in correspondence with those observed
in photoelectron spectroscopy although a quantitative correspondence
between the photoelectron spectra (modified for a linear threshold law
and thermal energy) and the unknown electron impact energy distribution
cannot be expected in general.

5.2 The Thermal Energy Distribution

The experimentally determined energy distribution (e.g. using the first
or second derivative method) contains both the energy deposited onto

the molecular ion during the ionization, E_{dep}, (described by the energy deposition function) and the thermal energy, E_{th}, present in the molecule before ionization. The probability that a molecule at a temperature T has an energy between E and E + dE is given by [6]

$$P_{th}(E)\ dE = \frac{\rho(E)\ \exp(-E/kT)}{\int_0^\infty \rho(E)\ \exp(-E/kT)\ dE} \qquad (I-5)$$

where k is the Boltzmann constant and $\rho(E)$ dE is the number of states of the molecule with energy between E and E + dE. Fig. I-11 represents the thermal energy distribution of propane calculated by Erhardt and

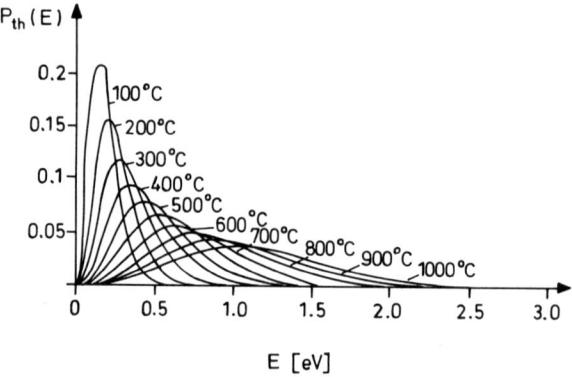

Fig. I-11. Thermal energy distribution of the propane molecular ion [49]. (By courtesy of Zeitschrift für Naturforschung.)

Osberghaus [49]. It is noteworthy that at 100 °C the average thermal energy of such a rather small molecule is already 0.2 eV and rises to more than 1 eV at 1000 °C. (The typical temperature of an electron impact source ranges from 150 - 250 °C). With increasing molecular weight the maximum of the thermal energy distribution is shifted to even higher energies [50].

If during the ionization process an energy E_{dep} is transferred to the molecular ion then at a total internal energy E the fraction $E - E_{dep}$ represents the thermal energy. The total internal energy distribution, P(E), can be obtained by convoluting the thermal energy distribution function $P_{th}(E - E_{dep})$ and the energy deposition function $P_{dep}(E_{dep})$.

$$P(E) = \int_0^E P_{th}(E - E_{dep}) P_{dep}(E_{dep}) \, dE \qquad (I-6)$$

6 Substituent Effects on Ionization Potentials

6.1 The Inductive Effect in Aliphatic Compounds

The ionization potentials within an homologous series of a monofunctional compound, RX, are a linear function of the polar substituent constants, σ^*, or the inductive substituent constants, σ_I [51], of the alkyl group, R, as shown in Fig. I-12 for thiols (X = SH) [52]. *Electron releasing alkyl groups* bonded to the S atom of the thiol molecule obvious-

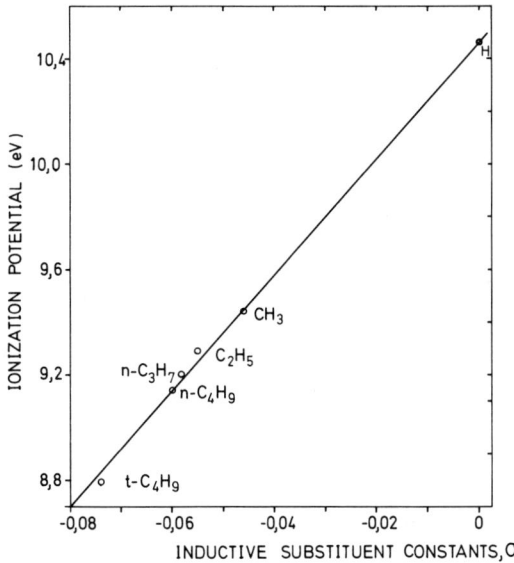

Fig. I-12. The ionization potential, IP, of thiols as a function of the inductive substituent constants, σ_I, of the corresponding alkyl groups [52].

ly *facilitate the electron removal and thereby lower the ionization potential*. It is interesting that the linear correlation even holds for hydrogen sulfide as the simplest thiol in the series. The equation for the correlation line in Fig. I-12 is given as

20   I   *The Ionization Process*

$$IP_{RX} = E_o + a_I \sigma_I \qquad (I-7)$$

With the intercept $E_o(H_2S) = 10.46$ and the slope $a_i = 22.2$ one obtains

$$IP_{RSH} = 10.46 + 22.2 \, \sigma_I$$

Similar equations have been obtained in a series of studies by Levitt et al. for X = OH [53-55], F, Cl, Br, I [56], COOH [57], NH_2 [58], CN [59] and NO_2 [59] and allow the ionization potentials of molecules in an homologous series to be calculated with high accuracy if experimental data are not available.

For aliphatic compounds of the general formula RXR' the ionization potentials are linear functions of the sum of the inductive (or polar) substituent constants of both alkyl groups R and R':

$$IP_{RXR'} = E_o + a_I[\sigma_I(R) + \sigma_I(R')] \qquad (I-8)$$

Such correlation lines have been determined for X = O [60], S [61], SS [62], CO [63], CO_2 [64], CH=CH [65], C≡C [66]. An excellent correlation has even been found for alkanes [67] using the (arbitrary) assumption that the electron is lost from the most central C-C-bond in the molecule. Moreover, a linear correlation was also observed for the ionization potentials of radicals, R˙ [68].

6.2 Mesomeric Effects in Aromatic Compounds

In monosubstituted aromatic compounds a linear correlation between the ionization potential and Brown's σ^+ constant is observed as shown in Fig. I-13 for monosubstituted benzenes, C_6H_5X. *Substituents with an electron-withdrawing effect* (NO_2, CN, CHO, COOH, F, Cl, Br, I) *increase the ionization potential* relative to benzene, whereas *substituents with electron-donating character* (OH, SH, OCH_3, NH_2, NR_2) *cause a reduction of the ionization potential*, as first shown by Price [69] and Crabble and Kearns [70]. Such linear correlations have since been observed for a wide variety of substituted aromatic compounds [71-83].

Photoelectron spectra of substituted benzenes demonstrate that electron-donating substituents split the first band in the benzene spectra into two parts*). One of the new bands is found to remain near its

---

*
  This band corresponds to ionization from the degenerate orbitals π_2 and π_3. When a substituent is introduced the degeneracy is removed. Only one of the two new orbitals can interact with the substituent, leading to a lowering of the ionization potential.

I-6 *Substituent Effects on Ionization Potentials*   21

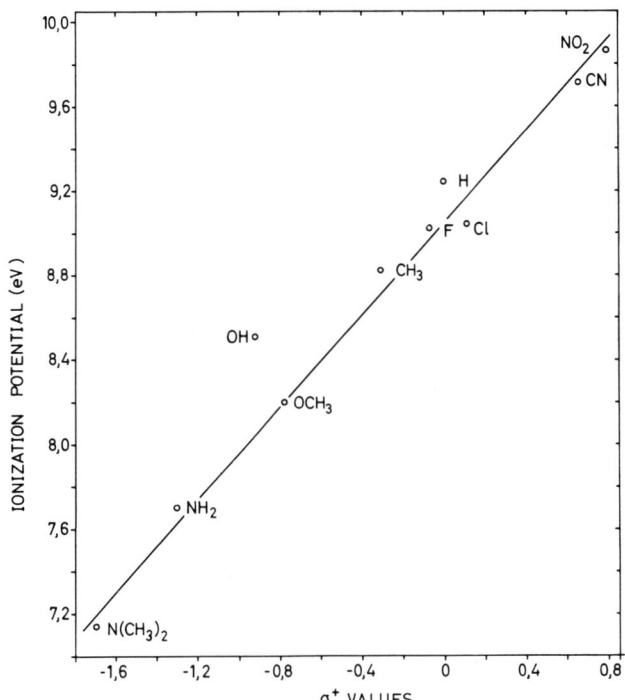

Fig. I-13. Substituent effect on the ionization potential of substituted benzenes.

original position, whereas the other is substantially shifted to lower ionization potentials. On the other hand, electron-withdrawing substituents do not produce detectable splittings, but shift the first band to higher ionization potentials.

In disubstituted aromatic compounds the effects of the substituents are additive, but not in a linear manner. Both empirical and theoretical equations have been derived for calculating the ionization potential of such disubstituted compounds from the individual σ^+ constants of both substituents [87-90].

7 References

1. D.R. Stull, E.F. Westrum and G.C. Sinke, "The Chemical Thermodynamics of Organic Compounds", Wiley, New York, 1969.
2. H.M. Rosenstock, K. Draxl, B.W. Steiner and J.T. Herron, J. Phys. Chem. Ref. Data, Vol. 6, 1977.
3. J.H. Beynon, R.A. Saunders and A.E. Williams, "The Mass Spectra of Organic Molecules", Elsevier, New York, 1968.
4. H.M. Rosenstock and R. Botter, "Recent Developments in Mass Spectrometry", Proceedings of Int. Conf. on Mass Spectrometry, Kyoto, Japan, University Press, Baltimore, 1970.
5. F.H. Field and J.L. Franklin, "Electron Impact Phenomena", Academic Press, New York, 1957.
6. M.L. Vestal in "Fundamental Processes in Radiation Chemistry" (P. Auloos, Ed.), Interscience Publisher, New York, 1968.
7. E.P. Wigner, Phys. Rev., $\underline{73}$, 1002 (1948).
8. S. Geltman, Phys. Rev., $\underline{102}$, 171 (1956).
9. V.H. Dibeler, R.M. Reese and M. Krauss, Advan. Mass Spectrom., $\underline{3}$, 471 (1966).
10. G.V. Marr, "Photoionization Process in Gases", Academic Press, New York, 1967, Chap. 7.
11. H.M. Rosenstock, Int. J. Mass Spectrom. Ion Phys., $\underline{20}$, 139 (1976).
12. W.A. Chupka, in "Ion-Molecule Reactions" (J.L. Franklin, Ed.), Butterworths, London, 1972, p. 45.
13. R. Stockbauer and M.G. Inghram, J. Chem. Phys., $\underline{54}$, 2242 (1971).
14. J.H. Beynon, R.G. Cooks, K.R. Jennings and A.J. Ferrer-Correia, Int. J. Mass Spectrom. Ion Phys., $\underline{18}$, 87 (1975).
15. R.W. Kiser, "Introduction to Mass Spectrometry", Prentice-Hall, Englewood Cliffs, Chapter 8.
16. J.D. Morrison in "Mass Spectrometry" (A. Maccoll, Ed.), MTP International Review of Science, Phys. Chemistry Series I, Volume 5, Butterworths, London, 1972, Chapter 2.
17. J.A. Simpson and C.E. Kuyatt, Rev. Sci. Instrum., $\underline{34}$, 265 (1963).
18. K. Maeda, G.P. Semeluk and F.P. Lossing, Int. J. Mass Spectrom. Ion Phys., $\underline{1}$, 395 (1968).
19. R.E. Fox, W.M. Hickam, T. Kjeldaas, Jr. and D.J. Grove, Phys. Rev., $\underline{84}$, 859 (1951).
20. R.E. Fox, W.M. Hickam, D.J. Grove and T. Kjeldaas, Jr., Rev. Sci. Instrum., $\underline{26}$, 1101 (1955).
21. R.E. Winters, J.H. Collins and W.L. Courchene, J. Chem. Phys., $\underline{45}$, 1931 (1966).
22. F.P. Lossing, A.W. Tickner and W.A. Bryce, J. Chem. Phys., $\underline{19}$, 1254 (1951).
23. J.W. Warren, Nature, $\underline{165}$, 810 (1950).
24. R.W. Kiser and E.J. Gallegos, J. Phys. Chem., $\underline{66}$, 947 (1962).
25. R.E. Honig, J. Chem. Phys., $\underline{16}$, 105 (1948).
26. J.D. Morrison, J. Chem. Phys., $\underline{21}$, 1767 (1953).
27. J.L. Occolowitz, B.J. Cerimele and P. Brown, Org. Mass Spectrom., $\underline{8}$, 61 (1974).
28. F.I. Vilesov, B.L. Kurbatov and A.N. Terenin Dokl. Akad. Nauk. SSSR, $\underline{138}$, 1329 (1961).
29. D.W. Turner and M.I. Al-Jobory, J. Chem. Phys., $\underline{37}$, 3007 (1962).

30. D.W. Turner, C. Baker, A.D. Baker and C.R. Brundle, "Molecular Photoelectron Spectroscopy", Wiley-Interscience, New York, 1970.
31. J.H.D. Eland, "Photoelectron Spectroscopy", Butterworths, London, 1974.
32. D. Villarejo, R.R. Herm and M.G. Inghram, J. Chem. Phys., 46, 4995 (1967).
33. D. Villarejo, J. Chem. Phys., 48, 4014 (1968).
34. W.B. Peatman, T.B. Borne and E.W. Schlag, Chem. Phys. Lett., 3, 492 (1969).
35. T. Baer, W.B. Peatman and E.W. Schlag, Chem. Phys. Lett., 4, 243 (1969).
36. R. Spohr, P.M. Guyon, W.A. Chupka and J. Berkowitz, Rev. Sci. Instrum. 42, 1872 (1971).
37. J.D. Morrison, J. Appl. Phys., 28, 1409 (1957).
38. W.A. Chupka and M. Kaminsky, J. Chem. Phys., 35, 1991 (1961).
39. W.A. Chupka, J. Chem. Phys., 30, 191 (1959).
40. W.A. Chupka and J. Berkowitz, J. Chem. Phys., 47, 2921 (1967).
41. K.M.A. Refaey and W.A. Chupka, J. Chem. Phys., 48, 5205 (1968).
42. G.G. Meisels, C.T. Chen, B.G. Giessner and R.H. Emmel, J.Chem. Phys., 56, 793 (1972).
43. W.C. Price, A.W. Potts and D.G. Streets, in "Electron Spectroscopy" (D.A. Shirley, Ed.), North-Holland, Amsterdam, 1972.
44. B. Brehm, J.H.D. Eland, R. Frey and H. Schulte, Int. J. Mass Spectrom. Ion Phys., 21, 373 (1976).
45. F.W. McLafferty, T. Wachs, C. Lifshitz, G. Innorta and P. Irving, J. Am. Chem. Soc., 92, 6867 (1970).
46. G. Innorta, S. Torroni and S. Pignataro, Org. Mass Spectrom., 6, 113 (1972).
47. B.N. McMaster in "Mass Spectrometry" Vol. 3 (R.A.W. Johnstone, Ed.), The Chemical Society, London 1975, Chapter 1.
48. H.M. Rosenstock and M. Krauss, Advan. Mass Spectrometry, 2, 251 (1963).
49. H. Ehrhardt and O. Osberghaus, Z. Naturforsch., 15a, 575 (1960).
50. B. Steiner, C.F. Giese and M.G. Inghram, J. Chem. Phys., 34, 189 (1961).
51. R.W. Taft and I.C. Lewis, Tetrahedron, 5, 210 (1959).
52. L.S. Levitt and B.W. Levitt, J. Org. Chem., 37, 332 (1972).
53. L.S. Levitt and B.W. Levitt, Chem. and Ind., 990 (1970).
54. L.S. Levitt and B.W. Levitt, J. Phys. Chem., 74, 1812 (1970).
55. L.S. Levitt and B.W. Levitt, Tetrahedron, 27, 3777 (1971).
56. L.S. Levitt and B.W. Levitt, Tetrahedron, 29, 941 (1973).
57. B.W. Levitt and L.S. Levitt, Chem. and Ind., 185 (1973).
58. B.W. Levitt and L.S. Levitt, Israel J. Chem., 9, 71 (1971).
59. B.W. Levitt, H.F. Widing and L.S. Levitt, Chem. and Ind., 793 (1973).
60. B.W. Levitt and L.S. Levitt, Experientia, 26, 1183 (1970).
61. B.W. Levitt and L.S. Levitt, Israel J. Chem., 9, 711 (1971).
62. L.S. Levitt and B.W. Levitt, Chem. and Ind., 132 (1973).
63. B.W. Levitt and L.S. Levitt, Chem. and Ind., 724 (1972).
64. L.S. Levitt, H.F. Widing and B.W. Levitt, Can. J. Chem., 51, 3963 (1973).
65. L.S. Levitt, B.W. Levitt and C. Párkányi, Tetrahedron, 28, 3369 (1972).
66. L.S. Levitt, Chem. and Ind., 637 (1973).
67. H.F. Widing and L.S. Levitt, Tetrahedron, 30, 611 (1974).

68. A. Streitwieser, Prog. Phys. Org. Chem., $\underline{1}$, 1 (1963).
69. W.C. Price, Chem. Revs., $\underline{41}$, 257 (1947).
70. G.F. Crable and G.L. Kearns, J. Phys. Chem., $\underline{66}$, 436 (1962).
71. P. Brown, Org. Mass Spectrom., $\underline{4}$, 519 (1970).
72. P. Brown, Org. Mass Spectrom., $\underline{4}$, 533 (1970).
73. M.S. Chin and A.G. Harrison, Org. Mass Spectrom., $\underline{2}$, 1073 (1969).
74. S. Pignataro, V. Mancini, G. Innorta and G. Distefano, Z. Naturforsch., $\underline{27a}$, 534 (1972).
75. G. Distefano, S. Pignataro, G. Innorta, F. Fringuelli, G. Marino and A. Taticchi, Chem. Phys. Lett., $\underline{22}$, 132 (1973).
76. S. Pignataro, A. Foffani, G. Innorta and G. Distefano, Z. Phys. Chem. Neue Folge, $\underline{49}$, 291 (1966).
77. A. Foffani, S. Pignataro, B. Cantone and F. Grasso, Z. Phys. Chemie, Neue Folge, $\underline{42}$, 221 (1964).
78. A.G. Harrison, P. Kebarle and F.P. Lossing, J. Am. Chem. Soc., $\underline{83}$, 777 (1961).
79. R.F. Pottie and F.P. Lossing, J. Am. Chem. Soc., $\underline{85}$, 269 (1963).
80. A. Buchs, G.P. Rosetti and B.P. Susz, Helv. Chim. Acta, $\underline{47}$, 1563 (1964).
81. F. Benoit, Org. Mass Spectrom., $\underline{7}$, 295 (1973).
82. I. Howe and D.H. Williams, J. Am. Chem. Soc., $\underline{91}$, 7137 (1969).
83. F.W. McLafferty and L.J. Schiff, Org. Mass Spectrom., $\underline{2}$, 757 (1969).
84. A.D. Baker, D.P. May and D.W. Turner, J. Chem. Soc. (B), 22 (1968).
85. E. Murad and M.G. Inghram, J. Chem. Phys., $\underline{40}$, 3263 (1964).
86. V.H. Dibeler and J.A. Walker, Int. J. Mass Spectrom. Ion Phys., $\underline{11}$, 49 (1973).
87. F. Benoit, Org. Mass Spectrom., $\underline{6}$, 1289 (1972).
88. F. Benoit, Org. Mass Spectrom., $\underline{9}$, 626 (1974).
89. T.W. Bentley and R.A.W. Johnstone, J. Chem. Soc. (B), 263 (1971).
90. J.M. Behan, R.A.W. Johnstone and T.W. Bentley, Org. Mass Spectrom., $\underline{11}$, 207 (1976).

Chapter II. The Quasi-Equilibrium Theory

1 The Theory

1.1 Introduction

The fragmentation of large organic ions produced by electron impact in a mass spectrometer has considerable similarity to the fragmentation processes associated with the thermal decomposition of organic molecules. This similarity consists of the large body of observations which indicate that ions undergo bond ruptures at weak bonds, loss of molecular species such as H_2, H_2O, CH_4 and molecular rearrangements, all of which are known to occur preferentially in thermal reactions. In addition, fragmentation threshold measurements on the ions indicate that in most instances the preferred processes are those which have the lowest energy requirement. The essential difference between thermal decomposition and electron impact ion fragmentation is that in the former class of processes the molecules are continuously energized and deactivated by collisions with the buffer gas and the resultant distribution of energy can be described by a *temperature*. In contrast, the ions produced in a conventional electron impact mass spectrometer are each energized by a single electron impact and undergo no subsequent collisions prior to detection. Thus each ion is formed with a specific amount of internal energy and angular momentum. These are conserved independently in all subsequent dissociations. During the ionization process, transitions to various electronic, vibrational and rotational states of the ion are possible, leading to a collection of ions with a distribution of internal energies. This distribution will be quite different from the thermal distribution and will differ with electron energy and with the nature of the molecular species.

The decomposition of diatomic ions is well understood. Many of the features of the fragmentation of these ions can be explained by examining the possible transitions to the various electronic states of the ions[*]. The potential curves for these states can be determined experimentally (e.g. by spectroscopic methods) or calculated with fair to good accuracy [251].

---

[*] In addition, transitions may occur to superexcited states of the molecule, which may dissociate to form a neutral pair or an ion pair or may autoionize to a molecular ion (see Section I-3.3).

On the other hand there are many experimental observations (e.g. the presence of metastable ions, the small amount of kinetic energy released upon fragmentation) which indicate that the simple decomposition model for diatomic ions, spontaneous dissociation, is not valid for polyatomic ions.

These observations imply rather that dissociation does not immediately follow ionization, but is slow enough to permit the transfer of energy into the various degrees of freedom involved in the observed dissociation. Inspired by these observations Rosenstock et al. [2] developed a statistical theory in 1952, the so called Quasi-Equilibrium Theory (QET).

1.2 Basic Assumptions

The Quasi-Equilibrium Theory (QET) is based upon the following assumptions [1,2,4-8]:

(1) *The time required for dissociation of the initial molecular ion is long compared with the time of interaction leading to its formation and excitation.* Ionization occurs roughly within 10^{-16} s. The shortest conceivable dissociation time would be of the order of one vibrational period ($\sim 10^{-14}$ s). The QET assumes that dissociation only occurs after at least several oscillations so that the molecular ion has lost all "memory" of how and where the ionization occurred, i.e. the rate of decomposition is independent of the ionization mode.

(2) *The rate of dissociation is slow relative to the rate of redistribution of the initial excitation energy over all degrees of freedom.* Possible mechanisms for this redistribution of excitation energy are discussed in Section II-2.1.1.

(3) *The fragmentation products are formed by a series of competing and consecutive unimolecular reactions.*

(4) *Ions generated in a mass spectrometer represent isolated systems in a state of "internal equilibrium"* [*]. Hence the rates for unimolecular decompositions can be calculated using the Absolute Reaction Rate Theory modified to apply to such non thermal situations, i.e. by integrating microcanonical rate expressions over the excitation energy distributions, as shown below.

---

[*] Isolated systems can be treated as microcanonical ensembles in which all systems within an energy interval between E and E + dE of the phase space are distributed uniformly over all states within this interval. Such an ensemble is in statistical equilibrium [3].

It follows from (4) that *the rate for a particular process is a function of the excitation energy only* and tends to a limiting value at high internal energies.

1.3 The Rate Expression

The ultimate goal of any kinetic theory is to calculate the rate of reaction. The rate expression of the QET was developed by Rosenstock et al. [2] by applying the Absolute Rate Theory of Eyring [9] to isolated systems. The model used in this treatment is illustrated schematically in Fig. II-1, where a slice through a potential surface is shown. In this model the dissociation is described as *a motion along a reaction coordinate separable from all other internal coordinates by a critical "activated complex configuration"*. Decomposition occurs if in the reaction coordinate

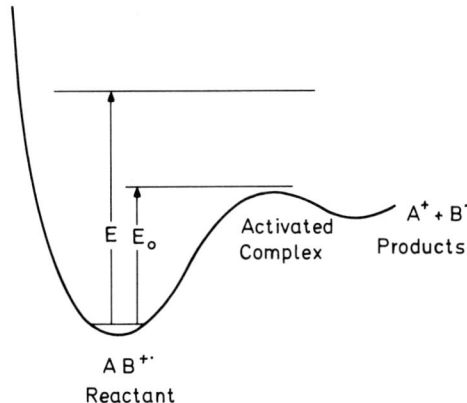

Fig. II-1: Schematic potential energy diagram of an ion.

a sufficient amount of internal energy has accumulated to overcome the energy barrier (activation energy, E_o) of the activated complex.

If an isolated system has an internal energy E, its rate of dissociation is given as

$$k(E) = \frac{\sigma}{h} \frac{W^{\neq}(E - E_o)}{\rho(E)} \qquad (II-1)$$

where h is Planck's constant, $W^{\neq}(E - E_o)$ is the number of vibrational and rotational states of the activated complex with an energy $\leq E - E_o$,

$\rho(E)$ is the density of states of the reactant ion with energy between E and $E + dE$ and the symmetry factor σ is the number of identical reactions i.e. reactions having the same activation energy and activated complex. (The original derivation of eq. II-1 by Rosenstock et al. [2] in 1952 is given in the Appendix of this chapter (Section II-3)). In this original treatment a continuous approximation for the expression $W^{\neq}(E - E_o)$ and $\rho(E)$ was used).

It is noteworthy that an equivalent equation was obtained for thermal reactions by Marcus and Rice [10] using a different approach consisting of a reformulation of the Rice-Ramsperger-Kassel theory, now termed RRKM-theory [10,11,252].

In the rate expression II-1 the dissociation is treated semi-classically. Decomposition is only possible if the excitation energy is equal to or larger than the activation energy (Fig. II-1); quantum mechanical reflection and tunnelling at the barrier are neglected.

The rate equation II-1 has a discontinuity at the activation energy. For $E < E_o$, $W^{\neq}(E - E_o) = 0$ and hence $k = 0$. Thus the semi-classical treatment predicts the existence of a minimum rate at $E = E_o$ ($W^{\neq} = 1$)*)

$$k_{min} = \frac{\sigma}{h\rho(E_o)} \qquad (II-2)$$

A quantum mechanical treatment including the reflection and tunnelling through the barrier removes this discontinuity [1]. Thus for the reaction $CD_4^{+\cdot} \rightarrow CD_3^+ + D^{\cdot}$ a minimum rate constant of $k_{min} = 4.4 \times 10^9 \text{ s}^{-1}$ has been calculated [12] using eq. II-2**) so that no metastable transition which would require a rate constant of $k \leq 10^6 \text{ s}^{-1}$ should be observed for this process. A metastable decomposition of $CD_4^{+\cdot}$ has, however, been found by Dibeler and Rosenstock [14] and may be explained as occurring via tunnelling [6,15]***).

---

*
At $E = E_o$ the only state available is the ground state of the activated complex.

**
$\rho(E_o)$ has been calculated using the approximation of Vestal, Wahrhaftig and Johnson [13].

***
Solka et al. [15] explained the metastable $H^{\cdot}$ loss from $CH_4^{+\cdot}$ as a tunnelling through the centrifugal barrier. Such a centrifugal barrier has been postulated also by Flamme et al. [16]. Ottinger [17] suggests that the discrepancy between the experimentally observed and theoretically calculated minimum rates for $H^{\cdot}$ elimination from $CH_4^{+\cdot}$ (based on the semi-classical treatment) can be resolved if rotational degrees of freedom are considered in the calculations of $\rho(E_o)$.

1.4 Enumeration of States

To calculate the rate according to eq. II-1 information on the density of states in the reactant and the number of states in the activated complex are necessary. For this purpose a variety of methods have been developed which allow a continuous approximation of the exact quantum mechanical density of states over the entire energy range of interest with high accuracy. Most of these methods for the enumeration of quantum states are based on two approximations: (1) only vibrational and rotational states are considered, i.e. it is assumed that excited electronic states do not take part in the decomposition; (2) the oscillators are assumed to be harmonic, the rotors assumed to be free. The various methods have been reviewed by Rosenstock [6] and in particular detail by Forst [253]. Only two of these approximations will be discussed here.

In the original formulation of the QET [2] the molecular ion was assumed to consist of N harmonic oscillators, the activated complex correspondingly of N-1 oscillators (one vibrational degree of freedom is converted into a translational degree of freedom in the activated complex, corresponding to the motion along the reaction coordinate).

Statistical mechanics [3] give the number of vibrational states in a system of energy E consisting of a collection of N identical harmonic oscillators as

$$\rho(E) = \frac{1}{h^N \prod_{i=1}^{N} \nu_i} \frac{E^{N-1}}{(N-1)!} \tag{II-3}$$

This is the so called *classical approximation* for the enumeration of vibration states which leads to the following expression for the rate equation

$$k(E) = \frac{\prod_{j=1}^{N} \nu_j}{\prod_{i=1}^{N-1} \nu_i} \left(\frac{E - E_o}{E} \right)^{N-1} \tag{II-4}$$

where ν_j are the vibrational frequencies of the reactant, ν_i those of the activated complex. The ratio of products of the vibrational frequencies in eq. II-4 is termed "frequency factor" (ν) so that eq. II-4 simplifies to

$$k = \nu \left(\frac{E - E_o}{E} \right)^{N-1}$$

If internal rotations are involved in the formation of the activated complex (as in the case of rearrangement reactions) the reactant is assumed to consist of a collection of N - L weakly coupled harmonic oscillators and L rigid internal rotors, and the activated complex of $N - L^{\neq} - 1$ harmonic oscillators and $L^{\neq}$ internal rotors. (The number of rotors in the activated complex needs not be identical with that in the reactant as in a rearrangement reaction several rotors will be stopped). In the case of internal rotations the classical approximation leads to the following rate expression

$$k = Z \left(\frac{E - E_o}{E} \right)^{N-(L/2)-1} (E - E_o)^{(L-L^{\neq})/2} \qquad (II-5)$$

where the factor Z contains the products of the moments of inertia and the vibrational frequencies of the reactant and the activated complex.

The rate equation based on the classical approximation has been used and repeatedly tested during the first decade of the QET and has been shown to be a very poor approximation particularly at low excitation energies [257]. The quantitative disagreement between theory and experiment, which will be discussed in Section II-2, prompted many mass spectrometrists to discard the entire theory, overlooking that its apparent failure did not result from incorrect basic assumptions but from inadequate mathematical approximations as well as lack of knowledge of the exact parameters used in the calculations.

In order to improve the agreement between the theoretical predictions of the classical approximation and the experimental results it was suggested to treat the number of oscillators in eq. II-4 as free parameter [177,179,193,194]. Using an "effective" number of oscillators which was of the order of 1/3 to 1/5 of the total degrees of freedom (and energy dependent) the agreement between theory and experiment was greatly improved.

However, a later analysis demonstrated that the use of better approximations for the enumeration of the density of states makes it unnecessary to invoke an "effective" number of oscillators. A variety of such methods has been described in the last two decades [18-29,254-256]. They have been discussed in a comprehensive review by Forst [253] who also tested the accuracy of the various approximations by comparison with direct state counting. While no method gives sufficiently accurate results near the threshold (< 0.3 eV), good agreement (i.e. results which are within 20 % of the direct count) is observed at higher energies with most methods. As mentioned above, most approximations rest upon the assumption that the oscillators are harmonic although corrections for the anharmonicity (in a system of classical Morse oscillators) have been reported [24,253]. Such corrections may be substantial at high internal energies [253].

Among the various approximations, that of Vestal et al. [29] has been most frequently used for QET calculations and will be briefly described*). The authors make an exact enumeration of the number of ways a given number of oscillators can remain completely unexcited and multiply this value by the number of ways the energy can be partitioned among the excited oscillators. The latter number is reasonably approximated by the semiclassical expression [6,29]

$$\frac{(E + E_o)^{N-1}}{(N-1)!} \tag{II-6}$$

leading to the following equation for the total number of vibrational states

$$W(E) = \int_o^E \rho(E)dE = \sum_{p=o}^{k} \binom{N}{p} \frac{1}{p!} \left[\frac{\sigma_p E}{h\bar{\nu}} - \frac{p-1}{2}\right]^p \tag{II-7}$$

where h is Planck's constant, $\bar{\nu}$ is the geometric mean frequency, N is the number of vibrational degrees of freedom, σ_p a frequency coefficient and p is an index varying from 1 to N. For example, if p oscillators are unexcited $\binom{N}{p}$ is the number of ways of doing this and the term in brackets is the probability of putting the energy E or less into the remaining N-p oscillators under the restriction that each holds at least one quantum. A similar equation has been developed for the case where free rotors are included in the model. Using the expression II-7 for the enumeration of vibrational states, the rate equation [29] becomes fairly complicated and its solution requires the use of a computer. Fig. II-2 compares the rate constant as a function of the internal energy calculated for 2,3-dithiabutane [30,31] using the classical approximation [2] (solid line) with the approximation by Vestal et al. [29] (dashed line). It is apparent that at higher energies the classical approximation leads to rate constants which are one order of magnitude smaller than those obtained by the method of Vestal et al.. The deviations become even more pronounced at low internal energies.

The accuracy of eq. II-7 for the enumeration of states has been tested by Vestal [1] for several simple models by calculating the num-

---

*) Although a variety of QET calculations are based on this approximation, it has been noted that this method gives inferior results to most other methods at a much larger expenditure of computer time [253].

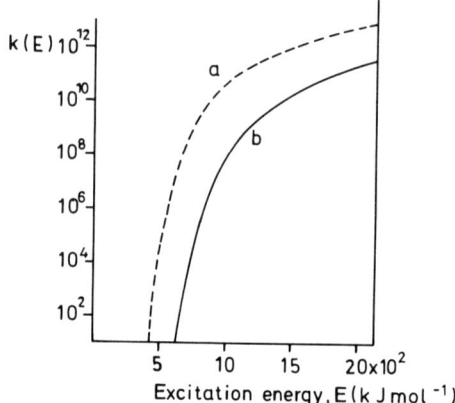

Fig. II-2. Calculated rate constant as function of the internal energy for the formation of CH_3S^+ from the 2,3-dithiabutane molecular ion using the classical approximation (b) and Vestal's approximation (a) [30,31].

ber of states using eq. II-7 or by direct counting. On average, a difference of only a factor two was found, demonstrating that the approximation by Vestal et al. [29] should give reliable quantitative data for rate constants. Experimental tests, described in Section II-2.2 support this conclusion.

It was one of the original goals of the QET to calculate complete mass spectra. Comparison of calculated and experimental mass spectra allow the validity of the theory to be tested although this is not a very rigorous test as will be shown later. However, the importance of the QET does not only lie in the possibility of calculating mass spectra. Its value also lies in the fact that it offers a theoretical concept for understanding the fragmentation of gaseous ions allowing a variety of phenomena in organic mass spectrometry, such as kinetic shifts, isotope effects, and the dependence of the fragment intensities on the internal energy and lifetime to be predicted and explained. These phenomena will be discussed in Section III.

1.5 Calculation of Rate Constants

Calculation of the rate constant of a given fragmentation reaction requires knowledge of at least four sets of data: the configuration of the molecular ion and the transition state, the vibrational and rotational frequencies in the molecular ion and the transition state, the symmetry factor and the activation energy.

1.5.1 Configuration of the Molecular Ion and the Transition State

There is growing evidence that with most organic molecules the constitution of the molecular ion is identical with that of the neutral molecule. The principal exceptions are small unsaturated hydrocarbons such as butenes [32-36] and butynes [1,37-39], where it has been demonstrated that isomerization of the molecular ion to a common structure or a mixture of interconverting structures occurs within a nanosecond after ionization [40]. Moreover, the observation of complete carbon scrambling in several aromatic hydrocarbon ions (see Section III-2.3) indicates that also in these cases the structure of the molecular ion may not be well defined. However, the molecular ions of alkanes which were the primary object of QET calculations, retain their constitution upon ionization [41] (in antithesis to a recent contention [42]).

On the other hand, removal of an electron from a σ-orbital will certainly lead to some change in the configuration, since the bond lengths and angles of the ions will not be identical with those of the neutral molecules. Most QET calculations are however, based on the assumption that the configuration of the ion is very similar to that of the neutral molecule (vide infra).

While the structure of the molecular ion is at least qualitatively known, less is understood about the configuration of the transition state, and this is the most severe handicap in QET calculations. The geometry of the activated complex is usually termed as either "loose" (no rotations are stopped) or "tight" (one or more internal rotations are stopped). Thus the transition state of a direct bond cleavage is classified as "loose" that of rearrangement reaction as "tight".

Isotopic labelling is the most important tool for gaining information on the geometry of the activated complex. Thus the **McLafferty** rearrangement [43] in aliphatic and aromatic ketones involves only the γ-hydrogen atom, pointing to a six-membered cyclic transition state (Scheme II-1). There is, however, considerable evidence that this reaction takes place stepwise [44,45] (at least at long ion lifetimes) as illustrated in the lower part of Scheme II-1. The overall reaction is therefore probably best described by two transition states both of which have to be taken into account for exact rate calculations. Moreover, hydrogen rearrangement reactions in general are non-specific, i.e. transition states of varying ring size may be involved. The elucidation of the structure of the activated complex by isotopic labelling is further complicated by possible label exchange ("scrambling") prior to decomposition, which may hamper identification of the transition state, if non-specific rearrangements are involved.

Finally, one may be tempted to conclude that the transition states

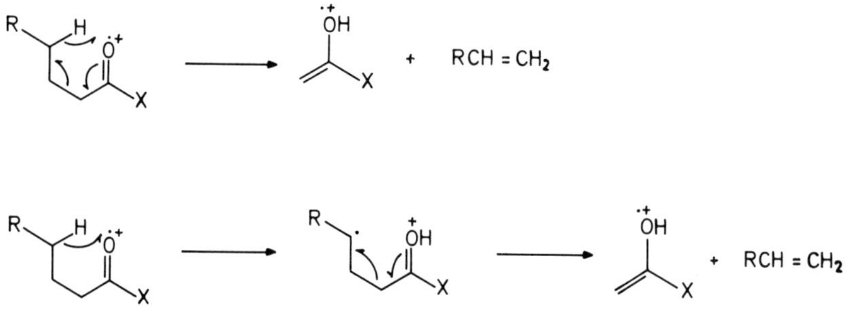

Scheme II-1

are well defined for those reactions which have been shown by isotopic labelling to be direct bond cleavages. However, there is a steadily increasing number of examples of formal direct bond cleavages which in reality are the result of complicated functional group interactions and thus have tight transition states [46].

In addition to isotopic labelling there are several methods which give further information on reaction mechanisms and thus on the transition state configuration. Such methods include the comparison of frag-

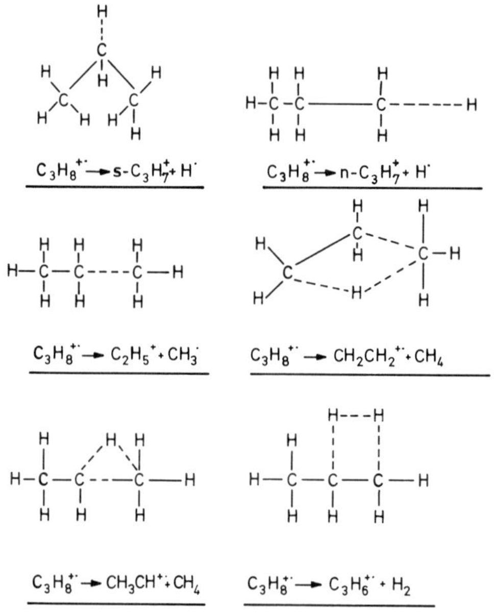

Scheme II-2

ment intensities [46], appearance potentials [47], the study of ion lifetimes [48] and the energy dependence [49], the detection of unusual isotope effects [50], and, especially important, the determination of the kinetic energy released upon metastable decomposition [51]. These are discussed in more detail in Section IV-2.

In summary, despite the various techniques available for elucidating the configuration of the transition state and numerous mechanistic studies knowledge of the transition state geometry is still largely the product of chemical intuition, and this is especially true in the case of hydrocarbon ions. Scheme II-2 shows the transition states used for the QET calculations of propane [1]. For the formation of both $C_3H_7^+$ and $C_2H_4^{+\cdot}$ two competing mechanisms with two distinct transition states have been assumed, although later deuterium labelling [52-54] revealed that the assumed 1,2-elimination for CH_4 loss does not take place and that for H elimination secondary hydrogens are predominantly involved, at least for the metastable decomposition.

The transition state is especially poorly defined if the reverse activation energy is small or negligible, i.e. if there is no pronounced maximum on the potential energy surface (save for a rotational barrier). For this case the transition state has been defined by Bunker and Pattengill [55] by the criterion of the "minimum local density of states", i.e. $d\rho^{\neq}/dq = 0$, if q is the reaction coordinate, although so far this criterion has been used only in neutral systems.

It is obvious that our poor knowledge of the transition state geometry severely limits the accuracy of QET calculations. Thus it is particularly interesting that a modification of the QET, the Phase Space Theory [56], discussed in Section II-1.6 does not require knowledge of the transition state geometry at all (except that it is assumed to be loose).

1.5.2 The Vibrational and Rotational Frequencies in the Molecular Ion and the Transition State

For small organic ions, the vibrational frequencies can be determined by photoelectron spectroscopy [57-59] or by high quality molecular orbital calculations. Moreover, high resolution photodissociation spectroscopy of ions trapped in an ion cyclotron resonance (ICR) cell as recently reported by Dunbar [60] may yield direct information about vibrational frequencies of gaseous ions in the near future.

While no significant change of the vibrational frequencies is expected if the electron is removed from a non or weakly bonding orbital, ionization of a bonding or antibonding orbital will lead to a decrease or increase of the frequencies. Changes in the vibrational frequencies are also expected if the geometry of the ion differs markedly from that of the neutral molecule. As far as data are available the change in frequencies is generally of the order of 10 - 30 %[*].

These rather modest changes justify equating the vibrational frequencies of the molecular ion with those of the neutral molecule in QET calculations, if experimental data are not available.

The vibrational and rotational frequencies in the transition state are usually estimated on the basis of the assumed configuration of the activated complex. Guidelines for the choice of "reasonably consistent" frequencies have been proposed [2,61,62,258]. For a loose activated complex, for instance, no vibrational frequency is assigned to the bond to be broken, while the vibrational frequencies in the vicinity of that bond are usually reduced by a factor of 2 - 4 and the rotation assumed to be free. On the other hand, for "tight" activated complexes, the skeletal frequencies are assumed to be rarely lower than those of the normal configuration. However, in this case some free rotations are stopped and treated as torsional vibrations. For instance, for the methyl elimination from the propane molecular ion (Scheme II-2) the C-C stretching is taken as the reaction coordinate (no frequency is assigned to this vibration), and both the CH_3 deformation frequencies (1200 cm^{-1}) and the C-C-C- bending frequency (400 cm^{-1}) are reduced by a factor of four while rotations of the methyl groups are assumed to be active [1].

It is evident that there is some arbitrariness in the assignment of the frequencies. It has been stated that it is rarely possible to choose the frequencies in a strictly a priori fashion. Some fitting to experimental data is nearly always required [1].

1.5.3 The Symmetry Factor

The symmetry factor, σ, in eq. II-1 is the number of identical reactions which lead to the formation of a given ion [1]. The total rate is the symmetry factor multiplied by the rate for one such reaction. For direct bond cleavages the symmetry factor can usually be derived simply. σ is the number of different bonds which, when broken, lead to the same fragment ion, e.g. for methyl loss from propane the symmetry factor is two [2]. For rearrangements a similar procedure is used based on the configuration of the activated complex and the assumed structure of the

---

[*] However, there are some instances where considerably larger variations in the frequencies are observed. Thus in ethylene the twisting vibration changes from 1027 cm^{-1} in the molecule to 430 cm^{-1} in the ion in its ground state [58].

products. (Examples are given in Ref. 1).

1.5.4 The Activation Energy

The activation energy of a given fragmentation reaction is determined from ionization and appearance potential measurements (see Section I-4 and III-4). The limited accuracy of the determination of ionization potentials has already been mentioned. The reliability of appearance potential measurements (i.e. the minimum energy to form a given fragment) is even poorer as on the one hand it is more difficult to detect the onset of fragmentation, and on the other hand the minimum energy determined this way is too large by the amount of the "kinetic shift" [63], the excess energy required to drive a fragmentation fast enough for the ion to decompose within the mass spectrometric time scale (usually 10^{-6} s) (see Section III-4.1). The experimental determination of this quantity is difficult as discussed in Section III-4.2. In addition the thermal and possibly also the competitive shift have to be taken into account (see Section III-4.3 and III-4.4).

In concluding this topic we see that it has become evident that reliable calculations of the rate constant do not only require adequate mathematical approximations for the enumeration of the density of states, but also demand knowledge of a variety of parameters which usually are not well known at all. Thus it has been stated that "the QET suffers unfortunately in virtually requiring the answer before it can calculate a mass spectrum" [42]. Hence it is admirable that in spite of the poorly defined parameters and sophisticated mathematical approximations not only complete mass spectra, but a variety of other mass spectrometric quantities have been calculated with results which are often in fair agreement with experimental data, as discussed in Section II-2.2, demonstrating that the exact choice of the parameters is less critical than might be expected.

1.6 Modified QET — Phase Space Theory

According to the microscopic reversibility principle (also termed the "principle of detailed balance") an activated complex configuration can be reached either by decomposition of the reactant ion or by recombination of the ionic and neutral products in an ion-molecule reaction, i.e. for the reaction $AB^{+\cdot} \rightarrow A^{+} + B^{\cdot}$ the rates, R, for the forward and backward passage over the activated complex are equal:

$$R_J(AB^{+\cdot} \rightarrow A^{+} + B^{\cdot}) = R_J(A^{+} + B^{\cdot} \rightarrow AB^{+\cdot})$$

(where R_J is the total rate with angular momentum J).

Using this principle it was first shown by Klots [56,64-69] that the problem of defining an unknown transition state configuration can be avoided by calculating instead the cross section for the formation of a collision complex from the separated fragments the thermodynamic properties of which are amenable to experimental observations. Another important advantage of this treatment in contrast to the original QET formulation is that the total angular momentum is conserved. It is important to recall that this theory, which is also referred to as "Phase Space Theory" [8], rests upon the basic assumption of the QET: the rate of dissociation is slow relative to the rate of redistribution of internal energy.

A disadvantage of this theory is that good collision cross section data are necessary for the calculations, and these are rarely available. Thus the Langevin collision model is applied to the calculation of these cross sections [259-261]. The model describes the attraction between an ion and a molecule polarized by the ion's field. The force of attraction accelerates the molecule towards the ion leading to a collision of the two particles even if their projected paths have a minimum distance which is larger than their combined radii. The cross section σ for a collision of a non-polar molecule of polarizability α and an ion of charge e, is, according to the Langevin model,

$$\sigma(E_{tr}) = \pi e (2\alpha/E_{tr})^{1/2}$$

where E_{tr} is the relative translational energy of the two-particle system. Hence the cross section depends inversely on the relative velocity of the molecule and the ion. (The Langevin model cannot be applied to molecules which have a permanent dipole moment).

Several rate equations based on this modified QET have been developed [56,70-74]. As example the rate equation given by Chesnavich and Bowers [73,74] is presented here as this equation uses a formalism which allows a direct comparison with the original QET - RRKM formulation in its integral form (eq. II-1 and Appendix).

$$k_J(E)_a = C \int_{E_{tr}^{\neq}}^{E-E_o} \frac{\rho_b(E - E_o - E_{tr})}{\rho_a(E - E_r^a)} \Gamma_b(E_{tr}, J) \, dE_{tr} \qquad (II-8)$$

a = reactant;
b = products;
$k_j(E)_a$ = unimolecular rate constant of a → b at total energy E and an-

gular momentum J;
C = factor containing the symmetry numbers as well as the degeneracies associated with rotational states;
$\rho_a(E - E_r^a)$ = vibrational density of states of the molecule "a" with total energy E and rotational energy E_r^a;
$\rho_b(E - E_o - E_{tr})$ = vibrational density of states of "b" at energy $E - E_o - E_{tr}$, where $E_{tr} = E_r^b + E_t$ is the rotation/translation energy sum.
$\Gamma_b(E_{tr}, J)$ = sum of angular momentum states at a given E_{tr} and J; $E_{tr}^{\neq}$ = minimum value of E_{tr} for which $\Gamma_b(E_{tr}, J) > 0$.

The Phase Space Theory has been used to calculate the rate constant, k(E), for a variety of systems ($CH_4^{+\cdot}$ [56], $C_6H_5CN^{+\cdot}$ [65,74], $C_6H_6^{+\cdot}$ [65], $SF_6^{+\cdot}$ [68], $C_4H_6^{+\cdot}$ [74]). It has been pointed out that rate constants predicted by this theory correspond to very loose transition states and hence represent upper limits [74] so that Phase Space Theory calculations, in general give rate constants which are too high if tight transitions states are involved. In Section II-2.2.3 some of these calculations will be compared with experimental data.

1.7 Calculation of Breakdown Graphs and Mass Spectra

The calculation of complete mass spectra is usually carried out in three steps:
(1) k(E) is determined for each fragmentation;
(2) a breakdown graph is constructed from the individual rate constants;
(3) convolution of the breakdown graph with the internal energy distribution yields the mass spectrum.
For such calculations all fragmentation routes, i.e. the so called breakdown scheme, must be known as illustrated in Scheme II-3 for propane, our model compound.

$$C_3H_8^{+\cdot} \rightarrow C_3H_7^+ \rightarrow C_3H_5^+ \rightarrow C_3H_3^+ \rightarrow C_3H^+$$
$$\rightarrow C_3H_6^{+\cdot} \rightarrow C_3H_4^{+\cdot} \rightarrow C_3H_2^{+\cdot} \rightarrow$$
$$\rightarrow C_2H_5^+ \rightarrow C_2H_3^+$$
$$\rightarrow C_2H_4^{+\cdot} \rightarrow C_2H_2^{+\cdot}$$

Scheme II-3

Rosenstock and Krauss [4] have summarized the methods which can be used to obtain information about the breakdown scheme:
(1) Energetic considerations
(2) Isotopic labelling
(3) Metastable transitions

(4) The dependence of the mass spectrum on the electron energy
(5) High resolution data
(6) Direct experimental determination of breakdown graphs.

Isotopic labelling and metastable transitions give the most valuable information on the breakdown scheme. It should, however, be kept in mind that not every fragmentation reaction is associated with a metastable peak.

The breakdown graph represents the relative intensity ratio of the molecular and all fragment ions as a function of the internal energy)* of the molecular ion, i.e. the graph shows the extent to which the intensity of the molecular ion is decreasing and the fragment ions are forming with increasing excitation energy within 10^{-5} s.

Once the rate constants have been determined the intensity of the molecular and fragment ions can be calculated using the laws for unimolecular decomposition [75]. For a consecutive reaction

$$M^{+\cdot} \xrightarrow{k_A} A^+ \xrightarrow{k_B} B^+$$

(where k_A and k_B are rate constants at a given energy E) the concentration of $M^{+\cdot}$, A^+ and B^+ at the time t is given as

$$[M^{+\cdot}] = \exp(-k_A t) \tag{II-9}$$

$$[A^+] = \frac{k_A}{k_B - k_A} [\exp(-k_A t) - \exp(-k_B t)] \tag{II-10}$$

$$[B^+] = 1 - \frac{k_B}{k_B - k_A} \exp(-k_A t) + \frac{k_A}{k_B - k_A} \exp(-k_B t) \tag{II-11}$$

if the initial concentration of $M^{+\cdot}$ at t = 0 is unity.

In the case of competing reactions of the general form

$$\begin{aligned} M^{+\cdot} &\to A_1^+ \\ &\to A_2^+ \\ &\to A_i^+ \end{aligned}$$

the concentration of $M^{+\cdot}$ and A_i^+ at the time t is given as

$$[M^{+\cdot}] = \exp\left(-\sum_{i=1}^{n} k_i t\right) \tag{II-12}$$

---

* This should not be confused with the energy of the bombarding particles.

$$[A_i^+] = \frac{k_i}{\sum_{i=1}^{n} k_i} \left[1 - \exp(-\sum_{i=1}^{n} k_i t)\right] \qquad (II-13)$$

The mass spectrometric time scale can be readily calculated for a given accelerating voltage, geometry of the instrument and mass of the ion. For a Nier type single focussing instrument the residence time of the ions in the source is usually 1 - 5 μs, they traverse the lens system within 1 μs, they remain for 5 μs in the field free region between entrance slit and magnet and arrive at the collector after 8 additional μs.

Finally convolution of the breakdown diagram with the internal energy distribution (Part I-5) leads to the actual mass spectrum, i.e. for competing reactions

$$[M^{+\cdot}] = \int_{E_{o1}}^{\infty} P(E) \exp(-\sum_{1}^{n} k_i t) \, dE \qquad (II-14)$$

$$[A_i^+] = \int_{E_{oi}}^{\infty} P(E) \frac{k_i}{\sum_{1}^{n} k_i} [1-\exp(-\sum_{1}^{n} k_i t)] \, dE \qquad (II-15)$$

2 Experimental Tests of the QET

2.1 Experimental Verifications of the Basic Assumptions of the QET

Since the QET was introduced in 1952, numerous experiments have been designed to test its basic assumptions and to check the quantitative agreement between QET calculations and experimental data. Tests of the basic assumptions and related experiments which allow a qualitative verification of these assumptions will be discussed first. These include studies on the energy randomization hypothesis, isolated electronic states and ion lifetimes. In a second section, quantitative tests of the QET such as a comparison of calculated and experimentally determined mass spectra, breakdown graphs, rate constants and kinetic energy release distributions will be discussed. It will be shown that the conception of such quantitative tests is no easy undertaking. This is not only a result of the limited accuracy of the experiments, but in particular due to the fact that the auxiliary information necessary for QET calculations (activation energies, frequencies, activated complex configurations and energy distribution functions) is in general, if available at all, of limited reliability.

2.1.1 Radiationless Transitions and Energy Randomization

The energy imparted to the molecular ion during ionization is redistributed in the ion mainly via radiationless transitions involving three mechanisms:

(1) *Vibrational relaxation*. Energy exchange within one electronic state (which may be the ground state or an excited state) results from the coupling of the oscillators and is determined by their anharmonicity.

(2) *Vibronic relaxation* (simultaneous electronic and vibrational relaxation). The energy exchange occurs by radiationless transitions via crossings or "avoided crossings" of the potential energy surfaces (see Fig. II-3). The relaxation rate depends on the electronic transition moments, the vibrational overlap integral (Franck-Condon factor, see Section I-1) and the final state degeneracy [76].

If the energy gap between two electronic states is large or the electronic transition is symmetry forbidden, then the vibronic relaxation rate may be slow, leading to "isolated electronic states".

(3) *Isomerization reactions*. Finally vibronic relaxation can also occur by isomerization of a molecular ion, a phenomenon which has to be taken into account when "isolated electronic states" are discussed.

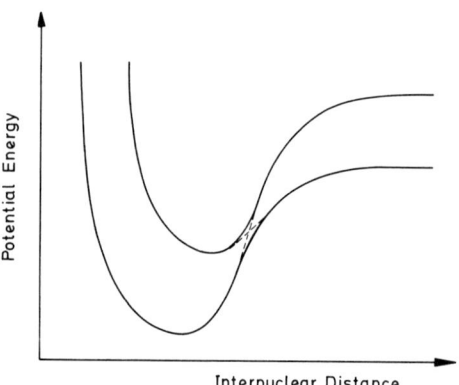

Fig. II-3. Avoided crossing of the potential energy curves of two electronic states of an ion.

Furthermore, energy relaxation may occur by radiative transitions. Whilst in the past[*] it was generally assumed that radiationless transitions in large organic ions are always much faster than radiative transitions[**], several examples of fluorescence from excited organic ions in the gas phase were recently reported by Maier et al. [78-82] including fluorobenzenes [78,82], polyacetylenes [79], 1,3,5-hexatrienes [80] and hexadiynes [81]. Although in most of these cases the ionization potential of the first excited electronic state (from which radiation is observed) is below the threshold for decomposition, in 1,3,5-hexatriene, 2,4-hexadiyne and 1,3-hexadiyne radical cations both the dissociative and radiative channels are in competition. Not only emission spectra but also lifetimes of the excited states have been reported. Surprisingly no fluorescence has been observed so far from excited benzene ions.

One of the basic assumptions of the QET (and RRKM) theory requires that the rate of dissociation is slow as compared to the rate of redistribution of the internal excitation energy. In its original version the QET assumes energy randomization between all electronic and vibrational states, e.g. the existence of isolated electronic states is not taken into account. In its later version energy randomization over all vibrational and rotational degrees of freedom is only assumed within one electronic state. The latter assumption will be discussed first. In recent years several tests of the energy randomization hypothesis have been reported, although in most cases for neutral systems. However, the results should apply qualitatively also to ionic systems. One experiment will be discussed in more detail.

Rynbrandt and Rabinovitch [83] studied the unimolecular decomposition of $[3,3-D_2]$-1-(1,2,2-trifluorocyclopropyl)-1,2,2-trifluorocyclopropane (2) chemically activated to 465 kJ mol^{-1} by addition of $^1CD_2$ to hexafluorovinyl cyclopropane (1) (see Scheme II-4).

Splitting off of CF_2 from the hot molecule from either ring gives tetrafluorovinyl cyclopropane (3). Upon activation, energy deposition occurred in the D-labelled nascent ring (signified by an asterisk). If energy randomization were complete prior to decomposition (3a) and (3b) should be formed in equal amounts. However a slightly preferential non-randomized decomposition of the D-labelled ring was observed to the extent of ∿ 3.5 %. Thus decomposition of the unrelaxed nascent ring com-

---

[*] Until 1975 only the emission spectrum of the diacetylene cation was known [77].

[**] For instance it has been concluded that in the benzene cation the non-radiative decay of the $\tilde{B}^2A_{2u}$ state proceeds with a rate constant of $\geq 10^{12}$ s^{-1} [81].

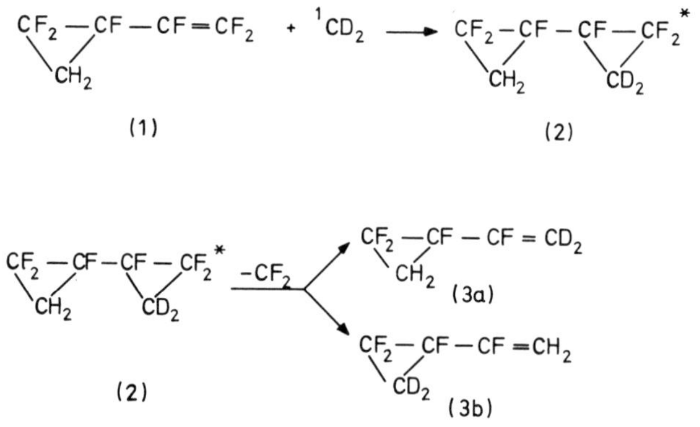

44   II *The Quasi-Equilibrium Theory*

$$CF_2-CF-CF=CF_2 \atop \diagdown CH_2 \qquad + \; ^1CD_2 \quad \longrightarrow \quad CF_2-CF-CF-CF_2^* \atop \diagdown CH_2 \quad CD_2$$

(1)   (2)

$$CF_2-CF-CF-CF_2^* \atop \diagdown CH_2 \quad CD_2 \quad \xrightarrow{-CF_2} \quad \begin{matrix} CF_2-CF-CF=CD_2 \\ \diagdown CH_2 \\ (3a) \\ \\ CF_2-CF-CF=CH_2 \\ \diagdown CD_2 \\ (3b) \end{matrix}$$

(2)

Scheme II-4

petes with intramolecular relaxation at short elapsed times. A rate for intramolecular energy relaxation of 1.1×10^{12} s^{-1} was calculated.

Incomplete energy randomization prior to unimolecular decomposition has also been reported by several other authors [84-93], e.g. using crossed beam studies. Of special interest is the often cited study by Lee et al. [84] of the system

$$C_2H_4^{+\cdot} + C_2H_4 \rightarrow C_4H_8^{+\cdot *} \rightarrow CH_3^{\cdot} + C_3H_5^{+} \rightarrow C_3H_3^{+} + H_2$$

Scheme II-5

However, the conclusion that energy randomization is incomplete in excited $C_4H_8^{+\cdot}$ prior to decomposition has been questioned as a result of recent theoretical calculations by Klots [69] (see Section II-2.2.4).

The above data suggest that complete energy equilibration is only

achieved several picoseconds after ionization*), demonstrating that the
QET assumption that energy randomization is fast as compared to decomposition, does not hold strictly. On the other hand, field ionization
kinetic measurements demonstrate [94] that many unimolecular decomposition reactions reach their maximum intensity only after about ten picoseconds, so that the large majority of ions seem to decompose after complete energy randomization in agreement with the quasi-equilibrium
hypothesis.

2.1.2 Isolated Electronic States

In its original version the QET assumes internal conversion of the initial excitation energy among whatever electronic states might be accessible. This hypothesis would be violated if "isolated electronic states" exist. Fragmentations from isolated electronic states would not be in competition with each other.

The existence of isolated states has been invoked repeatedly to explain unexpected effects in organic mass spectrometry although an unambiguous verification of the existence of such isolated states is difficult. In the following some criteria used in the past to identify isolated electronic states will be discussed critically. The experimental techniques employed in these studies are described in more detail in the following sections.

(1) *Ion lifetimes*. The apparent observation of a single rate constant over a range of ion lifetimes for the process $C_7H_8^{+\cdot} \rightarrow C_7H_7^+$ has been interpreted as proof for an isolated electronic state [99]. However, it has been pointed out that these and similar early ion lifetime measurements are clearly unreliable in view of the limited time range covered [8]. Thus a continuum of rate constants, as predicted by the QET, has been found for the same process in a later experiment [100] (see Section II-2.1.3).

(2) *The observation of metastable ions*. It has been argued [101] that intensive metastable decomposition of ions must occur from the ion ground state as electronically excited states have maximum lifetimes of 10^{-7} to 10^{-8} s. Where no metastable ion is observed an electronically excited state may or may not be involved. However, there is no evidence that metastable ions are representative of all the ions fragmenting at

---

* It has also been demonstrated in a variety of studies of neutral systems that radiationless transitions need not be fast. Rate constants for such vibronic relaxation of the order of several 10^8 s^{-1} have been reported, throwing a strong suspicion on the assumption of energy randomization [95-98].

faster rates [76].

(3) *Differences in heats of formation*. Unexpected differences (up to 1 eV) observed for the heats of formation of the benzoyl ion generated from compounds of the general type C_6H_5COX have been rationalized in terms of "isolated electronic states" [102]. It has however been pointed out that the apparent differences for $\Delta H_f(C_6H_5CO^+)$ (determined by AP measurements) result from variations of the excess energy [103].

(4) *Structure in photoionization yield curves*. Unusual structure in the photoionization yield curve has been tentatively interpreted as resulting from isolated electronic states [104]. A later photoelectron-photoion-coincidence (PEPICO) measurement did not show this structure, suggesting that it arises from autoionization [105].

(5) *Photo-electron-photoion-coincidence (PEPICO)*. A PEPICO study by Simm et al. [106] demonstrated that $C_2F_6^{+\cdot}$ ions in the ground state decompose to give only CF_3^+ while nearly all those in the first excited state give $C_2F_5^+$. The abrupt change in products between the ground state and the first excited state is evidence that the decomposition of the ion in the excited state does not involve prior internal conversion to the ground state, i.e. an isolated electronic state must exist as suggested earlier by Lifshitz and Long [107]. Similarly, isolated states were observed in $CH_3X^{+\cdot}$ (X = F,Cl,I) [108].

(6) *Unusual appearance potentials for metastable ions*. In some systems ($CH_4^{+\cdot}$ [109], $CD_2O^{+\cdot}$ [110]) the threshold energy for metastable decomposition was observed to be considerably in excess (~ 2 eV) of the thermochemical dissociation limit. The phenomenon may result from dissociation of an excited electronic state.

(7) *Direct determination of k(E)*. The direct determination of k(E) by ion lifetime measurements of energy-selected ions represents the most stringent test for the existence of isolated electronic states. Thus in a PEPICO study Baer et al. [111,112] observed several unimolecular reactions (loss of HCl from chloroethane and 1,2-dichloroethane and Cl loss from propargyl chloride) which proceed at two distinct rates (for a given internal energy) to form identical products whereas according to the QET a given fragmentation reaction at a certain internal energy should be described by a single rate constant. Their observation can be explained by assuming (1) two isolated long-lived states or (2) isomerization prior to decomposition. Although it is difficult to differentiate between these two cases, the authors assume that isolated states are involved in the HCl elimination from mono-and dichloro-alkanes while isomerization (to a cyclic structure) occurs prior to Cl loss from propargyl chloride.

Similarly, Andlauer and Ottinger [113,114] observed that the rate constants for loss of H and C_2H_2 from the benzene molecular ion differ by several orders of magnitude at higher internal energies, demonstrating that both reactions cannot be in competition with one another (see Section II-2.2.3). Their conclusion is supported by photo-ionization studies by Rosenstock et al. [115,116], PEPICO studies by Eland et al. [117,118] and MO considerations by Jonsson and Lindholm [119]. Also in the case of the benzene molecular ion it is difficult to distinguish between fragmentation from isolated electronic states and isomerization processes. Thus the molecular ion of benzene and hexadiene-[1,3]-yne-5 seem to isomerize to a mixture of interconverting structures prior to decomposition [120,121].

(8) *Fluorescence spectra*. The emission spectra observed by Maier et al. for several organic ions (see Section II-2.1.1) demonstrate that in some cases the first excited electronic state may be stable for several 10^{-8} s [78-82].

(9) *Photodissociation*. Photodissociation of the butyrophenone cation, $C_6H_5COC_4H_9^{+\cdot}$, leads to the exclusive formation of $C_4H_9CO^+$ at photon energies up to 3.3 eV, but to ethylene elimination (Norrish type II reaction) at higher energies pointing to a *state specific reaction* [122]. However, a later collisional activation study gave no evidence to support the existence of an isolated electronic state in the precursor ion [262].

Summarizing, the examples discussed demonstrate that many phenomena which where originally rationalized by assuming isolated electronic states can be explained without the intervention of such states. There is, however, growing evidence that isolated electronic states do exist in some systems such as $C_2F_6^{+\cdot}$ [106], $CH_3X^{+\cdot}$ (X = F,Cl,Br,I) [108] and probably also in $C_2H_5Cl^{+\cdot}$ [111], $C_2H_4Cl_2^{+\cdot}$ [111], $C_3H_3Cl^{+\cdot}$ [111] and $C_6H_6^{+\cdot}$ [113-119]. In many cases it is, however, difficult to differentiate unambiguously between isolated electronic states and isomeric forms of the same molecular ion.

Although in the original formulation of the QET internal conversion of all ions to the electronic ground state was assumed, it was later pointed out by Rosenstock [115] that this is not a central assumption of the QET. If isolated electronic states are present in a molecular ion each state can be described by the QET separately. This is a complication but not a failure of the theory.

2.1.3 Ion Lifetimes

Lifetime measurements of ions with a distribution of internal energies will be reported first. Such measurements have been carried out for more than three decades with the purpose of testing one of the basic assumptions of the QET, the existence of a continuum of rate constants.

Such a continuum of rate constants follows immediately from eq. II-1 if there is a continuous distribution of internal energies. On the other hand a few discrete ion lifetimes would indicate a few discrete reacting energy states (or, perhaps, a very sharply spiked energy deposition function).

2.1.3.1 The Principle of Ion Lifetime Measurements

Four principal approaches can be used to measure mass spectrometric fragmentation processes as a function of the ion lifetime:
(1) The ions are accelerated by a constant potential U_o to a final velocity

$$v_o = (2eU_o/m)^{1/2} \qquad (II-16)$$

before they enter a field free drift tube. By sampling the ions at various distances, Δx, within the drift tube or by using drift tubes of different lengths the residence time in the field free region can be varied.
(2) Alternatively, the flight path in the field free drift region may be kept constant while the acceleration voltage U_o and, thus, the maximum final velocity, v_o, are varied.

With both methods the field free drift region may be at ground potential or at a constant potential U. The decay is integrated over the time interval $\Delta t = \Delta x/v_o$. The shortest resolvable time with this method is that necessary to accelerate the ions to their final velocity, v_o, and will, in general, be of the order of 10^{-7} to 10^{-6} s.
(3) The time which elapses between the formation of an ion within the ionization zone and its entrance into the field free drift tube can further be varied by using a pulsed ion source where the electron beam and the acceleration voltage are pulsed alternately and the variable delay between both pulses, Δt, determines the lifetime of the ions before acceleration.
(4) Shorter ion lifetimes are resolvable if the decay within the acceleration field can be determined. This is simple as ions decomposing within the acceleration field have a deficit of translational energy compared to the non-decomposing ions, as illustrated by Fig. II-4. If an ion of mass m_1 is decomposing in the acceleration field at the position x (corresponding to a potential U_x) according to

$$m_1^+ \rightarrow m_2^+ + m_3$$

it has a translational energy $e(U_o - U_x)$ which is partitioned between the ionic and the neutral fragment according to their mass ratio, the

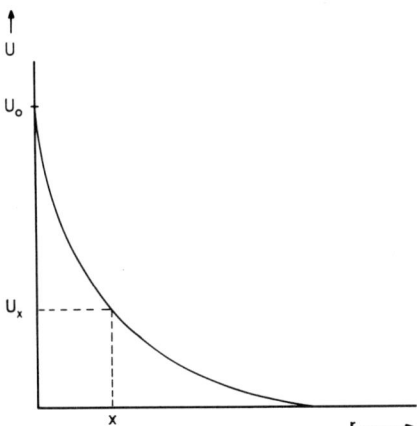

Fig. II-4. Schematic inhomogeneous potential distribution in an ion source used for lifetime measurements.

ionic fragment having an energy $e(U_o - U_x)m_2/m_1$. While passing through the remaining potential U_x the fragment ion reaches a final translational energy eU

$$eU = e(U_o - U_x)m_2/m_1 + eU_x \leq eU_o \qquad (II-17)$$

which is smaller than that of non-decomposed ions (eU_o). The energy deficit $e(U_o - U)$ can readily be measured with a magnetic, or preferably an electrostatic sector field, and allows the immediate calculation of the decomposition potential U_x and, if a homogeneous acceleration field is used, the determination of the decomposition time, t. If an inhomogeneous electric field is used the potential distribution within the field must be known. The decomposition time is given as

$$t = \int_o^x \frac{dx}{v} = \int_o^x \frac{dx}{\left[\frac{2e(U_o - U_x)}{m_1}\right]^{1/2}} \qquad (II-18)$$

The resolvable ion lifetimes are the shorter the higher the initial potential gradient of the acceleration field, the narrower the ionization zone and the better the energy resolution of the instrument.
While with the first three methods the decay is integrated over

relative large lifetimes*⁾ the fourth method samples the decomposition within extremely short time intervals and thus represents a differential method.

2.1.3.2 Long Ion Lifetimes (t > 10^{-7} s)

As outlined above the methods (1) to (3) permit only the study of relatively long ion lifetimes (> 10^{-7} s). The first lifetime measurements date back to Hipple in 1947 (i.e. before the QET was established). Since then various modifications of the methods (1) to (3) (outlined in the previous section) have been employed to study the mass spectrometric fragmentation as a function of time. Thus the variation of the repeller voltage (draw out field) in conventional single focussing mass spectrometers (method 2**⁾) has been used repeatedly [125-130], allowing in favorable cases the resolution of a time interval of 0.3 - 3 x 10^{-6} s [127]. This time interval can be extended to larger values by applying the pulse technique (method 3) which has been used in conjunction with electron impact [99,100,131,132] and photoionization [133-135] coupled to conventional magnetic type instruments [99,100,131,132], quadrupole mass filters [134,135] and time-of-flight mass spectrometers [136]. Using this technique a time interval of 1 - 20 μs was resolved by Lifshitz [100].

Cycloidal mass spectrometers have been used in two instances for ion lifetime measurements [137,138]. Here a delayed decomposition outside the ion source on a cycloidal orbit leads to an apparent mass, m*, which depends on the orbital position and thus the decomposition time. Such measurements are based on method (1) (variation of the flight distance) although a differential measurement of the fragment intensity as function of time is possible in this case.

Modified time-of-flight instruments can also be used for ion lifetime measurements [136,139,140]. Usually ions decomposing within the drift tube of such an instrument continue their flight path with constant velocity and cannot be differentiated from non-decomposing ions. Introduction of a flat-top potential barrier, however, allows the separation of parent ions, fragment ions and neutrals. The neutrals are not affected by the deceleration barrier and arrive at the detector first, followed by the parent and the fragment ions which are succes-

---

*
 With these methods the "metastable" ions are sampled as a function of the ion lifetime.
**
 Note that in this case, in contrast to method (2), not the total acceleration voltage, but only the repeller voltage is varied.

sively shifted to even longer flight times (see Fig. II-5). The reaction time can be varied by changing the location of the potential barrier [139,140] (method 1) or by delaying the pulse which extracts the ions from the source with respect to the ionizing pulse (method 3)[136].

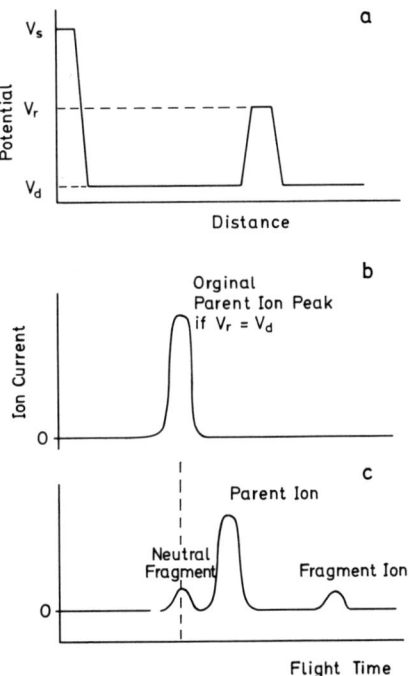

Fig. II-5. Potential distribution in a time-of-flight mass spectrometer modified for ion lifetime measurements (a) [139,140]. Without potential barrier parent ions, fragment ions and neutrals reach the detector after the same flight time (b). Introduction of a flat-top potential barrier allows the separation of parent ions, fragment ions and neutrals according to their flight time (c). (By Courtesy of the American Institute of Physics).

The most severe shortcoming of all these methods is that only a rather small time interval can be studied. A considerable extension of the time scale to longer times has been achieved by Tatarczyk and v. Zahn [141-143]. The instrument used by these authors is shown schematically in Fig. II-6. Ionization and mass selection occurs in a small double focussing mass spectrometer followed by a 175 cm long drift tube. A radial quadrupole d.c. field stabilizes the ion trajectory within this tube. Ions decomposing within this drift region (consisting of several segments) are analyzed by a quadrupole mass filter. The residence time can be either varied discontinuously by changing the length of the drift tube (method 1) or continuously by decelerating the ions in one

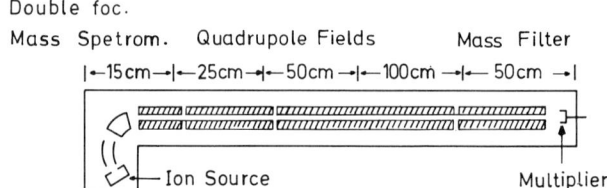

Fig. II-6: Tandem Dempster-quadrupole mass spectrometer with a 175 cm drift tube for the study of long ion lifetimes (t > 10^{-6} s) [141-143]. (By courtesy of Zeitschrift für Naturforschung.)

or several segments of the drift tube. Using this technique a lifetime window from 5 - 500 μs is covered. A similar experimental arrangement which, however, uses a tandem Dempster-ion cyclotron resonance (ICR) instrument has been employed recently by Smith and Futrell [144-147].

Results. Using the above techniques several fragmentation reactions mainly of the molecular ions of alkanes, toluene and benzonitrile have been studied repeatedly. Results are usually represented as semilog plots which should give curves with straight segments for one or a few distinct rate constants, but an exponentially declining slope, if a continuum of rate constants is involved. The actual results especially of the early lifetime measurements are confusing: While most authors who used variations of the draw out field, cycloidal mass spectrometers and pulsed ion sources observed one or several distinct rate constants [99,125-127,129,131,134,137], others using the same technique for the same fragmentations reported a continuum of rate constants [100,128, 136,138].

This demonstrates that most early ion lifetime measurements are clearly unreliable as a result of the finite time interval resolved in these measurements. Thus the measurements of Tatarczyk and v. Zahn [141-143] are of special importance as their experiments covered a relatively large time scale. Their results, represented in a double logarithmic plot in Fig. II-7, demonstrate unambiguously the existence of a continuum of rate constants [143]. Satisfactory agreement with the predictions of the QET has also been reported by other authors [100, 128]. In this context the study of Ryan et al. [136] is of special interest. In this work the calculated and measured metastable abundances are compared as a function of time. Although some discrepancies between experimental and calculated data were observed the authors concluded that the QET provides a satisfactory rationale for their results.

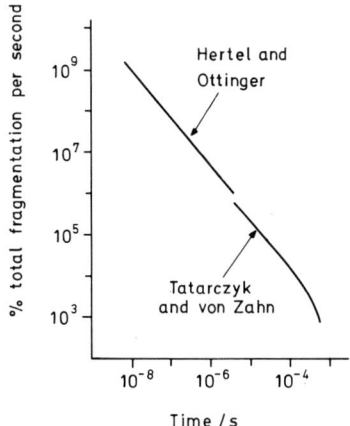

Fig. II-7. Comparison of rates for the reaction $C_5H_{11}^+ \to C_3H_7^+$ in n-heptane measured by Hertel and Ottinger [152] and Tatarczyk and von Zahn [143]. (By courtesy of Gordon and Breach.)

2.1.3.3 Short Ion Lifetimes ($t < 10^{-7}$ s)

An extension of the resolvable ion lifetime window to shorter times is especially interesting as one is approaching the ionization process itself. Shorter ion lifetimes ($t < 10^{-7}$ s) can only be resolved if the ions are accelerated by a high electric field immediately after their formation and if the decay within the acceleration field is determined as outlined in Section II-2.1.3.1. Such measurements have been first reported by Karachevtsev and Tal'rose [148] and Osberghaus and Ottinger [149] using electron impact ionization and by Beckey [150] using field ionization.

The main experimental difficulty does not arise from the high electric field necessary for a good time resolution but from the necessity of forming the ions in a very narrow region within this field so that the potential drop within the ionization zone is small compared with the total acceleration potential. These experimental difficulties have been overcome by Osberghaus and Ottinger [149], Ottinger [17,151, 153], and Hertel and Ottinger [152] using a molecular beam with a half width of only 11 μm. Their instrumental arrangement is shown in Fig. II-8a. By applying a potential of 10 kV to the anode (AC_+) with respect to the counterelectrode (AC_-) a homogeneous electric field of 10^4 V/cm is generated between AC_+ and AC_-. The molecular beam is formed by two concentric slits S_1 and S_2 and traverses the electric field on an equipotential surface while electrons are shot into the molecular beam

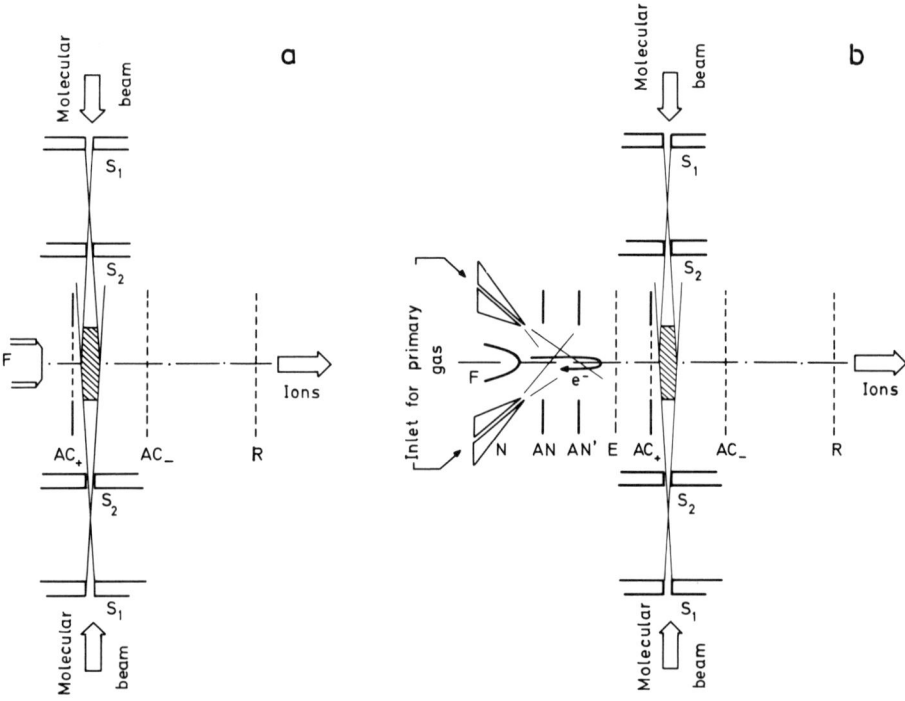

Fig. II-8. Electron impact ion source used by Ottinger [17,114] for ion lifetime measurements in the nanosecond range. (a) Ionization by electron impact (F = filament, AC_+, AC_- = ion acceleration electrodes, R = ion retardation electrode, S_1, S_2 = beam slits). (b) Ionization by charge exchange (AN, AN' = anodes, E = electron barrier electrode, N = primary gas inlet, other symbols as in (a). (By courtesy of Zeitschrift für Naturforschung.)

through fine holes in the anode. Fragment ions formed within the electric field have a deficit of translational energy compared with ions which do not decompose (eq. II-17 in Section II-2.1.3.1). This deficit is measured by an electrostatic sector field followed by a magnetic field for mass analysis. From the energy deficit the decomposition time can be determined using equation II-18. The actual experimental procedure corresponds to that used for the detection of metastable peaks in a double focussing mass spectrometer (Barber-Elliot technique [154]): after the electrostatic and magnetic sector have been adjusted to pass the desired fragment the acceleration voltage is increased stepwise until the metastable peak is observed at $U = m_1/m_2 U_o$ (m_1 = precursor ion, m_2 = fragment ion) leading to a peak profile shown in Fig. II-9. Fragment ions formed within the molecular beam contribute to a peak observed at the same potential as the non-decomposed molecular ion. Decomposition within the acceleration field leads to the peak tailing which

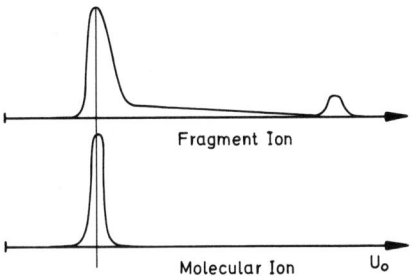

Fig. II-9. Typical peak shapes (ion intensity as a function of the acceleration potential) for a molecular and a fragment ion observed in Ottinger's experiments [17].

extends to the metastable peak. Using this technique a time range of 3×10^{-9} to 1×10^{-6} s can be covered. Fig. II-10 shows the normalized rates, di/dt, for methyl loss from the n-butane ion as a function of time in a semilog plot [17]. The exponentially declining slope shows unambiguously the existence of a continuum of rate constants predicted by the QET if there is a continuous distribution of internal energies.

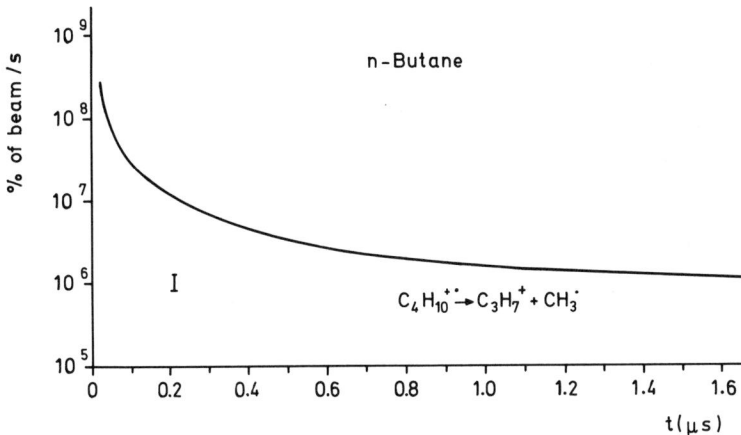

Fig. II-10. Rate for methyl loss from the n-butane molecular ion as a function of the ion lifetime according to Ottinger [17]. The error bar shows the reproducibility of the measurements.

Such continuum of ion lifetimes is also apparent from Fig. II-7 for the process $C_5H_{11}^+ \rightarrow C_3H_5^+$. In this figure the data of Hertel and Ottinger [152] and Tatarcyk and von Zahn [143] are contrasted demonstrating that these results covering different time ranges supplement each other with surprisingly good consistency.

Brief mention should be made of an ingenious, but little used method for ion lifetime determinations described by Karachevtsev and Tal'-rose [148,155]. The authors used an inhomogeneous electric field of 10^4 V/cm produced by applying a potential of 2 kV to a thin filament positioned in the center of a cylindrical counterelectrode. Only ions formed by the bombarding electrons in the vicinity of the filament are accelerated along radial trajectories towards the exit slit and thus provide the narrow ionization zone necessary for a good time resolution. Lifetimes of the order of a nanosecond have been resolved with this method.

Extension of the time scale to even shorter times necessitates the use of much higher initial field gradients and even better defined ionization zones. Both conditions are readily fulfilled under field ionization (FI) conditions where field strengths of the order of 10^7 to 10^8 V/cm are used to ionize organic molecules and the ionization zone is confined to a few Å by the ionization mechanism itself (quantum mechanical tunnelling). The basic principles of field ionization and field ionization mass spectrometry have been reviewed in detail in various publications [156-165] and will not be described here. The necessary field strengths are achieved by applying a potential of 10^3 to 10^4 V to a strongly curved metal surface (point, thin wire or blade). Using this technique ion lifetimes as short as 10^{-11} s*) can be resolved and the whole spectrum of ion lifetimes down to 10^{-5} s after ionization can be measured [164-165]. If a double focussing instrument is used the procedure for measuring ion lifetimes corresponds to that used by Ottinger (vide infra) i.e. the Barber-Elliott[154] or Major [166] technique for the detection of metastable ions in conventional double focussing instruments is employed. Unfortunately little energy is transferred during the ionization process, so that in general only processes with low activation energy ($\leq$ 2 eV) can be studied.

Fig. II-11 shows the relative rate, di/dt**), for ethyl loss from the n-heptane molecular ion as function of time from 2×10^{-11} to 2×10^{-6} s in a double logarithmic plot. The results give unambiguous evidence for the existence of a continuum of rate constants over the entire time range (a single rate constant would lead to a horizontal line). Fair agreement is observed with Ottinger's results (dashed line). The larger slope of Ottinger's decomposition curve results from differ-

---

* Even shorter ion lifetimes ($\leq 10^{-12}$ s) can be resolved if a point emitter is used.
** Normalized to the intensity of the molecuar ion.

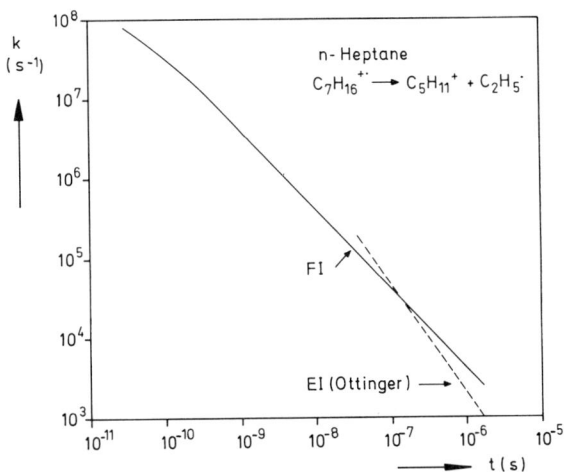

Fig. II-11. Rate for ethyl loss from the n-heptane molecular ion (solid line: field ionization [165], dashed line: electron impact ionization [151].

ences in the internal energy distribution: under EI conditions ions of high internal energies (E > 1 eV) prevail so that the fraction of ions with high rates is larger under EI than under FI conditions.

Summarizing, the ion lifetime measurements of Tatarczyk and v. Zahn [141-143], Ottinger [17,151-153] and Beckey and Tenschert [164-165] clearly demonstrate that there exists a continuum of rate constants and thus a continuous distribution of energies. The almost straight lines observed for the relative rates in the double logarithmic plots (Fig. II-7 and II-11) suggest a smooth energy deposition function. However, the limited reproducibility of such ion lifetime measurements does not allow the detection of a possible fine structure in the energy deposition function. In conclusion it should be stated that ion lifetime measurements constitute only a qualitative test of the QET. A quantitative test of the QET is possible with ion lifetime measurements of energy selected ions. Such measurements have recently been reported and are discussed in Section II-2.2.3.

2.2 Quantitative Tests of the QET

While the discussion of the experimental results in the previous section showed that in general the QET describes the decay of excited organic ions at least qualitatively correctly, it has been repeatedly

II The Quasi-Equilibrium Theory

questioned whether quantitative agreement can be achieved between theory and experiments. These doubts were based largely on the observed discrepancies between experimental and calculated breakdown curves where calculations employed the classical approximation of the QET. It will be shown in the following how far the use of better approximations for the enumeration of states removes this discrepancy. Propane will be used as model compound where data applying to it are available.

2.2.1 Mass Spectra

It was the original goal of the QET to calculate 70 eV mass spectra and to demonstrate that reasonable agreement with experimental data could be achieved. The mass spectrum of propane was first calculated by Rosenstock et al. [2] 1952 using the classical approximation. This molecule was considered large enough to represent a "real" organic molecule, while its physical properties (vibrational frequencies, activation energies, moments of inertia) were still well known. Although the original calculations by Rosenstock et al. [2] have been superceded by more refined ones (vide infra) they shall be briefly discussed as some important conclusions can be derived from these results. Table II-1 compares the

Table II-1. Experimental and Calculated Propane Spectrum (Using the Classical Approximation)[a]

Ion	$C_3H_8^{+\cdot}$	$C_3H_7^+$	$C_3H_5^+$ / $C_3H_3^+$	$C_3H_6^{+\cdot}$ / $C_3H_4^{+\cdot}$	C_3H^+ / $C_3H_2^{+\cdot}$	$C_2H_5^+$	$C_2H_4^{+\cdot}$	$C_2H_3^+$ / $C_2H_2^{+\cdot}$
Calculated	0.102	0.064	0.059	0.062	0.249	0.161	0.272	0.031
Experimental	0.090	0.071	0.103	0.042	0.310	0.183	0.122	0.027

[a] $P(E)$ = constant

experimental and calculated mass spectrum of propane, assuming a uniform energy distribution from 0 - 12.5 eV (i.e. $P(E)$ = constant). Although the calculations are based on the classical approximation of the density of states (equation II-4) which has been shown to be a poor approximation, and the assumed energy distribution is completely unrealistic the agreement between experiment and theory is surprisingly good: only for two fragments do the relative intensities deviate by more than $\pm$ 20 % which is the usual reproducibility of EI mass spectra obtained with different instruments. Two conclusions can be drawn from this result:

(1) The relative abundance of fragments in a 70 eV EI spectrum reflects variations in the internal energy distribution only to a minor extent as result of the integration over a large energy interval. (On the other hand, variation within the internal energy distribution may have a large effect on the relative fragment abundances if ions within a narrow energy interval (e.g. metastable ions) are sampled).

(2) The comparison of the experimental and calculated mass spectra does not represent a very stringent test of the QET.

The 50 eV spectrum of propane has been recalculated by Vestal [167] using eq. II-7 for the enumeration of states [29], a refined breakdown scheme, more reliable values for the activation energies, and an energy distribution function close to that reported by Chupka and Kaminsky [168] (Fig. I-9) and assuming a linear threshold law [169].

Table II-2. Experimental and Calculated Propane Spectrum (Using Vestal's Approximation)[a]

Ion	Calc.	Observed (API)[b]		
		No. 3	No. 61	No. 112
$C_2H_2^{+\cdot}$	2.9	2.7	1.7	3.0
$C_2H_3^+$	13.2	13.5	10.1	13.9
$C_2H_4^{+\cdot}$	19.9	19.9	19.7	19.5
$C_2H_5^+$	33.0	33.4	32.0	33.2
$C_3H_3^+$	3.5	5.8	5.3	6.3
$C_3H_4^{+\cdot}$	0.1	0.9	0.9	0.9
$C_3H_5^+$	5.2	4.3	4.8	4.4
$C_3H_6^{+\cdot}$	1.9	2.0	2.0	2.0
$C_3H_7^+$	11.5	7.7	10.8	7.7
$C_3H_8^{+\cdot}$	8.8	9.8	12.9	9.1

[a] Calculations are based on a temperature of 250 °C

[b] The numbers correspond to the API catalogue

Table II-2 contrasts his results with published propane spectra (API - catalogue [170]). Good agreement is observed with serial no. 3 and 112 except for the abundance of the propyl ion which is about 50% too high. However, the table also demonstrates the limited reproducibility of experimental data, which is a further reason why the calculation of complete mass spectra does not represent a good test of the theory.

A more stringent test of the QET is possible if the mass spectrum is calculated as a function of the electron energy, as such calculations allow verification of whether correct abundance ratios are obtained in the critical threshold region. Fig. II-12 compares the variation of the relative abundances in the EI spectrum of propane, as calculated by Vestal [167], with experimental data. The calculations are again based on Vestal's approximation for the density of states [29]. The agreement between theory and experiment is excellent except for the $C_3H_7^+$ ion the abundance of which is again too high especially at low

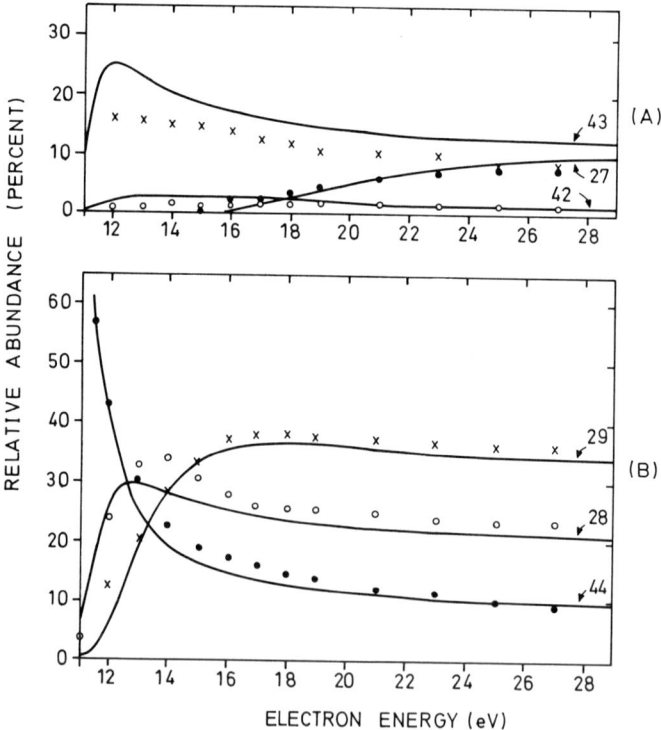

Fig. II-12. The variation of the propane mass spectrum with electron energy. Comparison of calculated values (solid line) with experimental results [167]. (By courtesy of the American Institute of Physics).

electron energies. The discrepancy observed for the propyl ion may result from a small error in the value of the appearance potential or from inaccurate assumptions about the activated complex configuration ($C_3H_7^+$ ions may have the structure of a protonated cyclopropane molecular ion instead of a linear structure [171,172,263]).

Finally Vestal also calculated the temperature dependence of the relative fragment intensity using the assumptions discussed above [167].

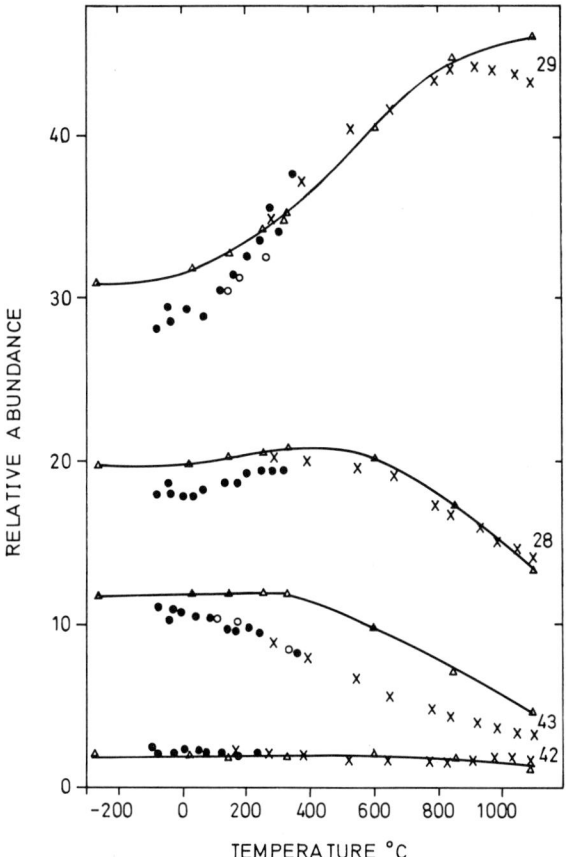

Fig. II-13. The variation of the 50 eV mass spectrum of propane with temperature [167]. Comparison of calculated data (solid line and triangles) with experimental values (other data points). (By courtesy of the American Institute of Physics.)

Fig. II-13 demonstrates that reasonable agreement is observed between calculated and experimental data [173-175] with the exception of the propyl ion where the trend is correctly represented while the absolute values of the calculated data are too high probably as a result of inaccuracies in the parameters used in the rate calculations. The result supports the assumption that the total thermal energy is transferred from the molecule to the ion and is effective in producing dissociation of the molecular ion [176].

The mass spectra of a variety of smaller molecules have been calculated by other authors [177-185]. Although the classical approximation for the density of states was used in most instances, in general fair agreement between experimental and theoretical data was observed. In spite of this apparent success it has, however, been questioned whether the QET can be applied equally well to more complex molecules for instance to those containing several functional groups [186]. The EI induced fragmentation of such compounds is usually rationalized by organic mass spectrometrists using the concept of charge (or radical site) localization [187] (see Section IV-1.4). Thus the failure of the QET might be expected for compounds in which the charge is predominantly localized. Therefore calculations by Yeo and Williams [188-191] and Heller et al. [192] are of interest. The authors chose a variety of mono- and bifunctional aromatic compounds, the molecular ions of which either decompose predominantly by two competing or by two consecutive reactions and calculated the intensity as a function of the electron energy. The authors used a modified version of the classical approximation in which the total number of oscillators was replaced by an effective number which increases linearly with rising electron energy. Thus the rate equation has the general form:

$$k = \nu \left(\frac{E - E_o}{E} \right)^{(N-1)x} \qquad (II-19)$$

where $x = 0.2 + 0.03 (E - E_o)$.

The same equation is used both for direct bond cleavages and rearrangement reactions. Furthermore an energy distribution similar to that determined by Chupka and Kaminsky for propane [168] has been used and equipartitioning of the energy during fragmentation assumed.

In a study of competing direct cleavages and rearrangement reactions [189] the frequency factor for the rearrangement reaction was fitted to give optimum agreement between theory and experiment. However, this procedure leads to unreasonably low frequency factors in some instances (e.g. $\nu = 3 \times 10^6$ s^{-1} for nitrobenzene). More convincing are calculations of consecutive fragmentations, where the frequency factor is chosen a priori, although arbitrarily ($\nu = 4 \times 10^{13}$ s^{-1} for direct bond cleavages and $1 \times 10^9 - 3 \times 10^{10}$ s^{-1} for rearrangement reactions)*). The calculated intensities for competing fragmentations from methylsalicylate [189] and from benzophenone [192] are contrasted with experimental data in Fig. II-14. In view of the arbitrary assumptions

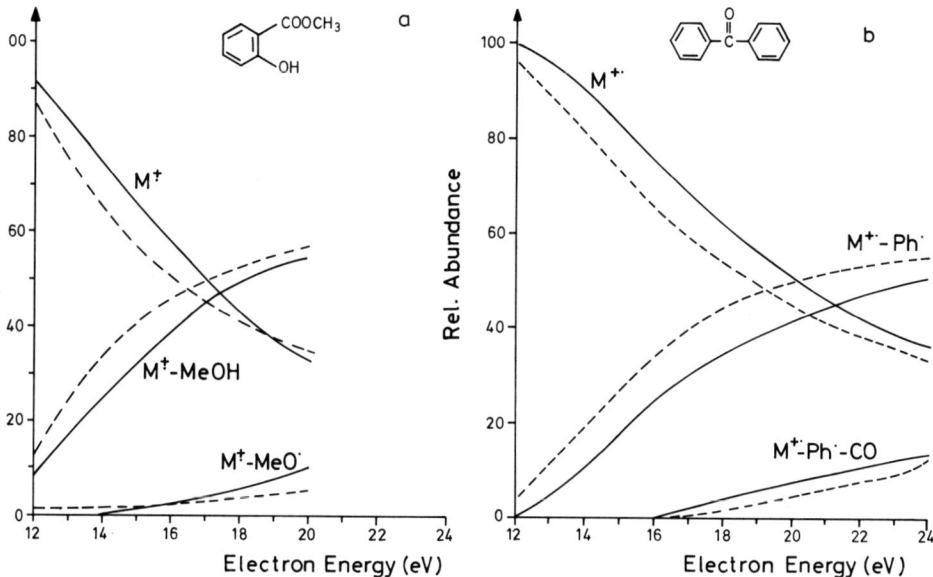

Fig. II-14. The variation with electron energy of the relative abundances for two competing decompositions from the methylsalicylate (a) [189] and benzophenone (b) molecular ion [192]. Comparison of calculated data (solid line) with experimental results (dashed line).

about the energy distribution function and the frequency factors and the poor approximation used for the density of states the agreement between calculated and experimental data is surprisingly good. Even the calculated intensities of the metastable ions are qualitatively consistent with measured values. Better agreement would certainly be possible if the molecular ion parameters were better known and if Vestal's approximations [29] were used for the calculations.

Although the results should be interpreted with caution in view of the arbitrary choice of the parameters they seem to support the premise that the fragmentation behavior of complex organic molecules with partial or predominant localization of the charge can also be readily explained within the framework of the QET.

---

* This procedure is justified as the calculated abundances are not very sensitive to the frequency factors chosen [190].

2.2.2 Breakdown Curves

In the calculation of the mass spectra the intensity of a given fragment is obtained by integration over a large range of energies. Such integrated values are relatively insensitive to the exact choice of the parameters (activation energies, frequencies) and the correctness of the applied rate equation. This is not the case if breakdown curves are determined. As outlined earlier breakdown graphs reflect the extent and variety of ion fragmentation as a function of the internal energy within 10^{-5} s after ionization. As such they give the exact branching ratios of all competing fragmentations at a given internal energy, and these depend sensitively on the parameters and rate equations employed.

Moreover, breakdown curves are especially suitable for comparison of experimental and calculated data, as a knowledge of the, in general, poorly defined internal energy distribution is not required. Some experiments, such as charge exchange and photoelectron-photoion-coincidence, allow the direct determination of breakdown graphs and emphasize the usefulness of such graphs for a quantitative test of the theory. Breakdown graphs can also be determined experimentally although indirectly from electron impact and photoionization measurements. In the following, breakdown graphs of the propane molecular ion obtained using the four techniques just mentioned will be compared with calculated ones.

2.2.2.1 Electron Impact

Breakdown graphs can be obtained from electron impact data by forming the second derivative of the ion current of each fragment with respect to the energy of the bombarding electrons and normalizing the individual curves.

Breakdown graphs from EI data were first determined by Chupka and Kaminsky [168] assuming that the threshold laws are valid (see Section I-3.1) and autoionization is unimportant. In Fig. II-15 their results (c) are contrasted with calculated graphs of Rosenstock et al. [2] (a) using the classical approximation, and Vestal [167] (b) using Vestal's approximation [29] for the enumeration of states. Although the breakdown curve calculated with the classical approximation [2] shows a qualitative similarity with that determined by electron impact, a quantitative comparison does not only reveal significant deviations in the shape of the individual curves, but also a striking discrepancy in the energy scales which differ by a factor of about two. A similar discrepancy in the energy scale between experimental and theoretical breakdown graphs was also observed with other compounds [177,193] and is one

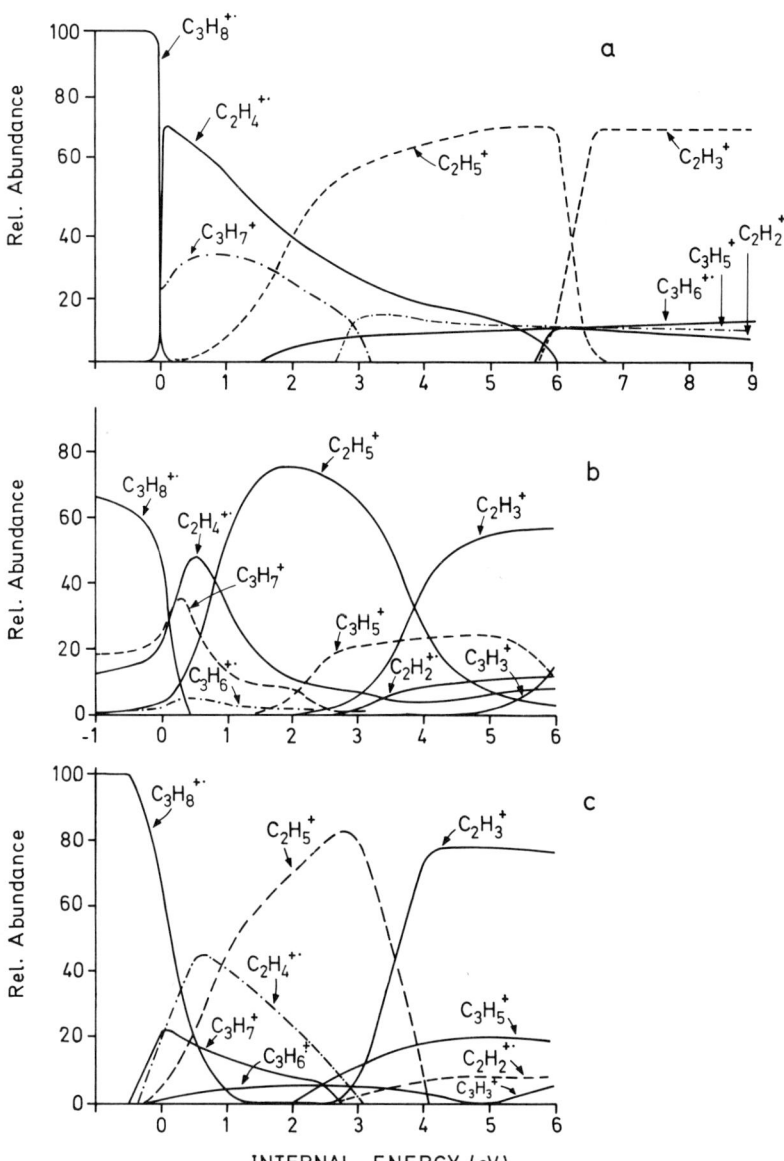

Fig. II-15. Breakdown graphs for the propane molecular ion. Comparison of calculated graphs using the classical approximation (a) [2] and Vestal's approximation (b) [29] with experimental results obtained by electron impact measurements (c) [168].

of the reasons why the validity of the QET was repeatedly questioned within the first decade after its introduction. That the failure of the theory to make quantitative predictions did not result from unsound basic assumptions, but rather from inadequate mathematical approximations was, however, overlooked by the critics.

Thus the breakdown graph of propane was recalculated by Rosenstock [5] and Vestal [167] using an improved enumeration of states and considering inter alia the fluctuation of energy between ionic and neutral fragments during decomposition [264]. The breakdown graph calculated by Vestal is shown in Fig. II-15b. The agreement, not only in the energy scale, but also in the shape of the individual curves is much better. The remaining differences are not necessarily due to inadequate calculations, but may in part reflect experimental errors, especially as the second derivative is formed from a quantity which can only be determined with limited reproducibility (non-monoenergetic electrons were used). Note however that the calculated intensity of the propyl ion is again too high, a phenomenon already discussed in the previous sections.

2.2.2.2 Photoionization

More reliable experimental breakdown curves should be obtainable using photoionization as on the one hand monoenergetic photons are used and on the other hand only the first derivative of the photoionization efficiency curve needs to be determined. Fig. II-16 contrasts the experimental breakdown curve determined by Chupka and Berkowitz [172] with Vestal's calculations [167] in the energy range from 11 to 14 eV. The major discrepancy is again found in the behavior of the curve for the $C_3H_7^+$ ion which, as already discussed in the previous section, obviously results from the use of inadequate parameters for the calculation of this fragment. Using photoionization, breakdown graphs for a large variety of alkanes have been determined by Steiner et al. [194] for energies up to 12 eV while theoretical calculations of some of these breakdown graphs based on Vestal's approximation [29] have been reported by Tou et al. [195]. Again good agreement between theory and experiment has been observed with the exception of metastable ions, the calculated abundance of which is often too high while the so called "missing metastables" (decay within the lens system and the magnetic sector) are too low in abundance.

Breakdown graphs obtained by photoionization have also been reported for ethane [172], butane [172], methanol [104], ethanol [104], propanol [104], isopropanol [104], dimethylether [196] and diethylether [196] as well as for ketones [197]. Where calculations using improved rate equations are available agreement between experiment and theory is good.

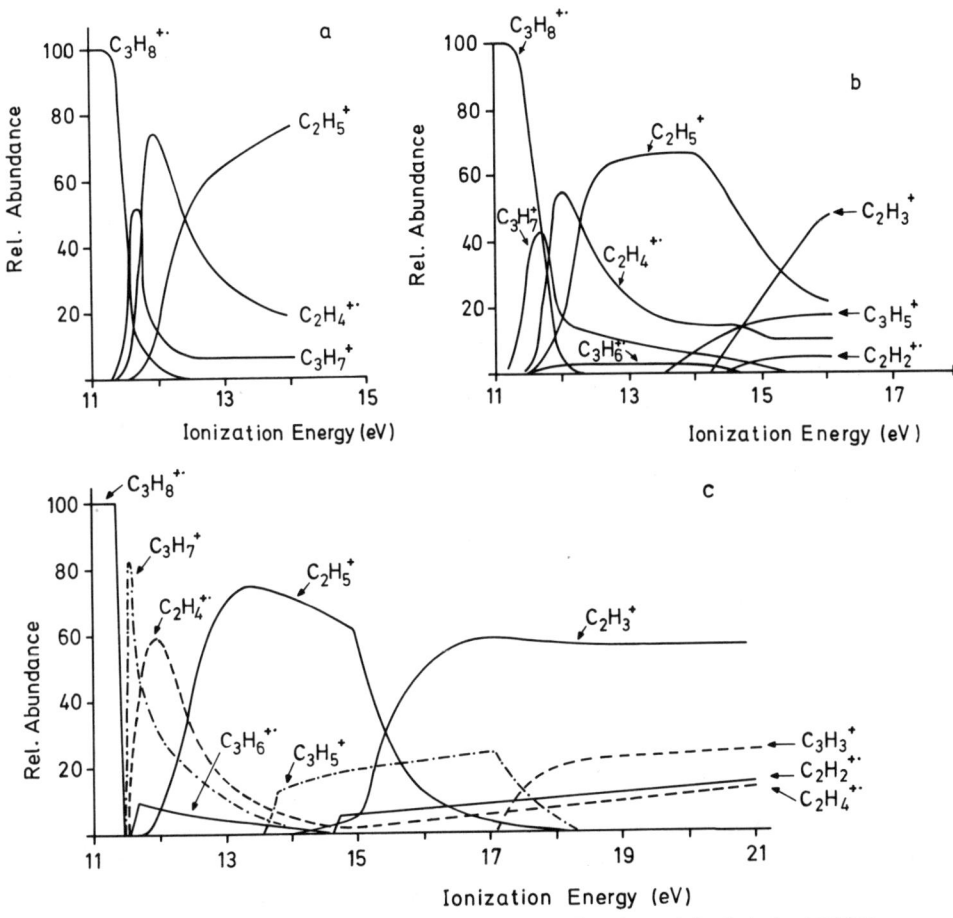

Fig. II-16. Breakdown graphs for the propane molecular ion; (a) photoionization [172]; (b) photoelectron-photoion-coincidence [222]; (c) calculations [167].

2.2.2.3 Charge Exchange

While breakdown graphs from electron and photon impact experiments can only be obtained indirectly via the first or second derivative of the individual ionization efficiency curves both charge exchange and photoelectron-photoion-coincidence experiments give direct information on the breakdown graph. If a molecule is ionized by charge transfer the difference between the recombination energy, RE, of the primary ion and the ionization potential, IP, of the molecule to be ionized, i.e. E = RE - IP, is deposited onto the molecular ion as excitation energy. Thus the internal energy of the molecular ion is known and can be varied by using different primary ions. Hence the charge exchange spectrum normalized to the total ion current immediately represents the breakdown diagram of the compound under study. Charge exchange spectra were first reported by Lindholm and coworkers [198-205]. They used a tandem mass spectrometer. The primary ions are mass selected in a magnetic sector field and retarded in a system of electrostatic lenses down to the desired translational energy (10 - 900 eV) before they enter a collision cell filled with the compound under study at a pressure of 1 x 10^{-4} torr. To ensure that only charge transfer processes with little or no transfer of kinetic energy are studied the secondary ions are extracted by a weak electric field perpendicular to the direction of the incident ions [205].The technique has recently been reviewed [205]. Charge exchange mass spectra can also be measured with conventional instruments having a slightly modified electron impact source [113,114].

Although the direct determination of breakdown curves from charge exchange spectra is of great advantage, several shortcomings of the method should be noted:
(1) Ionization cross sections for the various fragments cannot be measured as a continuous function of the internal energy.
(2) Only a limited number of primary ions are available.
(3) The primary ions often have several recombination energies, leading to ambiguous results. Moreover, the recombination energies of some primary ions are in dispute [167].

Figure II-17 contrasts the breakdown graph calculated by Vestal [167], with charge exchange measurements by Petterson and Lindholm [206]. Although the overall agreement is quite satisfactory two discrepancies are observed. First, the maximum in the propyl ion curve differs by 1 eV between the graphs and second, the abundance of the $C_3H_5^+$ does not decay to zero at higher internal energies as predicted by the theory. It is likely that the observed differences for the propyl ion which are also found if charge exchange and photoionization data are compared results from uncertainties in the assignment of recombination energies rather than from erroneous assumptions in the calculations. (The possibility of hydride ion transfer has been invoked to explain the discrepancy [167]). The fragmentation of propane following charge exchange has also been measured by Cermak and Herman [207] and Futrell and Tiernan [208]. The latter results are in particularly good agreement with Pettersons and Lindholms data. A careful comparison of breakdown curves obtained from charge exchange mass spectra of iso-butane [209], n-propanol and isopropanol [210] with QET calculations [211] was recently reported by Sunner and Szabo. The authors observed a good agreement between theoretical and experimental breakdown curves at low internal energies, whilst substantial deviations are found at higher internal energies (> 6 eV). They conclude that the discrepancies are mainly due to an over-simplified decomposition scheme used in the calculation. Breakdown curves obtained by charge exchange have also been reported by other authors [212-214].

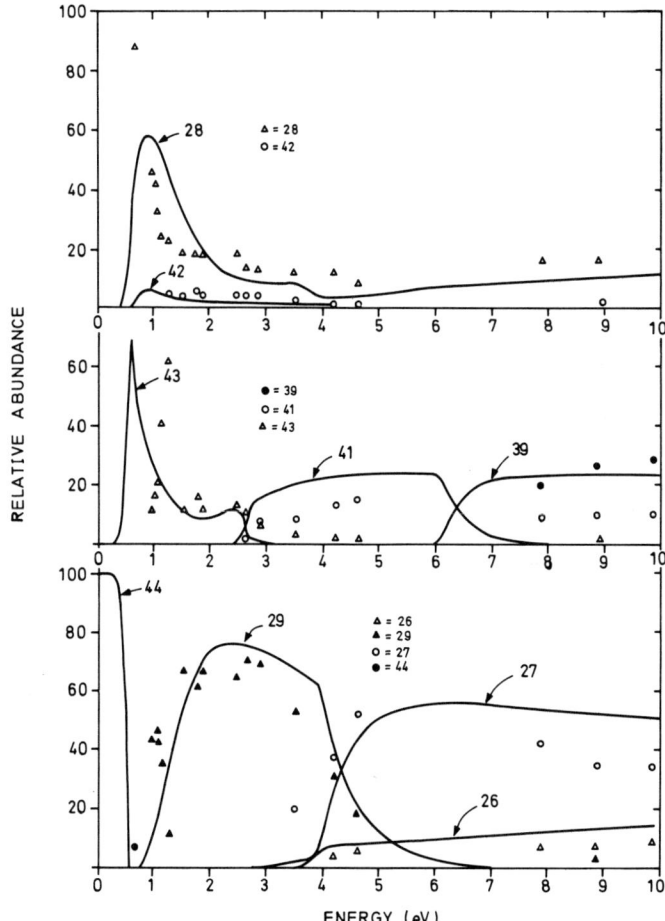

Fig. II-17. Comparison of the calculated breakdown graph [167] for the propane molecular ion (solid lines) with charge-exchange measurement [206]. (By courtesy of the American Institute of Physics.)

2.2.2.4 Photoelectron-photoion-coincidence (PEPICO)

In photoelectron-photoion-coincidence spectroscopy the translational energy of the electrons ejected during the ionization process is measured and allows the direct determination of the excitation energy of the ion to be made using eq. II-20:

$$E_{ion} = h\nu - IP - E_{electron} \qquad (II-20)$$

Photon sources of either variable [215-216] or constant energy [217-221] can be used (see Section I-4.3). If the mass of the ionized molecule or fragment is measured in time coincidence with the ejected electron the energy of which has been analyzed a direct construction of the breakdown diagram is possible. Ions and electrons are drawn out of the ion source in opposite directions. Energy analysis of the electrons can be achieved using a retardation system [217], a parallel plate [218, 219] or a 127 ° cylindrical plate [216] analyzer while a magnetic field [217] or a time-of-flight system [215,216,218,219] is applied for mass analysis. A time-of-flight spectrometer is advantageous as not only the mass, but also the kinetic energy released upon decomposition can be determined. Finally further modifications even allow the measurements of ion lifetimes as demonstrated in the following section. In spite of considerable experimental difficulties the PEPICO technique seems to be the method of choice for the experimental determination of breakdown diagrams as it does not suffer from the shortcomings mentioned for charge exchange experiments. The internal energy of the ions can be varied continuously and is always well defined.

Breakdown graphs were first reported by Brehm and Puttkamer [220, 221]. Fig. II-16b reveals that the experimental breakdown diagram of propane [222] is not only in good agreement with that obtained from photoionization data (Fig. II-16a) but also with that calculated by Vestal (Fig. II-16c). Minor deviations are observed in the width of the $C_3H_7^+$ curve and the height of the $C_2H_5^+$ curve, and these may be due to incorrect parameters used in the calculations.

Breakdown graphs (although not always normalized in the usual fashion) have also been determined using this technique for other organic compounds such as CH_4 [216,217,221], CD_4 [216,217], C_2H_2 [217], C_2H_4 [217,223], C_2H_6 [216,217], HCOOH [217], CH_3OH [105,217], C_2H_5OH [105], C_4H_6 [215], C_5H_{12} [222], C_6H_6 [118], CH_3X (X = F, Cl, Br, I, NH_2, SH) [108,210], D_2O [225], C_3H_6O [224], $Hg(CH_3)_2$ [224], C_2D_6 [216], C_3H_8 [222,226], $C_3H_6D_2$ [226], C_3D_8 [226] and allene (C_3H_4) [227]. Of special interest is a detailed study of unlabelled and deuterium labelled alkanes and alkenes by Stockbauer and Inghram [223] using the threshold [216] (or zero kinetic energy [215]) PEPICO technique (see Section I-4.3). The authors pointed out that a stringent test of the QET is its ability to predict the breakdown curve of a set of isotopically substituted compounds. The authors found good qualitative agreement with QET calculations and suggested that the existing slight discrepancies could be removed using other activated complex configura-

2.2.3 Determination of the Rate Constant, K(E)

Breakdown curves only give the ratio of rate constants in which case errors associated with the distribution of states in the precursor ion cancel out. Thus the direct experimental determination of rate constants as functions of the internal energy constitutes the most stringent test of the QET (and RRKM) theory. Direct determination of k(E) has recently become possible from lifetime measurements of energy selected ions. If ions of one distinct internal energy are sampled the decay can be described by a single rate constant. Thus the distribution of ion lifetimes is described by the laws for unimolecular reactions (eq. II-9 to II-13). If only one fragment is formed from the molecular ion and there is no competition from fluorescence or collisional stabilization the rate of formation is described by

$$\frac{di_F}{dt} = k i_M \qquad (II-21)$$

(i_F = fragment intensity, i_M = molecular ion intensity)

Ion lifetime measurements of energy selected ions have been carried out by Andlauer and Ottinger using charge exchange [113,114]. Their apparatus (Fig. II-8b) resembles that used for earlier lifetime measurements [149,151,152] (Fig. II-8a). However the original electron gun was replaced by an auxiliary ion source in which the primary ions are produced by electron impact. The primary ions are drawn by a small field gradient through the holes in the anode (AC_+) into the molecular beam, while a grid at a negative potential with respect to the filament (E) prevents electrons from passing through the anode. In practice the authors actually did not observe a single rate constant for a given reaction gas, but rather a distribution of rate constants with several maxima, which correspond to the various recombination energies of the primary ions. Moreover, both the initial thermal energy distribution as well as the kinetic energy transferred during collision lead to the formation of ions with a finite energy distribution thus giving a rate constant distribution. Hence rather than determining k(E) directly from the experimental results the authors chose a distribution of k values to fit the experimental data. K(E) was determined for HCN loss from the benzonitrile ion, loss of H and C_2H_2 from the benzene ion and loss of C_2H_2 from the thiophene ion. Fig. II-18 compares the experimentally determined rate constant as a function of the internal energy for HCN loss from the benzonitrile molecular ion with Phase Space and RRKM

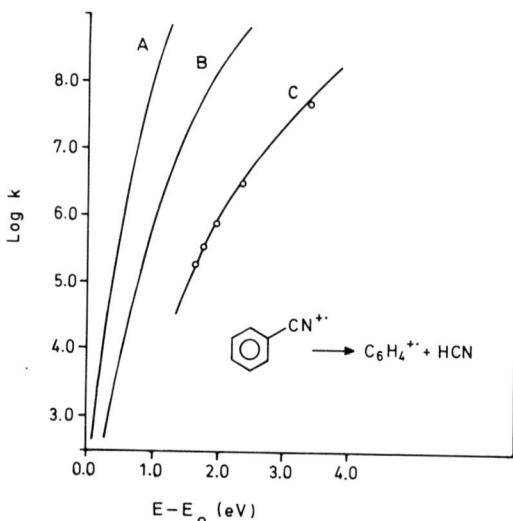

Fig. II-18. Rate constant for loss of HCN from the benzonitrile ion as a function of the internal energy. Comparison of experimental results (circles) [114] with calculations using the Phase Space Theory (curve A and B) and the RRKM Theory (curve C) [74]. (By courtesy of the American Chemical Society.)

calculations [65,74]. For the Phase Space calculations either a cyclic (curve A) or acyclic structure (curve B) has been assumed for the $C_6H_4^{+\cdot}$ fragment. Although the cyclic benzyne structure leads to a predicted k(E) curve which is much closer to experiment, the theoretical rate remains ca. 10^2 larger than the experimental rate over most of the energy range. Hence it is obvious that Phase Space calculations overestimate the rate constant by a substantial margin. On the other hand excellent agreement is observed if the experimental data are compared with RRKM calculations, assuming a four-centered transition state (curve C). The failure of the Phase Space Theory to predict the k(E) curve quantitatively apparently results from the fact that in these calculations a totally loose transition state is assumed which does not appear to correspond to reality.

Whilst charge exchange lifetime measurements still suffer from the drawback that the internal energy cannot be varied continuously, this shortcoming is overcome using the PEPICO technique in conjunction with ion lifetime measurements although only a smaller range of rate constants is accessible to this technique due to the poorer time resolution. Using this method ion lifetime measurements were reported simultaneously by Eland and Schulte [230], using a fixed energy photon source and by Baer et al. [39,231,232] using the threshold technique

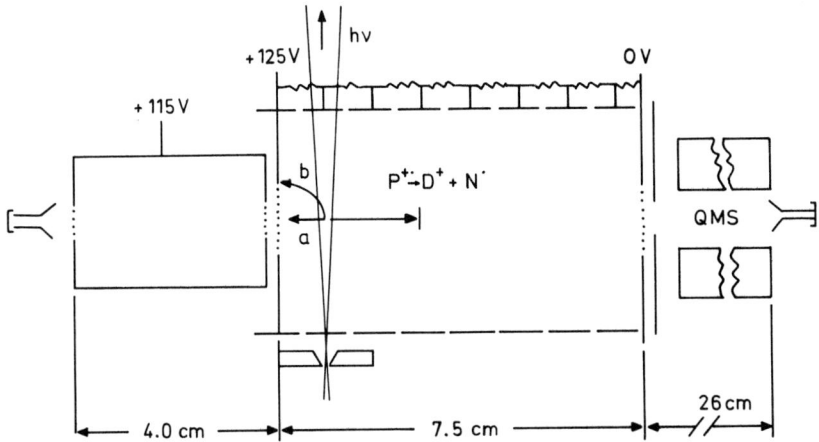

Fig. II-19. Instrumental set up used by Baer et al. [39] for ion lifetime measurement using the zero kinetic energy photoelectron-photoion-coincidence method. A series of resistors establishes a constant electric field throughout the acceleration region. A quadrupole mass spectrometer acts both as a mass filter and drift region. Arrows "a" and "b" are representative trajectories for zero kinetic energy and hot electrons. (By courtesy of the American Institute of Physics.)

Baer's apparatus is illustrated in Fig. II-19. Zero kinetic energy electrons are detected as described in Section I-4.3. Ion lifetimes are determined by a time-of-flight analysis. The positive ions fall through a uniform electric field established by a series of dropping resistors, before they enter a quadrupole mass filter. Ions decomposing within the electric field reach the mass filter with a deficit of translational energy compared with non-decomposing ions and will arrive at the detector after shorter flight times. Such delayed decompositions lead to an asymmetrical broadening of the time-of-flight curves in direct correspondence to the peak shapes observed in lifetime measurements by EI [17] and FI [150] (see Fig. II-9). For evaluation of the data, k(E) was again used as an adjustable parameter from which time-of-flight distributions were calculated to fit the experimental ones. While fair agreement was observed between experimental and calculated k(E) curves for X· loss (X = Cl, Br, I) from $C_6H_5X^{+\cdot}$ [231,232] (see Fig. II-20), considerable discrepancies between RRKM calculations and experiment were originally reported for $CH_3^{\cdot}$ loss from the molecular ion of C_4H_6 isomers [39].

However, recent recalculations by Chesnavich and Bowers [74] were able to reconcile the apparent disagreement. In their calculations the authors assume that in the $C_4H_6^{+\cdot}$ ion the isomerization reaction to a 2-methylcyclopropenium ion is the rate-determining step rather than the

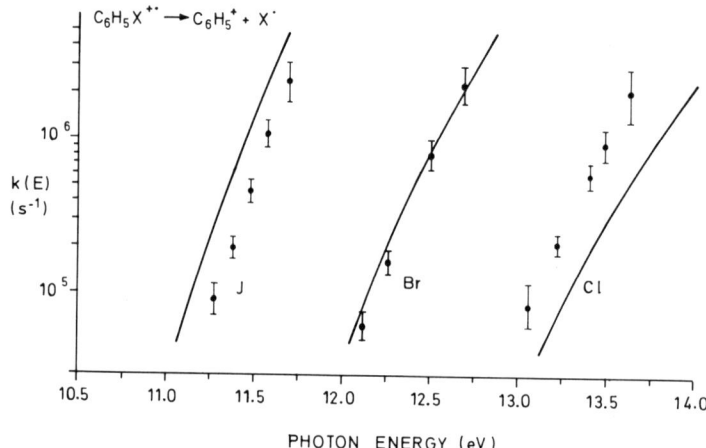

Fig. II-20. Rate constants for halogen loss from halobenzene molecular ions as a function of the internal energy. Comparison of calculated (solid lines) and experimental data [231].

fragmentation reaction, demonstrating again that discrepancies between theoretical and experimental data often arise from inadequate assumptions about the transition state, but not from a general failure of the theory.

In conclusion fair agreement between calculated and experimental rate constants was observed for several systems ($C_6H_5X^{+\cdot}$, $C_6H_5CN^{+\cdot}$, $C_4H_6^{+\cdot}$) in support of the general validity of the QET (RRKM Theory). The reasons for the remaining discrepancies observed for the benzene system remain open to question, but may again result from the fact that the Phase Space calculation employed lead to values for the rate constant which are too high.

2.2.4 Kinetic Energy Release Distribution

Decomposition of an excited organic ion occurs with a release of kinetic energy which stems from two sources:
(1) The energy in excess of the activation energy, E_o, termed "non fixed energy".
(2) The activation energy of the reverse reaction.

The kinetic energy released upon metastable decomposition of an ion will be discussed in detail in Section III-7.2. Here only its usefulness to test the QET will be discussed. The dissociation of a polyatomic ion does not occur with a single kinetic energy release because the available excess energy is partitioned into translational, rotational, and vibrational degrees of freedom leading to a distribution of kinetic energies. According to the QET (RRKM Theory) the probability that the activated complex has translational energy between E_t and $E_t + dE_t$ is

$$P(E_t) = \frac{\rho^{\neq}(E - E_o - E_t) \, dE_t}{W^{\neq}(E - E_o)} \qquad (II-22)$$

where $\rho^{\neq}$ is the density of vibrational and rotational states in the activated complex and $W^{\neq}$ the total number of states in the activated complex configuration with energy $\leq E - E_o$ [1]. According to eq. II-22 the distribution of kinetic energies should have a maximum at zero kinetic energy and decay exponentially with increasing energy. Equations for this distribution have also been derived using the modified QET (Phase Space Theory) where the exact mathematical expression depends on whether the potential surface of the collision complex is strongly or weakly attractive [69]. According to these calculations the most probable kinetic energy released is usually not zero.

Kinetic energy release distribution functions can now be determined experimentally with sufficient accuracy to allow a comparison with theoretical predictions. It has been pointed out that such comparisons represent tests of the theory which are as rigorous as and at the same time complementary to tests based on the experimental determination of k(E) [233]. While originally the deflection technique [234,235] was applied to obtain information on the kinetic energy release distribution, more recently the analysis of peak shapes of metastable ions in conventional double focussing instruments [228,229,236-239] or in time-of-flight instruments [233,240-249] has been employed successfully in such determinations. While $P(E_t)$ can be determined from metastable peak shapes even if the kinetic energy released is extremely small (≤ 1 meV), the time-of-flight method has the advantage that it can be used in conjunction with the PEPICO technique thus allowing $P(E_t)$ to be determined as function of the internal energy. $P(E_t)$ is usually obtained by calculating the theoretical peak shape using a set of discrete energy releases each with a characteristic weight. The weight factors are then fitted to give optimum correspondence between the experimental and theoretical peak shape.

Whilst QET (RRKM) calculations of the average kinetic energy release were in disagreement with experimental results [240,248], recent calculations by Klots [69,249] and Chesnavich and Bowers [74] based on the modified QET-Phase Space Theory, demonstrated very good agreement between theory and experiment for low internal energies*) (i.e. if kinetic energy release distributions derived from metastable ions were compared with the calculations). This is illustrated in Fig. II-21 for HCN loss from the benzonitrile ion. Such calculations have also been

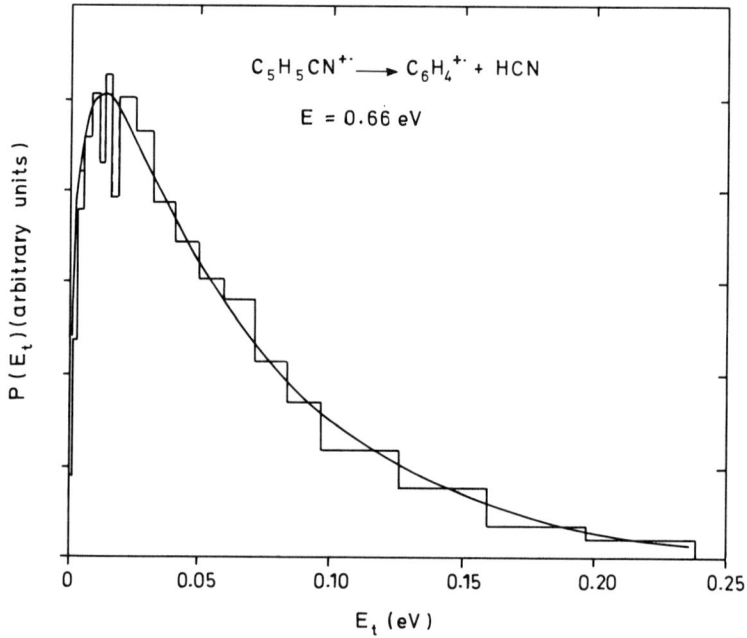

Fig. II-21. Comparison of calculated (smooth curve) and experimental (histogram) kinetic energy distribution of $C_6H_4^{+ \cdot}$ ions formed by HCN loss from benzonitrile ions [69]. (By courtesy of the American Institute of Physics.)

applied to the decomposition of short-lived collision complexes, e.g. for the system

$$C_2H_4^{+ \cdot} + C_2H_4 \rightarrow C_4H_8^{+ \cdot *} \rightarrow C_3H_5^+ + CH_3^{\cdot}$$

Scheme II-5

---

* Kinetic energy release distributions determined using the PEPICO technique [233, 246,249] revealed, however, that with increasing internal energies substantial discrepancies between theory and experiment are observed suggesting that the Langevin model used in Phase Space calculations to determine the collision cross section is not appropriate at such energies.

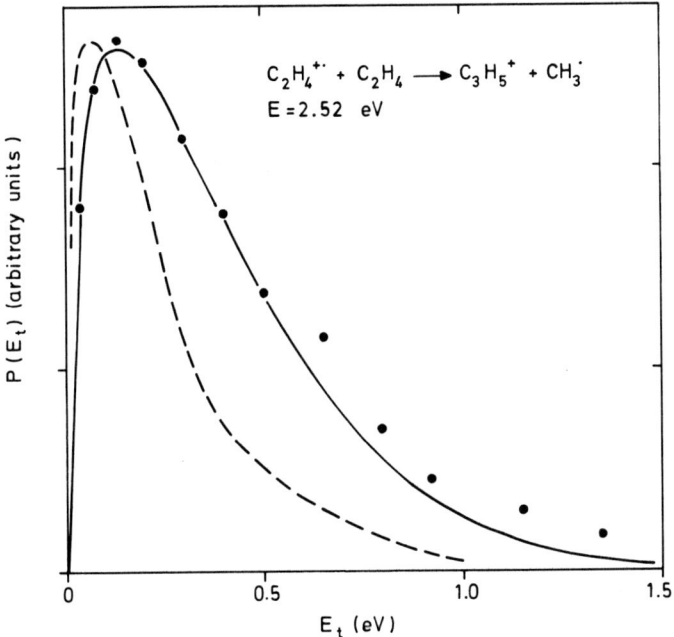

Fig. II-22. Kinetic energy distribution of $C_3H_5^+$ fragments formed by decomposition of the $C_4H_8^{+\cdot}$ collision complex. Comparison of experimental data (filled circles) with calculations using the RRKM Theory (dashed line) or the Phase Space Theory (solid line) [69]. (By courtesy of the American Institute of Physics.)

Original calculations of the kinetic energy distribution of the products of the $C_4H_8^{+\cdot}$ collision complex based on the RRKM theory were in severe disagreement with experimental results [84] (see Fig. II-22, dashed line). This was taken as evidence against the energy randomization hypothesis [84]. However a recalculation by Klots [69] based on the modified QET gave excellent agreement between theory and experiment even for this case (Fig. II-22).

The experimental determination of kinetic energy release distributions can also be used to demonstrate the role of angular momentum in unimolecular kinetics as recently shown by Klots et al. [249] for the reaction

$$C_4H_6^{+\cdot} \rightarrow C_3H_3^+ + CH_3^{\cdot}.$$

$C_4H_6^{+\cdot}$ ions with negligible angular momentum were prepared by electron impact of 1,3-butadiene. The kinetic energy release distribution was deduced from the metastable peak shape, i.e. ions with internal energies near the threshold for decomposition were sampled (Fig. II-23a). $C_4H_6^{+\cdot}$ ions with large angular momentum were prepared by collision of

$C_2H_2^{+\cdot}$ and C_2H_4 [250]. In the latter case the total energy is predetermined by the collision energy. The kinetic energy release distribution is shown in Fig. II-23b.

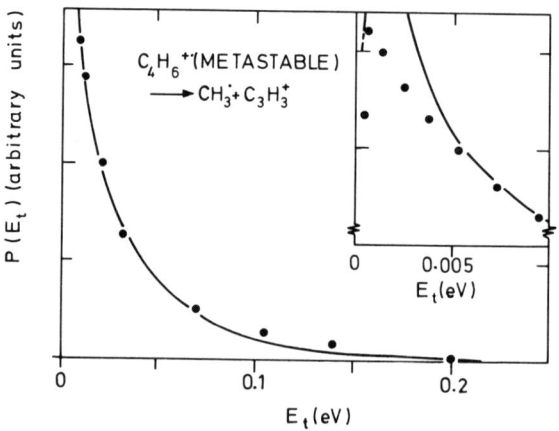

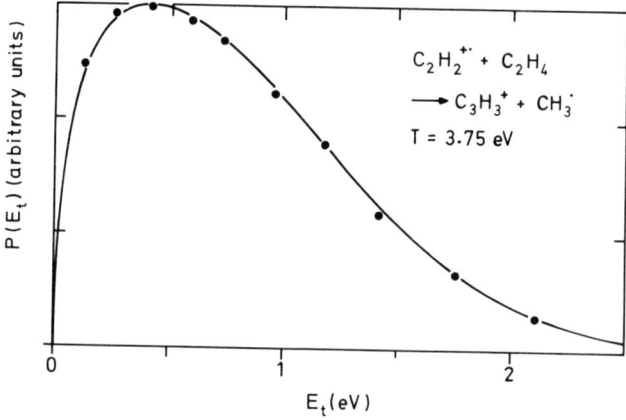

Fig. II-23. Comparison of observed kinetic energy release distributions (filled circles) from $C_4H_6^{+\cdot}$ ions with Phase Space calculations (solid lines) [249]. (a) "Metastable" $C_4H_6^{+\cdot}$ ions. (b) $C_4H_6^{+\cdot}$ collision complex [250]. (By courtesy of the American Institute of Physics.)

Comparison of Fig. II-23a and II-23b reveals substantial differences between the kinetic energy release distributions. Whilst the distribution of ions with low angular momentum shows a maximum at or close to zero kinetic energy, this maximum is shifted to considerably higher energies for ions of high angular momentum. The high angular momentum of the collision complex appears predominantly as orbital angular momentum giving rise to a centrifugal barrier and hence to large kinetic energy releases.

The experimental results in Fig. II-25 are compared with QET calculations using the Phase Space formalism (solid line). Again the agreement between experiment and theory is excellent.

The examples discussed in this and the previous section demonstrate that the RRKM formalism should be used to calculate rate constants while the Phase Space formalism (in which the conservation of angular momentum is taken into account) is more appropriate for the calculation of kinetic energy release distributions.

Summarizing, the various experiments described in this chapter have demonstrated that when an adequate mathematical treatment of the QET is used at least qualitative and often almost quantitative agreement between theory and experiments has been observed in support of the general validity of the theory. Further theoretical and experimental efforts are necessary in order to substantiate whether the remaining discrepancies between theory and experiment indeed result from a partial failure of the theory or not.

3 Appendix. The Rate Equation

Ions in a mass spectrometer are isolated systems which can be described in terms of microcanonical ensembles in which the individual systems are uniformly distributed over all states having energies in the range of E to E + dE. The dissociation of the ion is described as a motion along a reaction coordinate separable from all other internal coordinates by a critical "activated complex" configuration located at a saddle point of the potential energy surface (see Fig. II-1). Furthermore it is assumed that the transitions between the various states (i.e. the energy fluctuation) are fast compared with the dissociation, so that the dissociation itself has only a negligible influence on the relative number of ions in the activated complex. Thus there exists a quasiequilibrium between ions in the activated complex and the reactant. The number of ions in the transition state is proportional to the number of states in the activated complex. If the number of states is large this number can be represented by a density function, $\rho(E)\,dE$. A given ion with the activated complex configuration will have a total energy between E and E + dE, part of which is the activation energy, E_o, necessary to reach the saddle point, as well as the translational energy between E_t and $E_t + dE_t$ corresponding to the motion along the reaction coordinate. Thus the total number of states in the activated complex will be given by $\rho^{\neq}(E - E_o - E_t) \rho_t(E_t) \, dE \, dE_t$ (where ρ_t is the density of translational states per length, l, in the reaction coordinate).

Then *the rate constant, k, is given by the ratio of activated complex ions to reactant ions multiplied by one-half of the frequency of crossing the barrier and integrated over all possible values of the translational energy in the reaction coordinate.* (The factor one-half arises from equal probability for forward and back translations)*)

$$k = 1/2 \int_0^{E-E_o} \frac{\rho^{\neq}(E - E_o - E_t) \rho_t(E_t) \, v}{\rho(E)} \, dE_t$$

It remains to calculate the frequency, v, for crossing the barrier. For a potential barrier of length, l, the translational energy is given by

$$E_t = \frac{n_t^2 h^2}{8 \mu l^2}$$

(μ = reduced mass).

Then the density of translational states is given by

$$\rho(E_t) = \frac{dn_t}{dE_t} = \frac{1}{h} \left(\frac{2\mu}{E_t} \right)^{1/2}$$

$$\nu = \frac{v}{l}, \text{ where } v = \left(\frac{2E_t}{\mu} \right)^{1/2} \text{ and thus}$$

$$\nu = \frac{2}{h \cdot \rho_t}$$

which leads to

$$k = \frac{1}{h} \int_0^{E-E_o} \frac{\rho^{\neq}(E - E_o - E_t)}{\rho(E)} \, dE_t$$

This is the original rate equation derived by Rosenstock et al. [2] in which the integration of the density function leads to a continuous approximation of the number of states. In eq. II-1 this continuous approx-

---

* The transmission coefficient is assumed to be unity.

imation is replaced by the exact number of states, $W^{\neq}(E - E_o)$. Moreover, a symmetry factor for the number of identical reactions has been introduced.

4 References

1. M.V. Vestal in "Fundamental Processes in Radiation Chemistry" (P. Ausloos, Ed.), Interscience, New York, 1968, Chapter 2.
2. H.M. Rosenstock, M.B. Wallenstein, A.L. Wahrhaftig and H. Eyring, Proc. Nat. Acad. Sci., $\underline{38}$, 667 (1952).
3. R.C. Tolman "The Principles of Statistical Mechanics", Oxford University Press, London, 1938.
4. H.M. Rosenstock and M. Krauss, "Mass Spectrometry of Organic Ions" (F.W. McLafferty, Ed.), Academic Press, New York, 1963.
5. H.M. Rosenstock and M. Krauss in "Advances in Mass Spectrometry", Vol.2, Pergamon Press, Oxford, 1963, p. 251.
6. H.M. Rosenstock in "Advances in Mass Spectrometry", Vol. 4, The Institute of Petroleum, London, 1968, p. 523.
7. A.L. Wahrhaftig in "Mass Spectrometry" (MTP International Review of Sciences), Physical Chemistry, Series I, Vol. 5 (A. Maccoll, Ed.), Butterworths, London, 1972, p. 1.
8. C. Lifshitz in "Advances in Mass Spectrometry", Vol. 7, Heyden, London, 1977.
9. S. Glasstone, K.J. Laidler and H. Eyring, "Theory of Rate Processes", McGraw-Hill, New York, 1941.
10. R.A. Marcus and O.K. Rice, J. Phys. and Colloid Chem., $\underline{55}$, 894 (1951).
11. R.A. Marcus, J. Chem. Phys., $\underline{20}$, 359 (1952).
12. M.L. Vestal, J. Chem. Phys., $\underline{41}$, 3997 (1964).
13. M.L. Vestal, A.L. Wahrhaftig and W.H. Johnston, J. Chem. Phys., $\underline{37}$, 1276 (1962).
14. V.H. Dibeler and H.M. Rosenstock, J. Chem. Phys., $\underline{39}$, 1326 (1963).
15. B.H. Solka, J.H. Beynon and R.G. Cooks, J. Phys. Chem., $\underline{79}$, 859 (1975).
16. J.P. Flamme, J. Momigny and H. Wankenne, J. Am. Chem. Soc., $\underline{98}$, 1045 (1976).
17. Ch. Ottinger, Z. Naturforsch., $\underline{22a}$, 20 (1967).
18. G.M. Wieder and R.A. Marcus, J. Chem. Phys., $\underline{37}$, 1835 (1962).
19. B.S. Rabinovitch and R.W. Diesen, J. Chem. Phys., $\underline{30}$, 735 (1959).
20. B.S. Rabinovitch and J.H. Current, J. Chem. Phys., $\underline{35}$, 2250 (1961).
21. E.W. Schlag and R.A. Sandsmark, J. Chem. Phys., $\underline{37}$, 168 (1962).
22. G.Z. Whitten and B.S. Rabinovitch, J. Chem. Phys., $\underline{38}$, 2466 (1963).
23. P.C. Haarhoff, Molec. Phys., $\underline{6}$, 337 (1963).
24. P.C. Haarhoff, Molec. Phys., $\underline{7}$, 101 (1964).
25. E. Thiele, J. Chem. Phys., $\underline{39}$, 3258 (1963).
26. W. Forst, Z. Prasil and P. St. Laurent, J. Chem. Phys., $\underline{46}$, 3736 (1967).
27. S.H. Lin and H. Eyring, J. Chem. Phys., $\underline{43}$, 2153 (1965).
28. S.H. Lin and H. Eyring, J. Chem. Phys., $\underline{39}$, 1577 (1963).

29. M.L. Vestal, A.L. Wahrhaftig and W.H. Johnston, J. Chem. Phys., *37*, 1276 (1962).
30. R.W. Kiser, "Introduction to Mass Spectrometry", Prentice-Hall, Englewood Cliffs, 1965, Chapter 7.
31. B.G. Hobrock, Doctoral Dissertation, Kansas State University, Manhattan, 1963.
32. W.A. Bryce and P. Kebarle, Can. J. Res., *34*, 1249 (1956).
33. B.J. Millard and D.F. Shaw, J. Chem. Soc. (B), 664 (1966).
34. G.G. Meisels, J.Y. Park and B.G. Giessner, J. Am. Chem. Soc., *91*, 1555 (1969).
35. F.P. Lossing, Can. J. Chem., *50*, 3973 (1972).
36. G.A. Smith and D.H. Williams, J. Chem. Soc. (B), 1529 (1970).
37. Z. Dolejsek, V. Hanus and K. Vokac in "Advances in Mass Spectrometry", Vol. 3 (M.L. Mead, Ed.), Elsevier, Amsterdam, 1966, p. 503.
38. J. Collin and F.P. Lossing, J. Am. Chem. Soc., *79*, 5848 (1957).
39. A.S. Werner and T. Baer, J. Chem. Phys., *62*, 2900 (1975).
40. R.P. Morgan and P.J. Derrick, Org. Mass Spectrom., *10*, 563 (1975).
41. K. Levsen, H. Heimbach, G.J. Shaw and G.W.A. Milne, Org. Mass Spectrom., *12*, 663 (1977).
42. T.W. Bentley and R.A.W. Johnstone, Adv. Phys. Org. Chem., *8*, 151 (1970).
43. F.W. McLafferty, Anal. Chem., *31*, 82 (1959).
44. J.S. Smith and F.W. McLafferty, Org. Mass Spectrom., *5*, 483 (1971).
45. D.G.I. Kingston, J.T. Bursey, M.M. Bursey, Chem. Rev., *74*, 215 (1974).
46. H. Bosshardt and M. Hesse, Angew. Chem., *86*, 256 (1974).
47. H. Schwarz, Org. Mass Spectrom., *10*, 384 (1975).
48. F. Borchers, K. Levsen and H.D. Beckey, Int. J. Mass Spectrom. Ion Phys., *21*, 125 (1976).
49. A.N.H. Yeo and C. Djerassi, J. Am. Chem. Soc., *94*, 482 (1972).
50. N.A. Uccella, I. Howe and D.H. Williams, Org. Mass Spectrom., *6*, 229 (1972).
51. R.G. Cooks, M. Bertrand, J.H. Beynon, M.E. Rennekamp and D.W. Setser, J. Am. Chem. Soc., *95*, 1732 (1973).
52. C. Lifshitz and M. Shapiro, J. Chem. Phys., *46*, 4912 (1967).
53. M. Vestal and J.H. Futrell, J. Chem. Phys., *52*, 978 (1970).
54. Ch. Ottinger, J. Chem. Phys., *47*, 1452 (1967).
55. D.L. Bunker and M. Pattengill, J. Chem. Phys., *48*, 772 (1968).
56. C.E. Klots, J. Phys. Chem., *75*, 1526 (1971).
57. D.W. Turner, C. Baker, A.D. Baker and C.R. Brundle, "Molecular Photoelectron Spectroscopy", Wiley-Interscience, New York, 1970.
58. A.D. Baker, C. Baker, C.R. Brundle and D.W. Turner, Int. J. Mass Spectrom. Ion Phys., *1*, 285 (1968).
59. C. Sandorfy in "Chemical Spectroscopy and Photochemistry in the Vacuum Ultraviolet", D. Reidel, Dordrecht, 1974, p. 177.
60. R.C. Dunbar, J. Am. Chem. Soc., *98*, 4671 (1976).
61. A. Kropf, E.M. Eyring, A.L. Wahrhaftig and H. Eyring, J. Chem. Phys., *32*, 149 (1960).
62. E.M. Eyring and A.L. Wahrhaftig, J. Chem. Phys., *34*, 23 (1961).
63. W.A. Chupka, J. Chem. Phys., *30*, 191 (1959).
64. C.E. Klots, J. Chem. Phys., *41*, 117 (1964).
65. C.E. Klots, Z. Naturforsch., *27a*, 553 (1972).

66. C.E. Klots, J. Chem. Phys., 58, 5364 (1973).
67. C.E. Klots, Adv. Mass Spectrom., 6, 969 (1973).
68. C.E. Klots, Chem. Phys. Lett., 38, 61 (1976).
69. C.E. Klots, J. Chem. Phys., 64, 4269 (1976).
70. P.F. Knewstubb, Int. J. Mass Spectrom. Ion Phys., 6, 217 (1971).
71. P.F. Knewstubb, J. Chem. Soc., Faraday Trans. II., 68, 1196 (1972).
72. P.F. Knewstubb, Int. J. Mass Spectrom. Ion Phys., 10, 371 (1973).
73. W.J. Chesnavich and M.T. Bowers, J. Chem. Phys., 66, 2306 (1977).
74. W.J. Chesnavich and M.T. Bowers, J. Am. Chem. Soc., 99, 1705 (1977).
75. A.A. Frost and R.G. Pearson, "Kinetics and Mechanism", 2nd Ed., Wiley, 1961, p. 166.
76. B.N. McMaster, in "Mass Spectrometry", Vol. 3 (R.A.W. Johnstone, Ed.), The Chemical Society, London, 1975, Chapter 1.
77. J.H. Callomon, Can. J. Phys., 34, 1046 (1956).
78. M. Allan and J.P. Maier, Chem. Phys. Letters, 34, 442 (1975).
79. M. Allan, E. Kloster-Jensen and J.P. Maier, Chem. Phys., 7, 11 (1976).
80. M. Allan and J.P. Maier, Chem. Phys. Letters, 43, 94 (1976).
81. M. Allan, J.P. Maier, O. Marthaler and E. Kloster-Jensen, Chem. Phys., 29, 331(1978).
82. M. Allan, J.P. Maier and O. Marthaler, Chem. Phys., 26, 131 (1977).
83. J.D. Rynbrand and B.S. Rabinovitch, J. Phys. Chem., 75, 2164 (1971).
84. A. Lee, R.L. LeRoy, Z. Herman, R. Wolfgang and J.C. Tully, Chem. Phys. Lett., 12, 569 (1972).
85. R.D. Levine, Ber. Bunsengesellschaft Phys. Chem., 78, 111 (1974).
86. R.L. LeRoy, J. Chem. Phys., 53, 846 (1970).
87. R.L. LeRoy, J. Chem. Phys., 55, 1476 (1971).
88. J.M. Parson, K. Shobatake, Y.T. Lee and St.A. Rice, J. Chem. Phys., 59, 1402 (1973).
89. K. Shobatake, J.M. Parson, Y.T. Lee and St.A. Rice, J. Chem. Phys., 59, 1416 (1973).
90. K. Shobatake, J.M. Parson, Y.T. Lee and St.A. Rice, J. Chem. Phys., 59, 1427 (1973).
91. K. Shobatake, Y.T. Lee and St.A. Rice, J. Chem. Phys., 59, 1435 (1973).
92. M.V. Gur'ev, Dokl. Akad. Nauk SSSR, 136, 856 (1961).
93. M.V. Gur'ev, L.V. Sumin and S.A. Volkov, Khim. Vys. Energ., 1, 40 (1967).
94. K. Levsen and H.D. Beckey, Int. J. Mass Spectrom. Ion Phys., 9, 63 (1972).
95. C. Guttman and S.A. Rice, J. Chem. Phys., 61, 651 (1974).
96. C.-S. Huang and E.C. Lim, J. Chem. Phys., 62, 3826 (1975).
97. D.A. Hansen and E.K.C. Lee, J. Chem. Phys., 62, 183 (1975).
98. M.H. Hui and S.A. Rice, J. Chem. Phys., 61, 833 (1974).
99. F. Meyer and A.G. Harrison, J. Chem. Phys., 43, 1778 (1965).
100. C. Lifshitz, J. Chem. Phys., 47, 1870 (1967).
101. R.C. Dougherty, J. Am. Chem. Soc., 90, 5780 (1968).
102. F. Benoit, Org. Mass Spectrom., 7, 1407 (1973).
103. J.F. Elder, J.H. Beynon and R.G. Cooks, Org. Mass Spectrom., 11, 415 (1976).
104. K.M.A. Refaey and W.A. Chupka, J. Chem. Phys., 48, 5205 (1968).

105. B. Brehm, V. Fuchs and P. Kebarle, Int. J. Mass Spectrom. Ion Phys., 6, 279 (1971).
106. I.G. Simm, C.J. Danby and J.H.D. Eland, Int. J. Mass Spectrom. Ion Phys., 14, 285 (1974).
107. C. Lifshitz and F.A. Long, J. Phys. Chem., 69, 3746 (1965).
108. J.H.D. Eland, R. Frey, A. Küstler, H. Schulte and B. Brehm, Int. J. Mass Spectrom. Ion Phys., 22, 155 (1976).
109. J.P. Flamme, J. Momigny and H. Wankenne, J. Am. Chem. Soc., 98, 1045 (1976).
110. P.M. Guyon, W.A. Chupka and J. Berkowitz, J. Chem. Phys., 64, 1419 (1976).
111. T. Baer, A.S. Werner and B.P. Tsai, J. Chem. Phys., 62, 2497 (1975).
112. B.P. Tsai, A.S. Werner and T. Baer, J. Chem. Phys., 63, 4384 (1975).
113. B. Andlauer and Ch. Ottinger, J. Chem. Phys., 55, 1471 (1971).
114. B. Andlauer and Ch. Ottinger, Z. Naturforsch., 27a, 293 (1972).
115. H.M. Rosenstock, J.T. Larkins and J.A. Walker, Int. J. Mass Spectrom. Ion Phys., 11, 309 (1973).
116. H.M. Rosenstock, K.E. McCulloh and F.P. Lossing, Int. J. Mass Spectrom. Ion Phys., 25, 327 (1977).
117. J.H.D. Eland, Int. J. Mass Spectrom. Ion Phys., 13, 457 (1974).
118. J.H.D. Eland, R. Frey, H. Schulte and B. Brehm, Int. J. Mass Spectrom. Ion Phys., 21, 209 (1976).
119. B. Jonsson and E. Lindholm, Ark. Fys., 39, 65 (1968).
120. J. Momigny, L. Brakier and L. D'Or, Bull. Cl. Sci., Acad. Roy. Belg., 48, 1002 (1962).
121. C. Köppel, H. Schwarz, F. Borchers and K. Levsen, Int. J. Mass Spectrom. Ion Phys., 21, 15 (1976).
122. R. Gooden and J.I. Brauman, J. Am. Chem. Soc., 99, 1977 (1977).
123. K. Levsen, Habilitation Thesis, Bonn, 1974.
124. P.J. Derrick in "International Review of Science", Physical Chemistry Series II, Vol. 5 (A. Maccoll, Ed.), Butterworths, London, 1975, Chapter 1.
125. J.A. Hipple, Phys. Rev., 71, 594 (1947).
126. J. Momigny, Bull. Soc. Chim. Belg., 70, 291 (1961).
127. N.D. Coggeshall, J. Chem. Phys., 37, 2167 (1962).
128. J.C. Schug, J. Chem. Phys., 40, 1283 (1964).
129. G.A. Muccini, W.H. Hamill and R. Barker, J. Phys. Chem., 68, 261 (1964).
130. L.V. Sumin, M.V. Gur'ev and M.V. Tunitskii, Kinet. Katal., 5, 961 (1964).
131. C. Lifshitz and M. Shapiro, J. Chem. Phys., 45, 4242 (1966).
132. K. Meier and J. Seibl, Int. J. Mass Spectrom. Ion Phys., 14, 99 (1974).
133. P.F. Knewstubb and N.W. Reid, Int. J. Mass Spectrom. Ion Phys., 5, 361 (1970).
134. S.E. Buttrill, J. Chem. Phys., 61, 619 (1974).
135. S.E. Buttrill Jr., J. Chem. Phys., 62, 1603 (1975).
136. P.W. Ryan, J.H. Futrell and M.L. Vestal, Chem. Phys. Lett., 18, 329 (1973).
137. H. Benz and H.W. Brown, J. Chem. Phys., 48, 4308 (1968).
138. R.B. Fairweather and F.W. McLafferty, Org. Mass Spectrom., 4, 221 (1970).
139. W.W. Hunt, R.E. Huffman, J. Saari, G. Wassel, J.F. Betts, E.H. Paufve, W. Wyess and R.A. Fluegge, Rev. Sci. Instrum., 35, 88 (1964).
140. W.W. Hunt, R.E. Huffman and K.E. McGee, Rev. Sci. Instrum., 35, 82 (1964).

141. U. v. Zahn and H. Tatarczyk, Phys. Lett., 12, 190 (1964).
142. H. Tatarczyk and U. v. Zahn, Z. Naturforsch., 20a, 1708 (1965).
143. H. Tatarczyk and U. v. Zahn, Z. Naturforsch., 27a, 1646 (1972).
144. R.D. Smith and J.H. Futrell, Org. Mass Spectrom., 11, 445 (1976).
145. R.D. Smith and J.H. Futrell, Org. Mass Spectrom., 11, 309 (1976).
146. R.D. Smith and J.H. Futrell, Int. J. Mass Spectrom. Ion Phys., 17, 233 (1975).
147. R.D. Smith and J.H. Futrell, Int. J. Mass Spectrom. Ion Phys., 23, 75 (1977).
148. G.V. Karachevtsev and V.L. Tal'rose, Kinet. Katal., 4, 923 (1963).
149. O. Osberghaus and Ch. Ottinger, Phys. Lett., 16, 121 (1965).
150. H.D. Beckey, Z. Naturforsch., 16a, 505 (1961).
151. Ch. Ottinger, Max-Planck-Institut f. Strömungsforschung, Report 107 (1970).
152. I. Hertel and Ch. Ottinger, Z. Naturforsch., 22a, 1141 (1967).
153. Ch. Ottinger, Fourth Int. Congress Radiat. Res., Evian, Gordon and Breach, 1970.
154. M. Barber and R.M. Elliott, 12th Annual Conference on Mass Spectrometry and Allied Topics, Montreal, 1964.
155. G.V. Karachevtsev and V.L. Tal'rose, Kinet. Katal., 8, 1 (1967).
156. R. Gomer, "Field Emission and Field Ionization" Harvard University Press, Cambridge, 1961.
157. H.D. Beckey, "Principles of Field Ionization and Field Desorption Mass Spectrometry", Pergamon, Oxford, 1977.
158. A.J.B. Robertson in "Mass Spectrometry", MTP International Review of Science, Physical Chemistry, Series I, Vol. 5 (A. Maccoll, Ed.), Butterworths, London, 1972, p. 103.
159. M.G. Inghram and R. Gomer, J. Chem. Phys., 22, 1279 (1954).
160. J.P. Pfeifer, A.M. Falick and A.L. Burlingame, Int. J. Mass Spectrom. Ion Phys., 11, 345 (1973).
161. H.D. Beckey and H. Knöppel, Z. Naturforsch., 21a, 1920 (1966).
162. H.D. Beckey in "Advances in Mass Spectrometry", Vol. 2 (R.M. Elliott, Ed.), Pergamon, Oxford, 1963, p.1.
163. P.J. Derrick and A.J.B. Robertson, Int. J. Mass Spectrom. Ion Phys., 10, 315 (1973).
164. H.D. Beckey, H. Hey, K. Levsen and G. Tenschert, Int. J. Mass Spectrom. Ion Phys., 2, 101 (1969).
165. G. Tenschert and H.D. Beckey, Int. J. Mass Spectrom. Ion Phys., 7, 97 (1971).
166. F.W. McLafferty, J. Okamoto, H. Tsuyama, Y. Nakajima, T. Noda and H.W. Major, Org. Mass Spectrom., 2, 751 (1969).
167. M.L. Vestal, J. Chem. Phys., 43, 1356 (1965).
168. W.A. Chupka and M. Kaminsky, J. Chem. Phys., 35, 1991 (1961).
169. S. Geltman, Phys. Rev., 102, 171 (1956).
170. "Mass Spectral Data", American Petroleum Institute, Project 44, Pittsburgh, 1955.
171. H.M. Grubb and S. Meyerson, in "Mass Spectrometry of Organic Compounds" (F.W. McLafferty, Ed.), Academic Press, New York, 1963, p. 516.
172. W.A. Chupka and J. Berkowitz, J. Chem. Phys., 47, 2921 (1967).
173. O. Osberghaus and R. Taubert, Phys. Chem., 4, 264 (1955).
174. A. Cassuto in "Advances in Mass Spectrometry", Vol. 2 (R.M. Elliott, Ed.), Pergamon, Oxford, 1963, p. 296.
175. H. Ehrhardt, Thesis, Bonn, 1960.

176. W.A. Chupka, J. Chem. Phys., 54, 1936 (1971).
177. L. Friedman, F.A. Long and M. Wolfsberg, J. Chem. Phys., 27, 613 (1957).
178. A.B. King and F.A. Long, J. Chem. Phys., 29, 374 (1958).
179. A. Kropf, E.M. Eyring, A.L. Wahrhaftig and H. Eyring, J. Chem. Phys., 32, 149 (1960).
180. M. Krauss, A.L. Wahrhaftig and H. Eyring, Ann. Rev. Nucl. Sci., 5, 241 (1955).
181. J. Collin, Bull. Soc. Roy. Sci., Liège, 7, 520 (1956).
182. E.M. Eyring and A.L. Wahrhaftig, J. Chem. Phys., 34, 23 (1961).
183. A. Kropf, Dissertation, University of Utah, Salt Lake City, 1954.
184. H. Wincel and Z. Kecki, Neuklonikas, 8, 529 (1963).
185. P. Nounou in "Advances in Mass Spectrometry IV" (E. Kendrick Ed.), Institute of Petroleum, London, 1968, p. 551.
186. G. Spiteller and M. Spiteller-Friedmann, Justus Liebigs Ann. Chem., 690, 1 (1965).
187. F.W. McLafferty "Interpretation of Mass Spectra", Benjamin, Reading (1973).
188. A.N.H. Yeo and D.H. Williams, Chem. Commun., 956 (1969).
189. A.N.H. Yeo and D.H. Williams, J. Am. Chem. Soc., 92, 3984 (1970).
190. A.N.H. Yeo and D.H. Williams, Org. Mass Spectrom., 5, 135 (1971).
191. I. Howe and D.H. Williams, J. Am. Chem. Soc., 91, 7137 (1969).
192. R. Heller, P. Krenmayr and K. Varmuza, Org. Mass Spectrom., 9, 1134 (1974).
193. W.A. Chupka, J. Chem. Phys., 30, 191 (1959).
194. B. Steiner, C.F. Giese and M.G. Inghram, J. Chem. Phys., 34, 189 (1961).
195. J.C. Tou, L.P. Hills and A.L. Wahrhaftig, J. Chem. Phys., 45, 2129 (1966).
196. R. Botter, J.M. Pechine and H.M. Rosenstock, Int. J. Mass Spectrom. Ion Phys., 25, 7 (1977).
197. E. Murad and M.G. Inghram, J. Chem. Phys., 40, 3263 (1964).
198. E. Lindholm, Proc. Phys. Soc. London A 66, 1068 (1953).
199. E. Lindholm, Conference on Applied Mass Spectrometry, Institute of Petroleum, London 1953.
200. E. Lindholm, Z. Naturforsch., 9a, 535 (1954).
201. E. Lindholm, Ark. Fys., 8, 257 (1954).
202. E. Lindholm, Ark. Fys., 8, 433 (1954).
203. E. Gustafsson and E. Lindholm, Ark. Fys., 18, 219 (1960).
204. H. von Koch and E. Lindholm, Ark. Fys., 19, 123 (1961).
205. E. Lindholm, in "Ion-Molecule Reactions" (J.L. Franklin, Ed.), Plenum Press, New York, 1972, p. 457, and references herein.
206. E. Pettersson and E. Lindholm, Ark. Fys., 24, 49 (1963).
207. V. Cermak and Z. Herman, Nucleonics, 19, 106 (1961).
208. J.H. Futrell and T.O. Tiernan, J. Chem. Phys., 39, 2539 (1963).
209. J. Sunner and I. Szabo, Int. J. Mass Spectrom. Ion Phys., 25, 241 (1977).
210. J. Sunner and I. Szabo, Int. J. Mass Spectrom. Ion Phys., 25, 263 (1977).
211. M. Vestal and G. Lerner, Fundamental Studies Related to Radiation Chemistry of Small Organic Molecules, ARL 67-0114, 1967.
212. S. Ikuta, K. Yoshihara and T. Shiokawa, Bull. Chem. Soc., Japan, 48, 2134 (1975).
213. I. Szabo, Ark. Fys., 31, 287 (1965).

214. W.A. Chupka and E. Lindholm, Ark. Fys., $\underline{25}$, 349 (1963).
215. A.S. Werner and T. Baer, J. Chem. Phys., $\underline{62}$, 2900 (1975).
216. R. Stockbauer, J. Chem. Phys., $\underline{58}$, 3800 (1973).
217. E. v. Puttkamer, Z. Naturforsch., $\underline{25a}$, 1062 (1970).
218. J.H.D. Eland, Int. J. Mass Spectrom. Ion Phys., $\underline{8}$, 143 (1972).
219. C.J. Danby and J.H.D. Eland, Int. J. Mass Spectrom. Ion Phys., $\underline{8}$, 153 (1972).
220. B. Brehm and E. v. Puttkamer, Z. Naturforsch., $\underline{22a}$, 8 (1967).
221. B. Brehm and E. v. Puttkamer in "Advances in Mass Spectrometry", Vol. 4 (E. Kendrick Ed.), The Institute of Petroleum, London (1968), p. 591.
222. B. Brehm, J.H.D. Eland, R. Frey and H. Schulte, Int. J. Mass Spectrom. Ion Phys., $\underline{21}$, 373 (1976).
223. R. Stockbauer and M.G. Inghram, J. Chem. Phys., $\underline{62}$, 4862 (1975).
224. C.S.T. Cant, C.J. Danby and J.H.D. Eland, J. Chem. Soc. Farad. Trans. II, $\underline{71}$, 1015 (1975).
225. J.H.D. Eland, Chem. Phys., $\underline{11}$, 41 (1975).
226. R. Stockbauer and M.G. Inghram, J. Chem. Phys., $\underline{65}$, 4081 (1976).
227. A.C. Parr, A.J. Jason and R. Stockbauer, Int. J. Mass Spectrom. Ion Phys., $\underline{26}$, 23 (1978).
228. J.F. Elder, J.H. Beynon and R.G. Cooks, Org. Mass Spectrom., $\underline{10}$, 273 (1975).
229. J.L. Holmes and A.D. Osborne, Int. J. Mass Spectrom. Ion Phys., in press.
230. J.H.D. Eland and H. Schulte, J. Chem. Phys., $\underline{62}$, 3835 (1975).
231. T. Baer, B.P. Tsai, D. Smith and P.T. Murray, J. Chem. Phys., $\underline{64}$, 2460 (1976).
232. T. Baer, D. Smith, B.P. Tsai and A.S. Werner, Adv. Mass Spectrom., $\underline{7}$, Heyden, London, 1977.
233. D.M. Mintz and T. Baer, J. Chem. Phys., $\underline{65}$, 2407 (1976).
234. R. Taubert, Z. Naturforsch., $\underline{19a}$, 484 (1964).
235. C.G. Rowland, J.H.D. Eland and C.J. Danby, Int. J. Mass Spectrom. Ion Phys., $\underline{2}$, 457 (1969).
236. D.T. Terwilliger, J.H. Beynon and R.G. Cooks, Proc. R. Soc. London, A $\underline{341}$, 135 (1974).
237. D.T. Terwilliger, R.G. Cooks and J.H. Beynon, Int. J. Mass Spectrom. Ion Phys., $\underline{18}$, 43 (1975).
238. J.F. Elder, R.G. Cooks and J.H. Beynon, Org. Mass Spectrom., $\underline{11}$, 423 (1976).
239. J.L. Holmes, A.D. Osborne and G.M. Weese, Int. J. Mass Spectrom. Ion Phys., $\underline{19}$, 207 (1976).
240. M.A. Haney and J.L. Franklin, J. Chem. Phys., $\underline{48}$, 4093 (1968).
241. J.H.D. Eland, Int. J. Mass Spectrom. Ion Phys., $\underline{9}$, 397 (1972).
242. I.G. Simm and C.J. Danby, J. Chem. Soc. Farad. Trans. II, $\underline{72}$, 860 (1976).
243. I.G. Simm, C.J. Danby, J.H.D. Eland and P.I. Mansell, J. Chem. Soc. Farad. Trans. II, $\underline{72}$, 426 (1976).
244. B. Brehm, J.H.D. Eland, R. Frey and A. Küstler, Int. J. Mass Spectrom. Ion Phys., $\underline{12}$, 197 (1973).
245. B. Brehm, J.H.D. Eland, R. Frey and A. Küstler, Int. J. Mass Spectrom. Ion Phys., $\underline{12}$, 213 (1973).
246. D.M. Mintz and T. Baer, Int. J. Mass Spectrom. Ion Phys., $\underline{25}$, 39 (1977).
247. R. Stockbauer, Int. J. Mass Spectrom. Ion Phys., $\underline{25}$, 89 (1977).
248. R. Stockbauer, Int. J. Mass Spectrom. Ion Phys., $\underline{25}$, 401 (1977).

249. C.E. Klots, D. Mintz and T. Baer, J. Chem. Phys., 66, 5100 (1977).
250. Z. Herman and K. Birkinshaw, Ber. Bunseng. Phys. Chem., 77, 566 (1973).
251. H.D. Haystrum, Rev. Mod. Phys., 23, 185 (1955).
252. W. Forst, "Theory of Unimolecular Reactions", Academic Press, New York, 1973.
253. W. Forst, Chem. Rev., 71, 339 (1971).
254. W. Forst and Z. Prâsil, J. Chem. Phys., 51, 3006 (1969).
255. W. Forst and Z. Prâsil, J. Chem. Phys., 53, 3065 (1970).
256. M.R. Hoare and T.W. Ruijgrok, J. Chem. Phys., 52, 113 (1970).
257. H.M. Rosenstock, J. Chem. Phys., 34, 2182 (1961).
258. S.W. Benson "Thermochemical Kinetics", 2nd Ed., Wiley, New York, 1976.
259. P. Langevin, Ann. Chim. Phys., Ser. 7, 317 (1903).
260. P. Langevin, Ann. Chim. Phys., 5, 245 (1905).
261. T.A. Lehman and M.M. Bursey, Wiley-Interscience, New York, 1976, p. 132.
262. M.S. Kim and F.W. McLafferty, J. Phys. Chem., 82, 66 (1978).
263. M.B. Wallenstein, Thesis, University of Utah, 1951.
264. M.B. Wallenstein and M. Krauss, J. Chem. Phys., 34, 929 (1961).

Chapter III. Application of the Quasi-Equilibrium Theory to Organic Mass Spectrometry

1   K Versus E Curves

Once the QET had been accepted by organic mass spectrometrists it proved to be extremely valuable in predicting or explaining a variety of phenomena related to the fragmentation of organic ions [1,3].

For the interpretation of the relative abundance of molecular and fragment ions as function of the internal energy, the ion lifetime, the activated complex configuration and variety of other parameters the discussion of k versus E curves has been shown to be very useful [1,2,3]. Both theoretical calculations and experimental results have demonstrated that near the threshold the rate constant, k, of a given decomposition process rises rapidly with increasing energy, but levels off at higher energies reaching a constant value at the highest energies. This is to be expected as the shortest physically meaningful decomposition time of an excited organic ion should be of the order of one vibrational period (10^{-13} - 10^{-14} s). Thus the maximum rate constant of a simple direct bond cleavage will be of the order of $\sim 10^{14}$ s^{-1}. As the intensity of a fragment depends both on the k(E) function and the energy distribution, P(E), it is instructive to represent both functions in one diagram [1,3]. Fig. III-1 shows in the lower part a schematic semilog plot of a single k(E) curve, in the upper part an arbitrary P(E) function. Moreover the energy intervals contributing to the molecular ion, the decay within the ion source and the first field free region of a conventional single focussing instrument have been indicated. On the left hand of Fig. III-1 the intensities of the molecular ion, the metastable ion and the normal fragment are plotted as function of the rate constant. The definitions of the molecular ion, the fragment ion, and the metastable ion are based on a time scale, but a *range* of rate constants contributes (with varying weight) to a given decomposition time. Hence the ranges of rate constants giving rise to the different types of ions are not sharp. Rather there is some overlapping of the various ranges as illustrated in Fig. III-1. Thus metastable ions decompose predominantly but not exclusively with rate constants between 5×10^5 and 5×10^6 s^{-1} as indicated by horizontal dashed lines in Fig. III-1. The abundance ratios of the molecular ion, the metastable ion, and the normal fragment ion are approximately proportional to the ratios of the appropriate areas of the P(E) function cor-

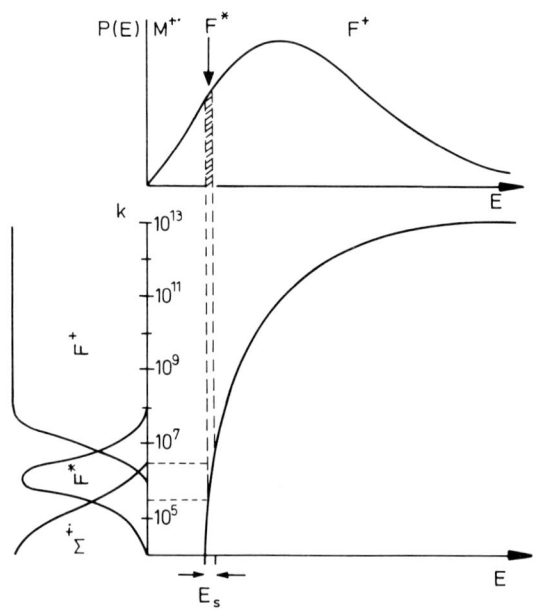

Fig. III-1. Schematic k(E) curve (lower part) and internal energy distribution (upper part). The energy range contributing to the molecular ion ($M^{+\cdot}$), the metastable ion (F^*) and the normal fragment (F^+) are indicated.

responding to each ion type. It is evident from Fig. III-1 that the intensity of the molecular ion is the larger the greater the difference between appearance and ionization potential (AP - IP). The relative abundance of the metastable ion of the assumed single decomposition process is the smaller the steeper the slope of the k(E) curve. As only a rather narrow range of energies contributes to the metastable ion, its abundance is also a sensitive function of the energy distribution of the molecular ion, especially as pronounced maxima and minima in the P(E) function are observed at low energies with many compounds. This critical influence of the energy distribution function is reflected, for example, in a strong temperature dependence of the metastable benzyl ion formation from 1,2-diphenylethane discussed in Section I-5.1 [4] and the abundance variations observed in the metastable decompositions of primary fragments of identical structures from different precursors (see Section V-3.4.3) [5-11].

The shape of the k(E) curves is predominantly determined by two factors: the geometry of the activated complex and the activation energy, E_o, of the process under study. This will be discussed qualitatively using the basic rate equation (see Section II-1.3)

$$k(E) = \frac{\sigma}{h} \frac{W^{\neq}(E - E_o)}{\rho(E)} \qquad (III-1)$$

(1) *Geometry of the activated complex.* As mentioned earlier the activated complex configuration of a rearrangement reaction is termed tight. This configuration is characterized by new bond formation and hence some vibrational frequencies will increase and some free internal rotations will be stopped leading to torsional vibrations. On the other hand, the transition state of a direct bond cleavage is termed loose and involves stretching of a bond along the reaction coordinate. Some vibrational frequencies will decrease and some torsional vibrations might change to internal rotations. Thus loose activated complexes have a larger number of low lying energy states than tight complexes. Therefore k increases much more rapidly with energy at low excitation energies for

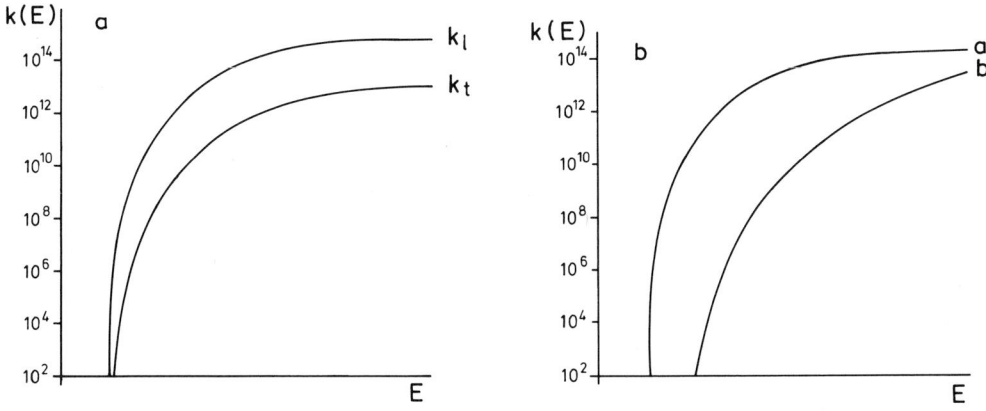

Fig. III-2. (a) Schematic k(E) curves for a reaction of either loose (l) or tight (t) transition state but identical activation energies. (b) Schematic k(E) curves for two reactions with identical transition state geometry but different activation energies.

the looser complex [12] (Fig. III-2a). At a given internal energy the intensity of a fragment is roughly proportional to the ratio of its rate constant relative to the sum of all rate constants. Thus if two decomposition processes of identical activation energy but different transition state geometry compete with each other, the process with the looser complex will lead to a more intense fragment at all internal energies, the effect becoming larger with increasing internal energy. However the dependence of the intensity ratio on the electron energy will be less pronounced than the dependence on the internal energy, as the fragment intensity at high electron energies results from an integration over a wide range of internal energies.

(2) *Activation energy*. While the energy dependence of the number of states in the activated complex, $W^{\neq}(E - E_o)$, is independent of the activation energy, the relative change of the density of states of the reactant, $\rho(E)$, with increasing excess energy $(E - E_o)$ becomes smaller as the activation energy is increased. Thus if $k(E)$ for two reactions with different activation energies but identical activated complex configuration is plotted as function of the excitation energy, E, the reaction with the higher activation energy shows a slower rise of $k(E)$ than that with lower activation energy [12]. This is illustrated schematically in Fig. III-2b. If two reactions proceeding via an identical transition state but differing in the activation energies are considered and a smooth energy distribution is assumed, the process with the lower activation energy will lead to the more abundant fragment. The effect will be especially pronounced at low internal energies (electron energies), i.e. for the metastable decomposition. If there is a large number of competing processes, only the processes with lowest activation energies will give rise to abundant metastable ions.

2 Internal Energy and Ion Lifetime Dependence of Competing Rearrangement Reactions and Direct Bond Cleavages

2.1 Energy Dependence

So far only the energy dependence of competing reactions of either identical activated complex configurations or identical activation energies have been considered. However this ideal situation will be rarely met in reality. Most competing reactions differ both in the geometry of their activated complexes and in the activation energy. Of special interest are systems where reactions having a loose activated complex and a high activation energy, such as direct bond cleavages, or a tight complex and a low activation energy, as assumed for rearrangement reactions,[*] are in competition as illustrated in Fig. III-3. Here the $k(E)$ curves intersect above the metastable "window" (log k $\sim$ 5.5 - 6.5).

---

[*] In rearrangement reactions bonds are not only broken, but new bonds are formed, which may explain the low activation energy.

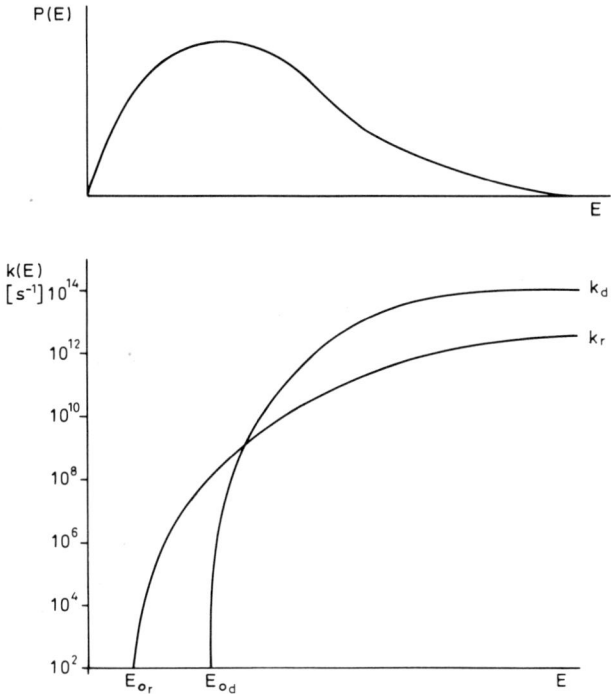

Fig. III-3. K(E) curves for competing rearrangement (r) and direct cleavage reactions (d). The upper part of the figure shows an arbitrary internal energy distribution function.

Assuming a smooth energy distribution function as shown in the figure the process with the lower activation energy and tighter transition state (rearrangement) will dominate at low internal energies, the process with higher activation energy and looser complex (direct cleavage) at high internal energies. Generally speaking the relative fragment intensities reflect the activation energies at low internal energies, but the transition state geometry at high internal energies. Experimental evidence for this postulated behavior was first reported by Williams and Cooks [13] who found for a variety of competing rearrangement reactions and (assumed) direct bond cleavages an increase of the rearrangement relative to the direct cleavage with decreasing electron energy. In support of their observation similar results were obtained by other authors [14-23]. Brown [22] studied eight compounds, the mass spectra of which show only one major direct cleavage and one major rearrangement. Scheme III-1 illustrates the schematic reaction mechanisms. In Table III-1 the difference between the appearance and the ionization potential, IP - AP, representing roughly the activation energy, is given for each reaction. It is evident that for all compounds the activation

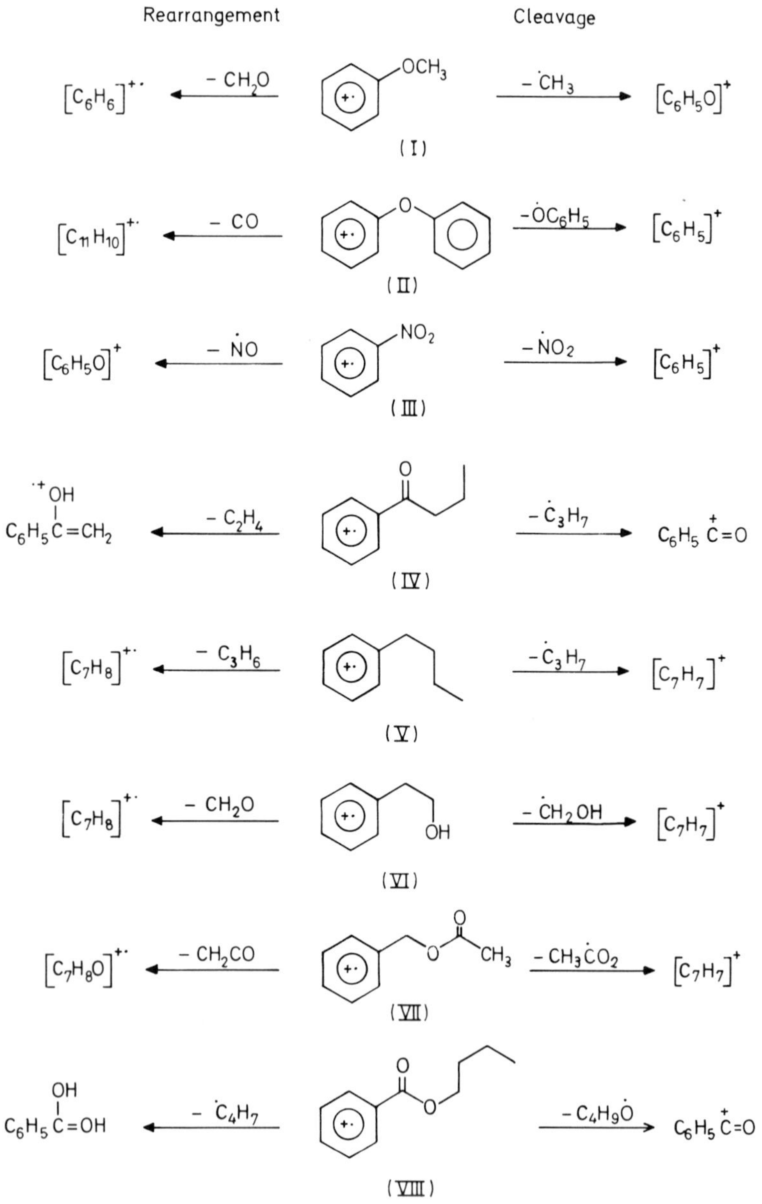

III Application of the QET

Scheme III-1 (Rearrangement vs. Cleavage pathways):

(I) Anisole (C₆H₅OCH₃):
- Rearrangement: $-CH_2O$ → $[C_6H_6]^{+\cdot}$
- Cleavage: $-\dot{C}H_3$ → $[C_6H_5O]^{+}$

(II) Diphenyl ether:
- Rearrangement: $-CO$ → $[C_{11}H_{10}]^{+\cdot}$
- Cleavage: $-\dot{O}C_6H_5$ → $[C_6H_5]^{+}$

(III) Nitrobenzene:
- Rearrangement: $-\dot{N}O$ → $[C_6H_5O]^{+}$
- Cleavage: $-\dot{N}O_2$ → $[C_6H_5]^{+}$

(IV) Butyrophenone:
- Rearrangement: $-C_2H_4$ → $C_6H_5\overset{+\cdot\,OH}{C}=CH_2$
- Cleavage: $-\dot{C}_3H_7$ → $C_6H_5\overset{+}{C}=O$

(V) Butylbenzene:
- Rearrangement: $-C_3H_6$ → $[C_7H_8]^{+\cdot}$
- Cleavage: $-\dot{C}_3H_7$ → $[C_7H_7]^{+}$

(VI) 2-Phenylethanol:
- Rearrangement: $-CH_2O$ → $[C_7H_8]^{+\cdot}$
- Cleavage: $-\dot{C}H_2OH$ → $[C_7H_7]^{+}$

(VII) Benzyl acetate:
- Rearrangement: $-CH_2CO$ → $[C_7H_8O]^{+\cdot}$
- Cleavage: $-CH_3\dot{C}O_2$ → $[C_7H_7]^{+}$

(VIII) Butyl benzoate:
- Rearrangement: $-\dot{C}_4H_7$ → $C_6H_5\overset{OH}{\underset{|}{C}}=OH$
- Cleavage: $-C_4H_9\dot{O}$ → $C_6H_5\overset{+}{C}=O$

Scheme III-1

energy for the rearrangement is indeed considerably lower than for the direct cleavage. Furthermore the relative abundance ratios of the rearrangement and direct cleavage fragment ions are shown in Table III-1 as functions of the electron energy both for the normal fragment and the metastable ion. For the normal fragment a dramatic increase of this

Table III-1. Competing Rearrangement and Direct Cleavage Process as Function of the Electron Energy.

			[Rearrangement]/[Cleavage] as Function of the Electron Energy[a]									
			Normal Ion Intensities					Metastable Ion Intensities				
Compound	Reaction	AP-IP[a]	70	15	13	11	9	70	15	13	11	9
anisole (I)	($M-CH_3^\cdot$) ($M-CH_2O$)	3.51 2.91	3.5	3.6	4.2	14	—	26	27	26	~23	—
phenyl ether (II)	($M-OC_6H_5^\cdot$) ($M-CO$)	6.31 4.12	0.82	6.5	42	~450	—	b	—	—	—	—
nitrobenzene (III)	($M-NO_2^\cdot$) ($M-NO^\cdot$)	1.64 0.39	0.095	0.23	0.47	2.0	—	7.6	5.0	5.4	4.0	—
butyrophenone (IV)	($M-C_3H_7^\cdot$) ($M-C_2H_4$)	1.46 0.41	0.081	0.22	0.33	0.69	~3	5.8	5.5	5.5	6.0	8.3
n-butylbenzene (V)	($M-C_3H_7^\cdot$) ($M-C_3H_6$)	2.57 1.39	0.48	1.1	1.6	4.1	—	6.2	9.8	11	14	—
2-phenylethanol (VI)	($M-CH_2OH^\cdot$) ($M-CH_2O$)	2.64 0.83	0.45	1.4	2.6	8.7	~45	30	38	41	45	—
benzyl acetate (VII)	($M-CH_3CO_2^\cdot$) ($M-CH_2CO$)	3.01 0.70	1.6	9.5	37	85	—	b	—	—	—	—
n-butyl benzoate (VIII)	($M-C_4H_9O^\cdot$) ($M-C_4H_7$)	2.25 0.11	0.70	4.7	25	59	—	b	—	—	—	—

[a] in eV.

[b] no cleavage metastable ion observed

ratio is observed while the electron energy is lowered as predicted by the QET. The same behavior is observed for most, but not all metastable ions. Nitrobenzene shows precisely the opposite behavior and it was concluded that in this case the k versus E curves intersect below the metastable rate region [22].

It is noteworthy that even at 70 eV electron energy in all instances the metastable ion for the rearrangement is more abundant than that for direct cleavage. (The metastable ion for the latter reaction could not be detected at all in some cases). Similar results have been reported by McLafferty and Fairweather [24] for a variety of competing direct bond cleavages and rearrangement reactions. This result is again a direct consequence of the low activation energy for the rearrangement reaction. However, not every rearrangement reaction must necessarily be accompanied by an abundant metastable ion. Thus it has been reported by Elder et al. [25] that many McLafferty rearrangements do not show a metastable peak as result of competing rearrangements with even lower activation energies.

Thus, both the energy dependence and the metastable ion ratios [26] may be used with confidence to identify reactions with tight transition states, provided there is no significant interference from competing reactions (see Section IV-2.5).

2.2 Ion Lifetime Dependence

As a given rate constant, k, contributes predominantly to the decomposition at time $t = 1/k$, Fig. III-3 also allows the dependence of the relative fragment abundance on the ion lifetime to be predicted. Assuming a uniform energy distribution, at short ion lifetimes (corresponding to high excitation energies) the process with loose transition state geometry and high activation energy will dominate, whilst at long ion lifetimes (corresponding to low internal energies) the process with low activation energy, but tight transition state will prevail. Thus the relative fragment abundance reflects the transition state geometry at short ion lifetimes, but the activation energy at long times.

Ion lifetime measurements using the field ionization method [27,28] are especially suited to test this postulate as this technique allows the time resolved study of the fragmentations of excited organic ions in the gas phase from 10^{-11} to 10^{-5} s after ionization. The principles of these measurements have been briefly outlined in Section II-2.1.3. Competing rearrangements and direct bond cleavages in a variety of compounds have been studied using this method [29-37]. The

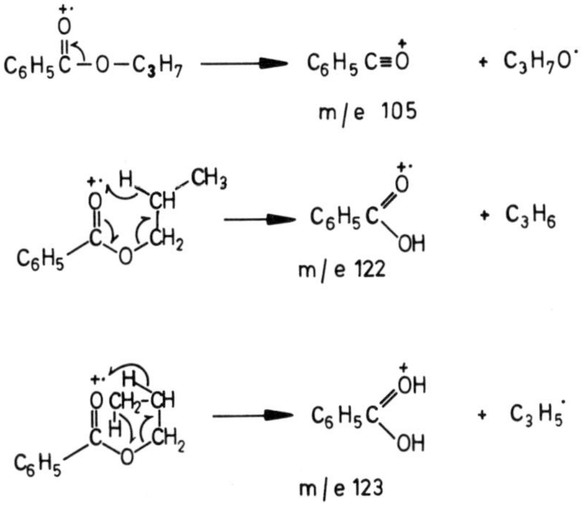

III-2 Energy and Lifetime Dependence

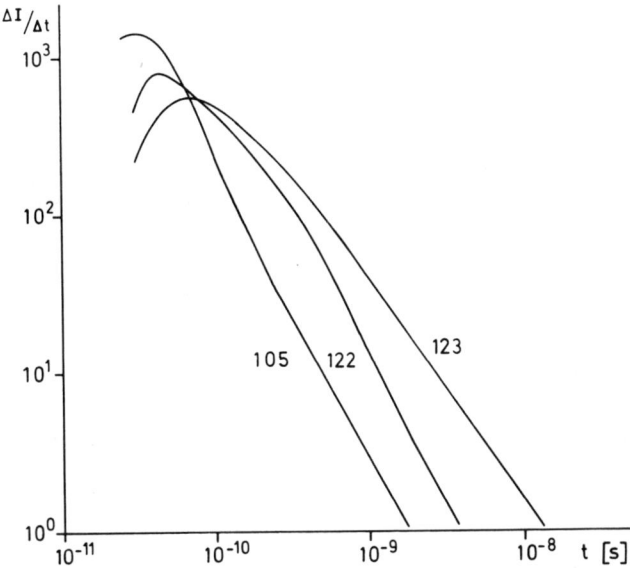

Scheme III-2

Fig. III-4. Benzoic acid propylester: Relative rates as a function of the ion lifetime [34] (double hydrogen rearrangement: m/e 123; single hydrogen rearrangement: m/e 122; direct cleavage: m/e 105).

results will be discussed with n-propylbenzoate as example [34]. The electron impact mass spectrum of this compound shows three intense fragments which may be formed according to Scheme III-2. The benzoyl ion at m/e 105 is formed by direct cleavage (AP - IP = 2.7 eV), while transfer of a single hydrogen gives rise to a fragment at m/e 122 (AP - IP = 1.1 eV) and migration of two hydrogens leads to m/e 123 (AP - IP = 0.3 eV). Fig. III-4 shows the relative rate of formation, di/dt, for these three processes as function of the ion lifetime on a double logarithmic scale. As a result of its low activation energy the double hydrogen rearrangement (m/e 123) is by far the dominant process at long ion lifetimes while the slowest rate is observed for the benzoyl ion having the highest activation energy, thus demonstrating that at long times the relative rates or fragment intensities reflect the activation energies. The situation is reversed at the shortest resolvable time ($t = 2 \times 10^{-11}$ s). As result of its loose transition state geometry it is now the direct cleavage process which has the highest rate in spite of its high activation energy, followed by the single and the double hydrogen rearrangement which have successively tighter transition state configurations. It is apparent that at the shortest decomposition time the geometry of the activated complex is rate determining. Finally two observations are of interest:

(1) The rates of formation not only for the hydrogen rearrangements, but also for the direct cleavage pass through a maximum between 10^{-11} and 10^{-10} s after ionization.

(2) Hydrogen rearrangements are often surprisingly fast reaching maximum rates between 10^{-11} and 10^{-10} s after ionization. Numerous examples for such fast hydrogen rearrangements have been reported [30-35,37].

2.3 Atom Scrambling

The term "atom scrambling" or "atom randomization" is normally used to characterize reversible atom exchange reactions in an ion prior to decomposition in which some or all atoms of the ion take part. If only hydrogen atoms are involved in this scrambling process the skeletal constitution of the ion is normally retained prior to decomposition. Such exchange reactions which usually proceed via specific mechanisms (see Section IV-2.6.1) may lead to a completely statistical distribution of atoms within the ion or within part of it. The extent of atom scrambling can only be studied if the ion decomposes. Thus any kinetic investigation of a scrambling reaction always has to deal both with the kinetics of the actual scrambling process and those of the succes-

III-2 *Energy and Lifetime Dependence*   99

sive decomposition which leads to the detection of scrambling, i.e. one is dealing with a consecutive reaction, the first step of which is a rearrangement or a series of reversible rearrangements with a relatively tight transition state. If the successive decomposition occurs by simple bond split then the consecutive reaction (scrambling and decomposition) is competing with a direct cleavage (without prior scrambling). Thus both the transition state geometry and the presence of a consecutive reaction should lead to a reduction of scrambling at high internal energies (short ion lifetimes). The same is true if the overall scrambling process competes with a rearrangement reaction. However, as result of the tighter transition state of the latter the energy dependence of the scrambling should be less pronounced [38]. It has been shown by a variety of authors that the extent of atom scrambling indeed depends sensitively on both the internal energy [38-48] and the ion lifetime [37,49-60]. The hydrogen scrambling in the benzoic acid molecular ion is one of the best studied examples [39,43,49,61-67]: Deuterium labelling revealed that only the hydroxylic hydrogen and the two hydrogens in the ortho positions take part in the scrambling process [61,66]. A mechanism involving hydrogen transfer via a five-membered transition state and an intermediate in which both oxygens become equivalent has been proposed by Shapiro et al. [62,63]. (Scheme III-3). An equilibration between the hydroxylic hydrogen and both ortho hydrogens is achieved by revolution of the side chain [62,63]. The authors conclude that at least eight such half revolutions must have occurred to account for their labelling data. This mechanism already implies that the H/D exchange depends on the ion lifetime and thus on the internal energy.

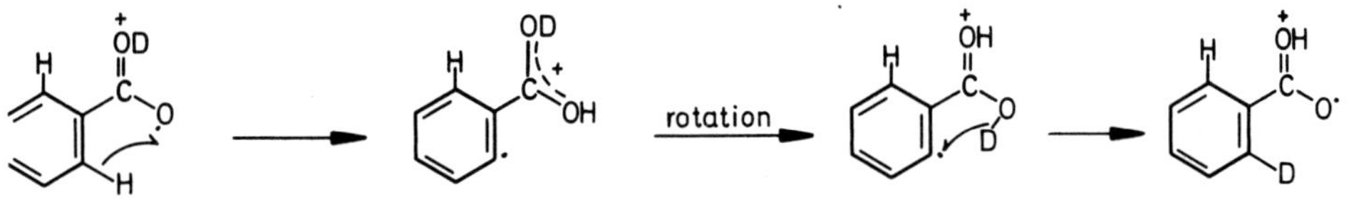

Scheme III-3

Hence Beynon et al. [43] noticed that the extent of H/D exchange observed in the (M-OH)$^+$ fragment formed in the ion source was considerably smaller than that in the metastable ion, pointing at a dependence both on the internal energy and on ion lifetime. A more detailed study on the energy dependence of this scrambling reaction was reported by Howe and McLafferty [39]. The basic principle of their method is the following: By studying the scrambling in fragments of successively higher activation energy one is sampling molecular ions with increasing average internal energy.[*]) Thus by measuring the extent of hydrogen scrambling prior to OH· and CO + OH· loss for both the metastable ion and the normal fragment and by including collision induced decompositions the

Table III-2. Relative Ratio of OH and OD Loss as Function of the Internal Energy in Benzoic Acid-O-d [39]

Energy (eV)	$[M^{+\cdot} - OH] / [M^{+\cdot} - OD]$	Process
∼12.7	1.5	$[M^{+\cdot}\text{-OH}]/[M^{+\cdot}\text{-OD}]$ meta
>12.7	1.04	$[M^{+\cdot}\text{-OH}]/[M^{+\cdot}\text{-OD}]$ coll
12.7 - 16.0	0.34	$[M^{+\cdot}\text{-OH}]/[M^{+\cdot}\text{-OD}]$ source
∼16.0	0.083	$[M^{+\cdot}\text{-OH-CO}]/[M^{+\cdot}\text{-OD-CO}]$ meta
>16.0	0.05	$[M^{+\cdot}\text{-OH-CO}]/[M^{+\cdot}\text{-OD-CO}]$ source

meta = metastable ions
coll = collision induced fragments
source = ion source fragmentation

authors were able to determine the hydrogen scrambling in benzoic acid-O-d molecular ions of five distinct energy ranges (Table III-2). Whilst at the lowest internal energy (∼3 eV) and a lifetime of ∼10^{-6} s the ratio for OH/OD loss from benzoic acid-O-d is close to that expected for complete equilibration of the three hydrogens involved (= 2.0 in the absence of an isotope effect), the scrambling is considerably reduced in molecular ions with an internal energy of more than 4.3 eV. The same method has been employed to study hydrogen scrambling as function of the internal energy in toluene [40], cycloheptatriene [40] and 2-phenyl-ethanol molecular ions [41].

The hydrogen scrambling in benzoic acid has also been studied as function of the ion lifetime using the field ionization technique [49] (Table III-3). While no scrambling is detectable at $t < 6 \times 10^{-11}$ s within the reproducibility of the measurements, 88% of the molecular ions show the statistical distribution of the hydrogens at 10^{-5} s after ionization.

---

[*] For a given fragmentation process $M^{+\cdot} \to F^+$ the molecular ion will predominantly have internal energies between the activation energy for F^+ and that for the process of next higher activation energy.

Table III-3. Degree of Scrambling (α) as Function of the Ion Lifetime in the Benzoic Acid Molecular Ion [49]

Lifetime (s)	α
$<6 \times 10^{-11}$	<0.03
2×10^{-10}	0.13 ± 0.01
3.5×10^{-10}	0.17 ± 0.01
1.7×10^{-9}	0.27 ± 0.03
8.0×10^{-9}	0.34 ± 0.03
$2.2 - 7.6 \times 10^{-8}$	0.39 ± 0.07
$0.76 - 1.9 \times 10^{-7}$	0.51 ± 0.03
$0.63 - 4.1 \times 10^{-6}$	0.74 ± 0.03
$0.57 - 1.1 \times 10^{-5}$	0.88 ± 0.04

The dependence of hydrogen-deuterium exchange on the ion lifetime has also been reported for cyclohexene [51], 2-methylpropene [56], 1-butene [50], toluene [49], aliphatic aldehydes [55] and ketones [37, 52,53], 3-phenylpropanal [59], benzyl cyanide [57] and n-pentylbenzene [60]. Some mechanistic aspects of these scrambling processes will be discussed in Section IV-2.6.1. The reported energy and ion lifetime dependence requires that the activation energy for the scrambling reaction should be lower than or comparable with that of the decomposition process in which the scrambling is detected. The data reported on energy barriers for H/D exchange support this conclusion. Thus ab initio calculations demonstrated that the activation energy for H-migration in $C_2H_5^+$ (~ 0.5 eV) [68] and $C_3H_7^+$ (<0.3 eV) [69] is far less than that for any dissociation of the ion. Similarly an energy barrier of 0.8 eV was calculated for H/D exchange in $C_2H_4^{+\cdot}$ [70] while AP measurements suggest a value of 1.6 eV [71,72]. Both values are considerably lower than the lowest threshold for decomposition (2.62 eV for H_2 loss) [73]. Finally, the calculated activation energy for H-scrambling in the benzene molecular ion (1.9 eV) [74] is considerably lower than that for fragmentation (4.5 eV).

However it has been suggested that the hydrogen scrambling in 2-methylpropane has both a high activation energy and a loose transition state as this H/D exchange is only detected in the 70 eV spectrum in those fragments having a high activation energy (e.g. CH_3^+) [75]. An unusual energy dependence of a H/D-exchange reaction has also been observed for decomposing $C_4H_8O^{+\cdot}$ ions [76].

Whilst so far only H/D-exchange reactions prior to decomposition have been discussed, such scrambling processes are also observed with other atoms. Numerous examples have been reported for $^{12}C/^{13}C$-exchange reactions in organic ions. For instance, complete carbon scrambling has been observed in the molecular ions of isomeric butenes [77], benzene [78,79], cyclopropabenzene [80], naphthalene [81], thiophene [82] and pyridine [83]. As expected such scrambling processes are reduced at shorter ion lifetimes [58,84].

3 Distribution of Internal Energy between Fragments

3.1 Internal Energy of Product Ions

If a fragment A^+ is formed by the reaction $M^{+\cdot} \rightarrow A^+ + N^{\cdot}$ its internal energy depends on the kinetic energy of the fragments and the internal energy of the neutral. Under the quasi-equilibrium hypothesis the probability that an energy remains as internal energy in the fragment ion A^+ is proportional to the number of ways of putting this energy into A^+ and the remainder of the available energy into the neutral fragment $N^{\cdot}$ [2]. Thus the probability that the ion A^+ has an energy between ε and $\varepsilon + d\varepsilon$ is given by

$$\rho(E,\varepsilon) \, d\varepsilon = \frac{\rho_A(\varepsilon) \, \rho_N(E - E_o - \varepsilon) \, d\varepsilon}{\int_0^{E-E_o} \rho_A(\varepsilon') \, \rho_N(E - E_o - \varepsilon') \, d\varepsilon'} \qquad (III-2)$$

where E is the internal energy of the reactant ion $M^{+\cdot}$, E_o is the activation energy, $\rho_A(\varepsilon) \, d\varepsilon$ is the number of states of the ion A^+ with energy between ε and $\varepsilon + d\varepsilon$, $\rho_N(\varepsilon) \, d\varepsilon$ the corresponding number of states in the neutral fragment N.

The major factors influencing the distribution of internal energy of the primary fragment have been discussed qualitatively by McAdoo et al. [85]. They conclude that the energy distribution $P(E)_{A^+}$ of the primary product A^+ as initially formed will be determined by (1) $P(E)_{M^{+\cdot}}$ for the molecular ion, (2) k(E) for the reaction forming A^+ ($M^{+\cdot} \rightarrow A^+ + N^{\cdot}$), (3) k(E) for other reactions competitive with A^+ formation and (4) partitioning of excess energy between the fragment ion and the neutral.

The situation is illustrated in a simplified schematic diagram in Fig. III-5. In the upper part of this figure an arbitrary P(E) function with a pronounced minimum is shown for the molecular ion.

III-3 *Distribution of Internal Energy between Fragments* 103

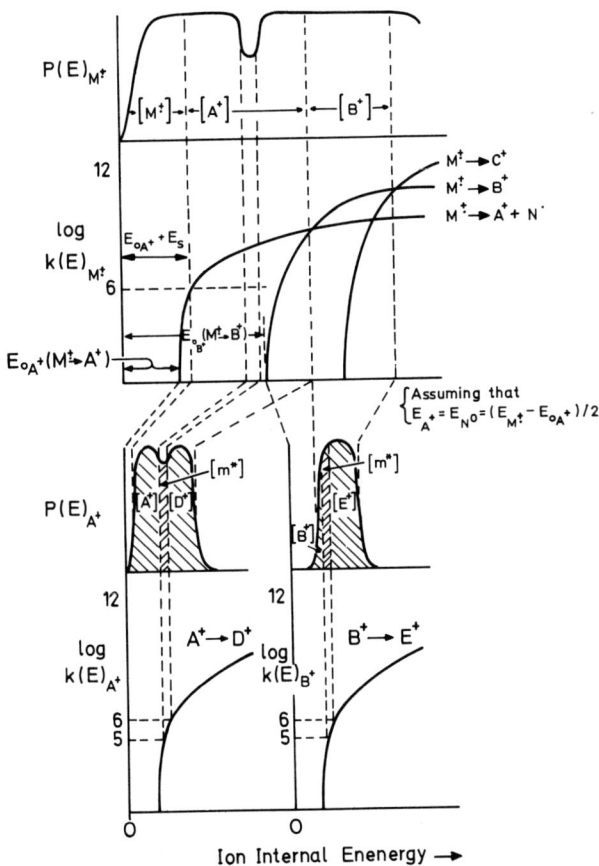

Fig. III-5. Relationship of P(E) and k(E) for molecular and product ions in mass spectral reactions; see text for definitions [85]. (By courtesy of Heyden & Son Ltd.)

The three primary decomposition reactions ($M^{+\cdot} \longrightarrow A^+$, $M^{+\cdot} \longrightarrow B^+$ and $M^{+\cdot} \longrightarrow C^+$) differ considerably in their activation energies and transition state geometries so that within a given energy interval only one process is of significant abundance. It is further assumed that the reverse activation energy can be neglected and that the excess internal energy of $M^{+\cdot}$ is divided equally between the ionic fragment A^+ and the neutral fragment $N^{\cdot}$ *). A^+ ions can be formed from molecular ions having

---

* This means that the molecular ion fragments in two halves and equipartitioning of the energy is assumed.

energies $E \geq E_{OA^+}$. As a result of the kinetic shift, E_s, (see Section III-4.1) there is, however, a very low probability of forming A^+ ions in the ion source with zero energy as illustrated in Fig. III-5. At higher energies the initial abundance of the molecular ion determines the abundance of the corresponding product ions. Thus the pronounced valley in $P(E)_{M^{+\cdot}}$ in Fig. III-5 is reflected in $P(E)_{A^+}$. However, as a result of the energy partitioning assumed above, the energy scale is reduced by a factor of two. At even higher internal energies the formation of B^+ from the molecular ion becomes competitive with the formation of A^+ as shown by the respective $k(E)$ functions. Where the rate constants for the reaction $M^{+\cdot} \to A^+$ and $M^{+\cdot} \to B^+$ become equal, half of the molecular ions of the corresponding internal energy decompose to yield A^+ and the remainder to yield B^+. Using this reasoning the energy distributions of the fragments A^+ and B^+ can be derived qualitatively as illustrated in the lower part of Fig. III-5. It is of special interest that the competition from A^+ leads to an energy distribution $P(E)_{B^+}$ with fewer low energy ions than in $P(E)_{A^+}$ explaining the operation of a "competitive shift" encountered in appearance potential determinations (see Section III-4.3).

The $P(E)_{A^+}$ function shown in Fig. III-5 would be strongly affected by a further competing decomposition process of similar activation energy to A^+ depleting the number of molecular ions which will decompose to give A^+ in the energy range shown in the figure. Thus competition from a process with a slower increase of $k(E)$ would take away more $M^{+\cdot}$ ions leading to low energy A^+ ions and thus would shift $P(E)_{A^+}$ to higher average energies while competition from a process with a steeper $k(E)$ function would shift $P(E)_{A^+}$ to lower average energies.

So far the discussion was based on the assumption of equipartitioning of the energy i.e. it was assumed that the energy retained by the primary fragment ion was directly proportional to the fraction of internal degrees of freedom of the daughter ion compared to those of the daughter ion and the neutral fragment combined. However, Wallenstein and Krauss [86,87] pointed out that while this was true on the average, there are very large fluctuations about this average. Applying this "fluctuation effect" to Fig. III-5 means that molecular ions of internal energy corresponding to the "valley" in $P(E)_{M^{+\cdot}}$ will produce A^+ ions of average energy $(E_{M^{+\cdot}} - E_{OA^+})/2$. However, this is only the most probable value. There are significant probabilities that the resulting ion A^+ will have a wide range of internal energies below and above this average value. The energy fluctuation thus has the effect of smoothening any structure in $P(E)_{A^+}$ (such as the valley discussed above) resulting from structure in $P(E)_{M^{+\cdot}}$.

A variety of experimental tests have been devised by McAdoo et al. [85] in support of the above discussion. Only one of these experiments will be mentioned: In the $P(E)_{A^+}$ distribution in Fig. III-5 the energy range contributing to metastable decomposition of A^+ to give D^+ is indicated. Decreasing the electron energy shifts $P(E)_{M^{+\cdot}}$ and thus $P(E)_{A^+}$ to lower energies, depleting the energy range for the metastable decay much faster than the energy range corresponding to the undecomposed ion A^+. Thus $[m^*]/[A^+]$ decreases to zero if the electron energy is lowered to the appearance potential of the metastable ion as illustrated in Fig. III-6 for the reaction $C_2H_5O^+ \xrightarrow{*} H_3O^+ + C_2H_2$ [85] (2-butanol) and $C_6H_{13}^+ \xrightarrow{*} C_3H_7^+ + C_3H_6$ (n-hexadecane).

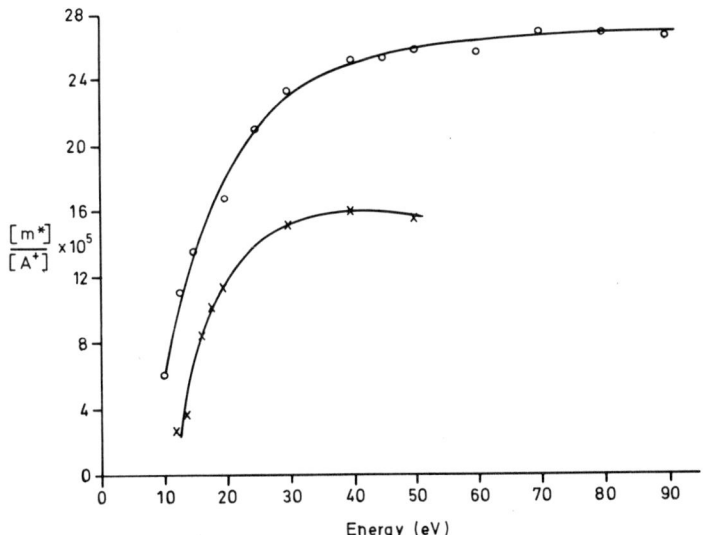

Fig. III-6. Abundance ratio of metastable to precursor ion, $[m^*]/[A^+]$, as a function of the electron energy [85]. o: 2-butanol ($C_2H_5O^+ \xrightarrow{*} H_3O^+ + C_2H_2$); x: n-hexadecane ($C_6H_{13}^+ \xrightarrow{*} C_3H_7^+ + C_3H_6$).

3.2 Degree of Freedom Effect

It has been shown in the previous section for the reaction $M^{+\cdot} \to A^+ \to D^+$ (m^*) that the metastable to precursor intensity ratio is a sensitive function of the internal energy distribution of the primary fragment, $P(E)_{A^+}$, which in return is influenced by the energy distribution of the molecular ion. Instead of varying the electron energy as shown in Fig. III-6 $P(E)_{A^+}$ can also be influenced by varying the number of degrees of freedom (DOF) in the molecular ion i.e. by studying an homologous series of compounds [5,6,88]. It was first shown by McLafferty and Pike [5,6] that the logarithm of the metastable to precursor ion abundance ratio $\log[m*(A^+ \to D^+)]/[A^+]$ is a linear function of the reciprocal of the number of internal degrees of freedom of the molecular ion (DOF = 3n-6, where n is the number of atoms in the molecular ion). This is illustrated in Fig. III-7 for the process $M^{+\cdot} \to C_3H_6O^{+\cdot} \to C_2H_3O^+$ (m*) for 16 2-alkanones ranging from $C_6H_{12}O$ to $C_{31}H_{62}O$ [89]. Similar

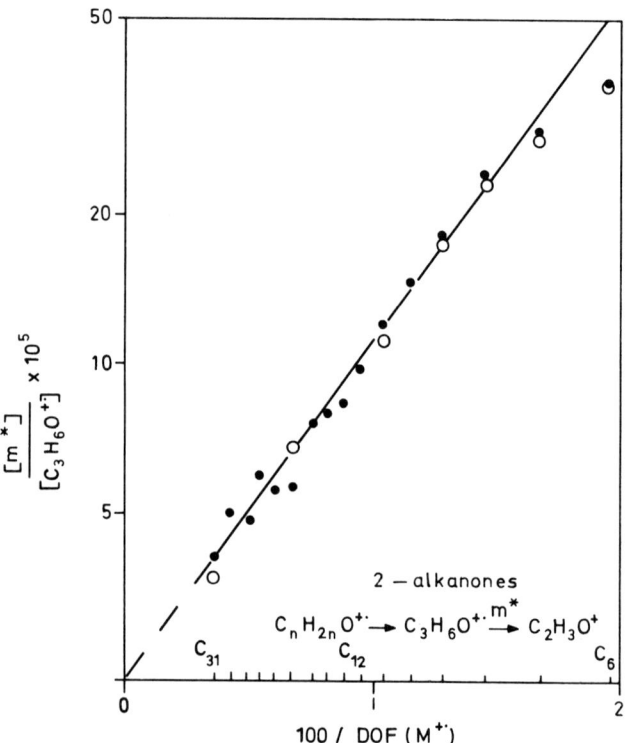

Fig. III-7. Degree of freedom effect for 2-alkanones (filled circles = experimental values, open circles = calculated values) [89]. (By courtesy of the American Chemical Society.)

linear correlations were observed for 30 other homologous series [89].

A qualitative explanation of this degree of freedom effect is readily on hand: With increasing size of the molecular ion increasingly larger neutral fragments carry away a larger fraction of the molecular ions internal energy leading to a decrease of the average internal energy of the primary fragment A^+. As discussed in the previous section (see Fig. III-5) this shift of $P(E)_{A^+}$ to lower energies depletes the energy range for the metastable decay much faster than the energy range corresponding to the undecomposed fragment A^+ as illustrated in Fig. III-8. For a given energy distribution of the molecular ion the QET allows the energy distribution of a primary fragment and thus the

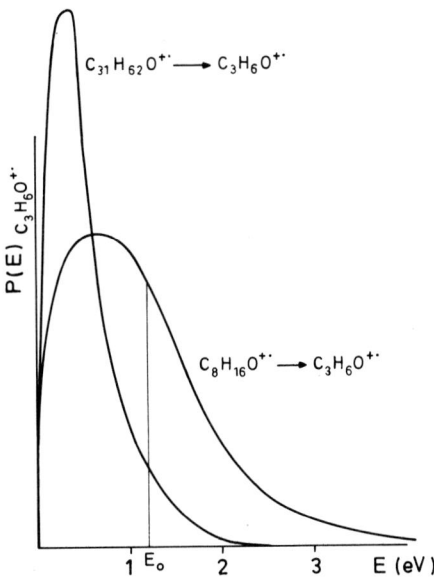

Fig. III-8. Calculated internal energy distribution function for $C_3H_6O^{+\cdot}$ fragment ions generated from $C_8H_{16}O^{+\cdot}$ and $C_{31}H_{62}O^{+\cdot}$ (semischematic representation). The vertical line represents the "metastable window" for $C_3H_6O^{+\cdot} \rightarrow C_2H_3O^+$ [89].

metastable to precursor abundance ratio to be calculated (eq. III-2). Such calculations of the DOF effect for $C_2H_5O^+$ ions from homologous 1-alkanols by Lin and Rabinovitch [90] were in qualitative agreement with experimental results, predicted however a non-linear DOF effect ($\log[m^*]/[A^+]$ approaches $-\infty$ as 1/DOF approaches zero). The calculations were repeated by Bente et al. [89] using photoelectron spectra to determine $P(E)_{M^{+\cdot}}$ and taking into account an increasing kinetic shift with increasing size of the molecular ion as result of competing reactions with low activation energy and tight transition state. Their calculations show a linear DOF plot in good agreement with the experimental data (Fig. III-7).

Lowering the electron energy may increase the slope of the DOF plot [85]. At higher molecular weights the A^+ ions yielding $m^*(A^+ \rightarrow D^+)$ must come from higher energy molecular ions whose relative abundances

108   III *Application of the QET*

should be reduced more by the reduction in electron energy. Finally, if the primary fragment ion A^+ does not decompose unimolecularly but collision-induced [91], the degree of freedom effect is considerably reduced or even negligible as the abundance of collision-induced fragments is relatively insensitive to internal energy variation.

4 Determination of Appearance Potentials

4.1 The Kinetic Shift

The appearance potential, AP, of a fragment ion represents the minimum energy necessary to form this ion within the mass spectrometric time scale. Appearance potential determinations are of great importance for the determination of activation energies and heats of formation of ions. High quality appearance potentials are usually determined by threshold measurements using mono- or quasimonoenergetic electrons or photons [92] as discussed in Section I-4. Although the accuracy of such AP measurements is somewhat lower than that of ionization potential (IP) measurements as result of the more gradual increase of the ionization efficiency curve at onset, reproducibilities ranging between 0.01 and 0.1 eV can be achieved with these techniques [92]. However, to date most appearance potentials have been determined with non-monoenergetic ions using one of the various extrapolation methods described in detail by Kiser [93]. With these techniques the accuracy ranges from less than 0.1 to more than 0.5 eV in an unpredictable manner [92].

The difference between appearance and ionization potential, AP - IP, cannot necessarily be equated with the activation energy. This difference is too high by the amount of the *kinetic shift*, the excess energy required to drive the fragmentation fast enough for the ion to decompose in the mass spectrometer source. Thus the existence of a kinetic shift, which was first discussed in detail by Chupka [94], is a direct consequence of the QET. If, for instance, an ion is formed in the ion source of a conventional magnetic type instrument, an energy E_s in excess of the activation energy, E_o, is necessary to obtain a rate of $\sim 10^6$ s^{-1} corresponding to the maximum ion lifetime in the source[*]) (see Fig. III-1).

---

[*]
   This is the most probable rate leading to decomposition after 10^{-6} s. However, considerably smaller rates ($\sim 10^4$ s^{-1}) also contribute with low probability to this decomposition time. This is of importance for threshold measurments.

It is apparent from Fig. III-1 that the kinetic shift is the larger the shallower the rise of k with E. Thus keeping the discussion in Section III-1 in mind, *the kinetic shift will be the larger the larger the molecule, the higher the activation energy and the tighter the activated complex.* The excess energies required to obtain rate constants of 10^5, 10^6, and 10^7 s^{-1} were calculated by Vestal [2] for a variety of decomposition processes. The results are summarized in Table III-4. It is evident

Table III-4. Calculation of Excess Energies Required for Selected Reactions to Attain Rates in the Range Appropriate to the Mass Spectrometer [2]

Reaction	E_o, eV	N^a	\multicolumn{3}{c}{Excess energy ($E' - E_o$) for indicated reaction rate}		
			10^5 s^{-1}	10^6 s^{-1}	10^7 s^{-1}
$CH_4^+ \to CH_3^+ + H$	1.70	9	<0.011	<0.011	<0.011
$C_2H_6^+ \to C_2H_4^+ + H_2$	1.00	18	<0.01	<0.01	0.06
$C_3H_8^+ \to C_3H_7^+ + H$	0.94	27	<0.01	0.02	0.17
$i\text{-}C_4H_{10}^+ \to sec\text{-}C_3H_7^+ + CH_3$	0.69	36	0.03	0.17	0.41
$n\text{-}C_4H_{10}^+ \to i\text{-}C_4H_{10}^+$	1.0	36	0.02	0.11	0.48
$C_2H_2^+ \to C_2H^+ + H$	5.8	7	<0.01	0.01	0.04
$C_2H_4^+ \to C_2H_2^+ + H_2$	2.63	12	<0.01	<0.01	0.01
$C_3H_6^+ \to C_3H_5^+ + H$	2.07	21	0.05	0.19	0.38
$C_4H_6^+ \to C_3H_3^+ + CH_3$	2.28	24	0.27	0.45	0.70
$C_6H_6^+ \to C_6H_5^+ + H$	3.9	30	1.50	2.00	2.60
$C_7H_8^+ \to C_7H_7^+ + H$	1.95	39	0.60	0.85	1.25
$CH_3OH^+ \to CH_2OH^+ + H$	1.15	12	<0.01	<0.01	0.01
$C_2H_5OH^+ \to C_2H_4OH^+ + H$	0.63	21	<0.01	<0.01	0.01
$i\text{-}C_3H_7OH^+ \to C_2H_4O^+ + CH_4$	0.47	30	<0.01	0.01	0.03
$n\text{-}C_3H_7OH^+ \to C_3H_6^+ + H_2O$	0.76	30	<0.01	0.08	0.24
$CH_3COCH_3^+ \to CH_3CO^+ + CH_3$	0.70	30	<0.01	0.01	0.01
$C_2H_5COCH_3^+ \to C_2H_5CO^+ + CH_3$	0.77	33	<0.01	0.02	0.06

[a] Number of internal degrees of freedom for reactant ion.

that the kinetic shift is negligible for most fragmentation processes, but reaches considerable values for reactions in aromatic hydrocarbons as result of the high activation energies of these processes. Thus for the loss of hydrogen from the benzene ion a kinetic shift of the order of 1.5 eV may be expected. Using the RRKM Theory kinetic shifts have also been calculated by McLafferty et al. [4] for substituted 1,2-diphenylethanes and by Gilbert and Stace [95] for substituted acetanilides leading to values ranging from 0.2 - 1.2 eV.

4.2 Experimental Determination of the Kinetic Shift

In view of the great importance of an accurate determination of activation energies a direct measurement of the kinetic shift is desirable. Originally it was assumed that a comparison of appearance potentials of ions formed in the ion source and in one of the field free regions (metastable ions) would allow the determination of at least part of the kinetic shift [22,96-98,104] (the so called "measurable" part [97]). Indeed earlier AP determinations using the semilog plot method showed considerable differences between the APs of "normal" and metastable ions [22,96-98], e.g. 1.3 eV for HCN loss from benzonitrile [96]. However, earlier photoionization studies as well as more recent investigations employing the energy distribution difference (EDD) technique demonstrate that the APs of metastable ions appear at or near to the APs of normal fragments [99-103]. Obviously the maximum lifetimes of normal and metastable ions are too close to show up as large kinetic shifts. It has recently been shown that the differences in the AP values observed with the semilog plot method for normal and metastable ions were indeed an artefact of the normalization procedure used in this method [105]. Since metastable ions arise from ions with energies near the threshold, but normal ions from a wide range of internal energies, a decrease of the electron energy will result in the loss of a larger fraction of ion current due to the normal ion. Thus in a semilog plot the curve of the normal ion will be displaced below that of the metastable ion leading to an apparent difference, Δ eV, in the range of 0.1 - 1 % of the ion current at 50 eV as illustrated in Fig. III-9 [106].

Thus in order to detect kinetic shifts experimentally it is desirable to study ions with considerably longer ion lifetimes (t > 10^{-3} s). Hence the ion cyclotron resonance technique (see Section V-3.5.1) was used by Gross [107] to determine the appearance potential of ions with an average ion lifetime of 10^{-2} s. Comparison with published AP values led to the conclusion that a kinetic shift of several tenths of an eV exists for HCN loss from aniline and benzonitrile and H

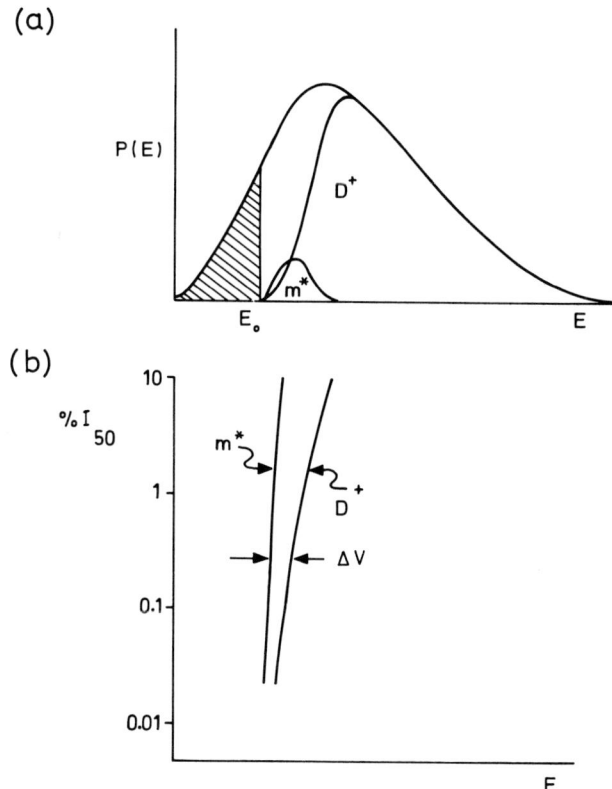

Fig. III-9. Schematic semilogarithmic plot of the normalized intensity of a fragment ion (D^+) and the corresponding metastable ion (m^*) as a function of the electron energy [106]. (By courtesy of Heyden & Son Ltd.)

loss from benzene*). An even better estimate of the kinetic shift may be possible using the photodissociation technique described by Dunbar [108,109], where ions are trapped in an ICR cell for seconds.

Ion trapping in the space charge of a pulsed electron impact source has been used by Lifshitz et al. [110] and Gordon and Reid [111] to determine the appearance potential as function of the ion lifetime from 1 μs up to 1.2 ms. Thus for H loss from the benzene molecular ion (a process for which a considerable kinetic shift has been predicted by Vestal, see Table III-4) the appearance potential varied from 14.2 eV

---

*) It should be noted that the energy scale in this experiment is consistently shifted to higher values by 0.3 eV. No explanation for this shift has been given by the authors.

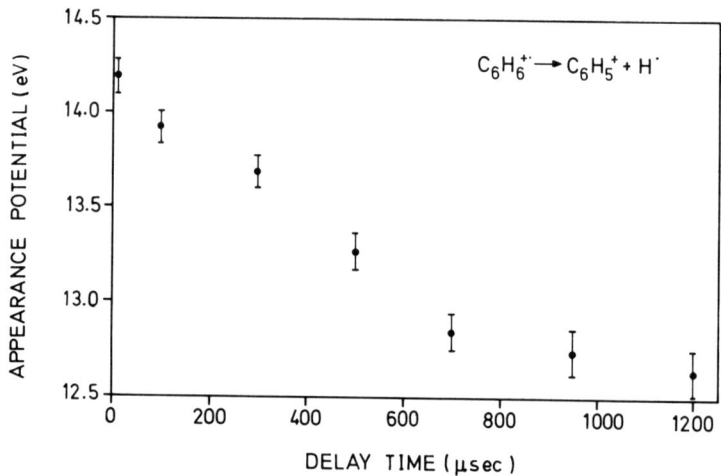

Fig. III-10. Experimental dependence of the appearance potential for $C_6H_5^+$ ion formation from benzene on the ion lifetime [111]. (By courtesy of Elsevier Scientific Publishing Company.)

at 1 μs to 12.7 eV at 1.2 ms (see Fig. III-10) suggesting a kinetic shift of at least 1.5 eV [111]. However, in all instances the appearance potentials were determined using the semilogarithmic plot method. As shown above this normalization procedure may lead to erroneous results although the error should be less pronounced than with metastable ions.

Finally it has recently been proposed that time-dependent breakdown graphs give reliable information on kinetic shifts [112]. Thus Lifshitz et al. [112] determined the normalized second derivative electron impact ionization efficiency curves of chlorobenzene as functions of time again using the method of ion trapping in the space charge of a pulsed electron impact source. If the ion lifetime is varied from 10^{-5} to 10^{-3} s a shift of 0.5 eV to lower energies is observed for the crossing point between parent and fragment ion curves, in good agreement with QET calculations. High quality time-dependent breakdown graphs of allene have also been reported by Stockbauer and Rosenstock [113] using the PEPICO technique although the ion residence time could only be varied from 0.7 - 4.7 μs causing a shift of the crossover between parent and fragment curve of 0.04 eV.

4.3 The Competitive Shift

The determination of activation energies by appearance potential measurements is further complicated by the occurrence of a competitive shift [4,85,114-116] as first pointed out by Lifshitz and Long [114]. This effect is observed if two k versus E curves intersect (see Fig. III-5: A^+ and B^+). Thus the reaction of higher activation energy (leading to B^+) must proceed at a rate considerably faster than $10^6 \, s^{-1}$ to compete effectively with the other reaction (leading to A^+) to be observed in the ion source [85]. As result of this effect the observed appearance potentials tend to be too high. This should be especially pronounced if appearance potentials are determined by the semilog plot method.

4.4 The Thermal Shift

As a result of the thermal energy present in the molecule before ionization less energy is required to produce the ion at threshold. For hot ion sources ($\sim 200 \, °C$) as normally used in electron impact studies and large molecules the average thermal energy may reach considerable values, e.g. an average thermal energy of 0.8 eV was calculated for 1,2-diphenylethane at 200 °C [4] (see also Fig. I-11). The effect of thermal energy on the ionization efficiency curves has been studied in detail by Chupka [117]. For a series of alkanes the thermal shifts of the ionization efficiency curves range from 0.10 eV to 0.24 eV if the temperature is raised from 28 °C up to 142 °C. Thus as expected the thermal shift increases with increasing size of the molecule as result of the increasing number of degrees of freedom over which the thermal energy can be distributed. Calculated thermal shifts were in all cases but one in good or even excellent agreement with experimental values, demonstrating that the thermal energy is fully effective in the dissociation of the molecular ion, as predicted by the QET. The thermal shift leads to apparently lower appearance potentials and therefore partially offsets the kinetic shift.

5 Kinetic Isotope Effects

Kinetic isotope effects are encountered in organic mass spectrometry in the study of isotopically labelled compounds. They are only significant in the case of deuterium labelling.

Intramolecular isotope effects are observed if two equivalent bonds in *one* ion, i.e. in partially deuterated compounds, are broken. Cleavage of the isotopically substituted bond (C-H, O-H, N-H) leads to a *primary*

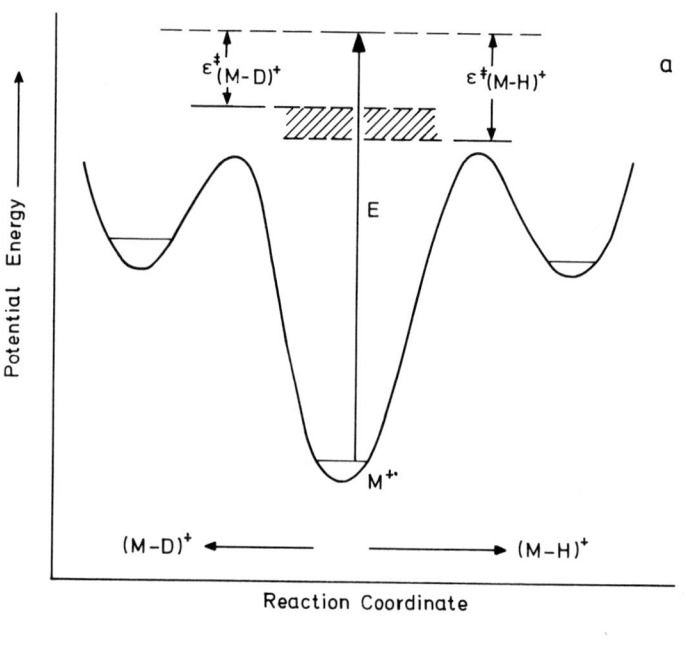

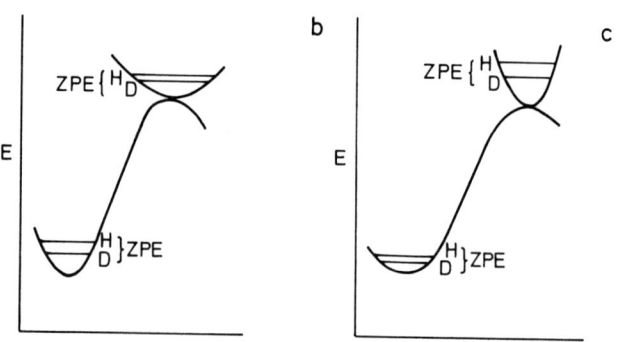

Fig. III-11. Potential energy diagram [104] illustrating the occurrence of a kinetic isotope effect; (a) intramolecular isotope effect; (b) normal intermolecular isotope effect; (c) reverse intermolecular isotope effect ($\varepsilon^{\neq}$ = non-fixed energy in the activated complex, ZPE = zero point energy). (By courtesy of Elsevier Scientific Publishing Company.)

isotope effect defined as $i = k_H/k_D$ whilst a *secondary isotope effect* is observed if other bonds, e.g. the C-C bond in CH_3CD_3, are involved. The primary isotope effect is caused by the slightly smaller activation

III-5 *Isotope Effects*

energy for X-H cleavage as compared to X-D cleavage (ΔE_o is usually $\leq$ 0.1 eV) leading to $i = k_H/k_D > 1$, i.e. to a faster elimination of H$^{\cdot}$ than D$^{\cdot}$ *). The differences in activation energies result from the *different zero point energies of the activated complexes* as illustrated in Fig. III-11a in which a schematic potential energy diagram for H$^{\cdot}$ and D$^{\cdot}$ loss from equivalent positions of the molecular ion is shown. As discussed by Cooks et al. [104] the activated complex corresponding to H$^{\cdot}$ loss must have a lower zero point energy since it possesses a C-D bond not involved in the fragmentation and vice versa**).

Intermolecular isotope effects are encountered if one compares labelled and unlabelled compounds, i.e. the rate constants for X-H and X-D cleavage in two different ions such as

$$C_2H_5 - H^{+\cdot} \rightarrow C_2H_5^+ + H^{\cdot}$$
$$C_2H_5 - D^{+\cdot} \rightarrow C_2H_5^+ + D^{\cdot}$$

In this case the isotope effect results from differences in zero point energies of the reactant and the activated complex as shown in Fig. III-11b and III-11c. Hence the activation energy for X-D cleavage in the labelled compound is given as

$$E_{oD} = E_{oH} + 1/2h \left[\sum (\nu_{iH} - \nu_{iD}) - \sum (\nu_{iH}^{\neq} - \nu_{iD}^{\neq}) \right]$$

where E_{oH} is the activation energy for the corresponding reaction in the unlabelled compound, ν_{iH} and ν_{iD} the vibrational frequencies of the labelled and unlabelled reactant, $\nu_{iH}^{\neq}$ and $\nu_{iD}^{\neq}$ those of the activated complexes.

A *normal isotope effect* is observed if the difference in zero point energies in the activated complex is smaller than that in the reactant ($E_{oD} > E_{oH}$ and $k_H/k_D > 1$) as shown in Fig. III-11b whilst a reverse isotope effect is encountered if the difference in zero point energies is larger in the activated complex than in the reactant ($E_{oD} < E_{oH}$ and $k_H/k_D < 1$, see Fig. III-11c). Reverse isotope effects are, however, rarely encountered in organic mass spectrometry. According to the QET

---

*
Moreover the difference in bond lengths between the X-H and X-D bond may contribute to the observed isotope effect.

**
As the two k(E) curves differ only by the small amount of ΔE_o, approximately the same energy range of P(E) contributes to both fragments. Thus the relative abundance of $[M-H]^+ / [M-D]^+$ can be approximately equated with the isotope effect.

the primary isotope effect is given as

$$i = \frac{k_H}{k_D} = \frac{W^{\neq}(E - E_{OH})}{W^{\neq}(E - E_{OD})} \qquad (III-3)$$

(For a given internal energy the density of states, $\rho(E)$, is identical).

It is obvious that at energies above E_{OH}, but below E_{OD}, $W_D^{\neq} = 0$ and thus the isotope effect becomes infinite. With increasing internal energy the ratio $(E - E_{OH}) / (E - E_{OD})$ decreases, rapidly approaching unity. Correspondingly the number of states of the two activated complexes and thus the rate constants become comparable (but not identical) at high internal energies.

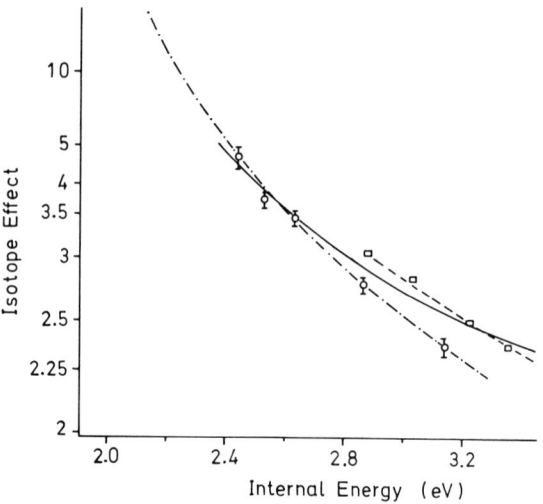

Fig. III-12. Isotope effect for H/D elimination from partially deuterated toluene (solid line = QET calculations [118]; circles = photodissociation ICR [108]; squares = electron impact [40]). (By courtesy of the American Chemical Society.)

The strong dependence of the isotope effect on the internal energy suggested by this qualitative discussion has indeed been verified both by QET calculations and experiments. One of the most thoroughly studied systems is that of toluene. Fig. III-12 compares the energy dependence of the isotope effect calculated by Vestal and Lerner [118] for H/D loss from partially labelled toluene with experimental data by Howe and McLafferty [40] (using electron impact) and by Dunbar [108] (using the photodissociation-ICR technique). The experimental data agree well with the calculated values showing the strong increase of the isotope effect with decreasing internal energy. Good agreement between the calculated and measured energy dependence of the isotope effect has also been reported by Corval and Masclet [119,120] for various labelled methanols and ethanols (using the modified classical approximation of the QET) and by Gordon et al. [121] for ethylene (using Haarhoff's approximation for the density of states [122]). Experimental evidence for a strong dependence of the isotope effect on the internal energy has been reported by a variety of other authors e.g. for $C_7H_8^{+\cdot}$ from cycloheptatriene [40] and 2-phenylethanol [41] as well as for nicotinic [44] and isonicotinic acid [44].

In agreement with the above-discussed energy dependence, the isotope effect is generally considerably larger in metastable ions than in fragment ions formed in the ion source. The magnitude of the isotope effect for metastable ions depends on the slope of the k(E) function: A steep k(E) curve leads to metastable ions with little non-fixed energy thus giving rise to a large isotope effect. In support of this assumption large isotope effects are observed for the metastable decay of small alkane ions where a steep k(E) function is expected. Thus an isotope effect of i = 600 was observed for metastable decomposition of ethane [123] and an isotope effect of > 1000 for isobutane [124]. Secondary isotope effects are expected to be considerably smaller than primary isotope effects as no C-H or C-D bonds are broken. Even for metastable ions such secondary isotope effects rarely exceed i = 1.2 [125]. However, some examples are known where relatively large secondary isotope effects have been observed for metastable ions. Thus secondary isotope effects ranging from 1.5 to 1.8 have been reported for methyl loss from t-butylbenzene and 4-t-butylpyridine molecular ions with a lifetime of 10^{-5} s [126] and an isotope effect of at least 3 for methyl loss from the n-butane molecular ion with a lifetime of 10^{-3} s [127]. These secondary isotope effects show the expected energy dependence.

Secondary deuterium isotope effects have recently been reported by Eadon and Zawalski [128] for iodine loss from the molecular ions of 4-t-butylcyclohexyl iodide and 5-iodononane. Interestingly, the authors observed an apparent inverse isotope effect (increased loss of iodine in the deuterated compound) if positions in the molecule were labelled which were remote from the iodine atom.

As the ion lifetime is a direct function of the internal energy, an increasing isotope effect with increasing ion lifetime is to be expected and has been observed in benzoic acid [49] and t-butylbenzene [129] using the field ionization kinetic technique.

Isotope effects have also been observed in the kinetic energy, T, released upon metastable decomposition [41,130-134] (see Section III-7.2). For example, the reaction $H-\overset{\cdot}{C} = \overset{+}{O}H \rightarrow CHO^+ + H^{\cdot}$ in methanol is accompanied by an energy release of 0.19 eV while the corresponding loss of $D^{\cdot}$ from $H-\overset{\cdot}{C} = \overset{+}{O}D$ has an energy release of 0.39 eV, i.e.

$T_H/T_D = 0.48$ [130]. The rationalization of this isotope effect depends on whether T includes contributions from the reverse activation energy or not (see Section III-7.3.1). If the reverse activation energy is negligible then the excess energy (non-fixed energy) for loss of H˙ should be greater than that for loss of D˙ (see Fig. III-11a) and hence $T_H > T_D$. In the presence of a considerable reverse activation energy one might anticipate from Fig. III-11a that there is no isotope effect on T as both the activated complex for H˙ loss and the resulting product, $(M-H)^+$, have a lower zero point energy than the activated complex and product ion formed by D˙ loss. It has, however, been pointed out by Bertrand et al. [134] that the C-H or C-D bond may be a little stronger or weaker in the activated complex than in the products so that the differences in the zero point energies of the activated complexes and the products may be different for H˙ or D˙ loss leading to a small isotope effect on the kinetic energy release. Relatively small isotope effects on D˙ loss are indeed observed in most cases [134].

As deuterium labelling is still the most powerful technique used for the elucidation of reaction mechanisms, the occurrence of isotope effects has to be taken into account in the interpretation of the data. It will, however, be shown later (Section IV-2.4) that the isotope effect does not always complicate the evaluation of labelling data, but may give valuable mechanistic information.

6 Substituent Effects on Fragmentation Reactions

6.1 Substituent Effects on Appearance Potentials

An electron donating substituent on an aromatic ring lowers the ionization potential while the opposite effect is observed for electron withdrawing groups as has been discussed in Section I-6. For most compounds there is an almost linear correlation between the IP and the σ^+ value of the substituent.

Linear correlations have also been observed between the appearance potentials [4,137-142] and Hammett's σ [135] and Brown's σ^+ [136] constants in several series of substituted compounds although the correlation is generally poorer than for ionization potentials, i.e. there is a more pronounced scatter of the AP values. Thus the dependence of the appearance potential on σ or σ^+ reflects the influence of the substituent on the ionization potential, the activation energy and to a smaller extent the kinetic shift. Introduction of an electron donating substituent leads in most cases to an increase in AP - IP and thus to an increased activation energy although several instances are known where the opposite behavior has been observed [4,138]. There is at present no theoretical concept to explain these opposite effects of the substituent on the activation energy.

6.2 Substituent Effects on Fragment Abundances

The strong influence of substituents on the reactivity of aromatic molecules and ions in solution as described by the Hammett equation stimulated Bursey and McLafferty [143] to look for similar correlations in organic mass spectrometry. For a variety of systems where a common fragment A^+ is formed either from a precursor M_Y^+ containing a substituent Y in meta or para position or from a precursor M_H^+ lacking the substituent they observed a correlation between fragment ion abundances and σ or σ^+ values which can be described by the general equation

$$\log \frac{[A^+]_Y / [M_Y^+]}{[A^+]_H / [M_H^+]} = \log \frac{Z}{Z_o} = \rho\sigma \qquad (III-4)$$

However, for other compound series poor or none correlations at all were observed. Since then various authors have studied substituent effects for a large variety of compound series using electron impact, chemical ionization [144] and field ionization [145,146] and including negative ions [147]. The literature on substituent effects on ion abundances has been extensively reviewed [148-152] and will not be repeated here. As an example of a good correlation, $\log Z/Z_o = \rho\sigma$ for acetyl ion formation from substituted acetophenones [137] is shown in Fig. III-13 whilst a poor correlation is observed for $CF_3C\equiv\overset{+}{N}CH_3$ formation from $CF_3CON(CH_3)C_6H_4Y^{+\cdot}$ [141] (Fig. III-14).

The formal analogy between equation III-4 and the Hammett equation might suggest that the abundance ratio is solely determined by the rate constant ratio. Unfortunately this simple relation does not exist. Rather, the abundance ratio is controlled by a variety of factors which can be explained within the framework of the QET [1,137,152-155].

(1) *Factors influencing the molecular ion intensity.*

As illustrated in Fig. III-1 the intensity of the undecomposed molecular ion, $[M_Y^+]$, is predominantly determined by (a) the activation energy of the process with lowest threshold and (b) the energy distribution, P(E), of the molecular ion.
a. *Activation energy.* It has been argued that in general the effect of a substituent on the ionization potential is more pronounced than on the

120   III *Application of the QET*

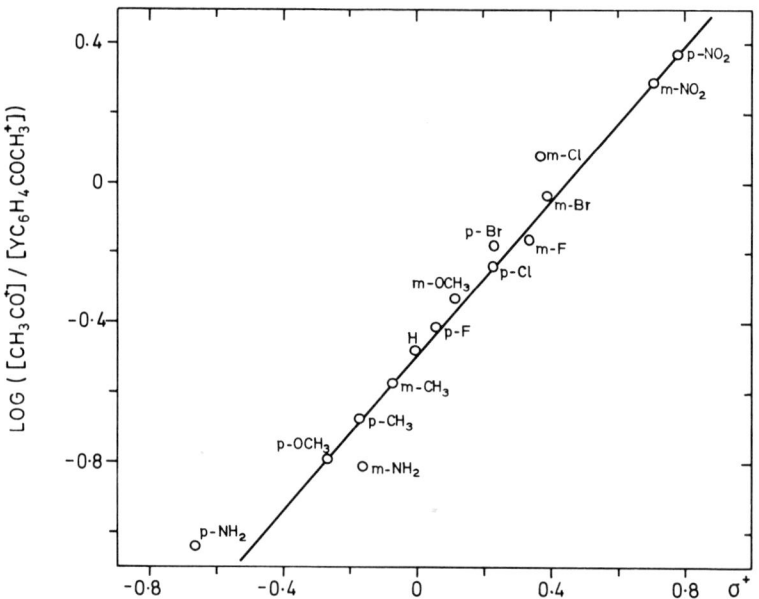

Fig. III-13. Substituent effect on decomposing acetophenone molecular ions. Log ($[CH_3CO^+]/[YC_6H_4COCH_3^{+\cdot}]$) versus Brown's σ^+ constants [137] (By courtesy of Heyden & Son.)

appearance potential. (The frequent exceptions from this rule have been mentioned above.) Thus, electron donating substituents often lead to an increase of the difference AP - IP (vide supra) and hence to an increase of the fraction of molecular ions with insufficient energy to decompose. This is, for instance, the case for $C_7H_7^+$ ion formation from substituted phenyl benzyl ethers (Fig. III-15) [138]. If this were the only effect influencing the abundance ratio, log Z/Z_o versus σ plots should show a positive slope*. Experimental evidence for a direct correlation between the fraction of undecomposed molecular ions and σ has been reported by Chin and Harrison [137] for the acetyl ion formation from acetophenones as shown in Fig. III-16. The figure reveals a considerable scatter of the data points which is to be expected, since not only AP - IP, but also P(E) determine the fraction of undecomposed mol-

---

*
 However, exceptions from this rule are known. For instance, AP - IP increases with electron withdrawing substituents for NO loss from substituted nitrobenzenes whilst the log Z/Z_o versus σ^+ plot has a negative slope [139], demonstrating that there are other factors influencing the relative fragment abundance. Moreover, although a very good linear correlation between AP - IP and σ^+ has been observed for $CF_3C\equiv N^+CH_3$ formation from $CF_3CON(CH_3)C_6H_4Y^{+\cdot}$ the corresponding correlation for the fragment abundances was poor [141] (see Fig. III-14).

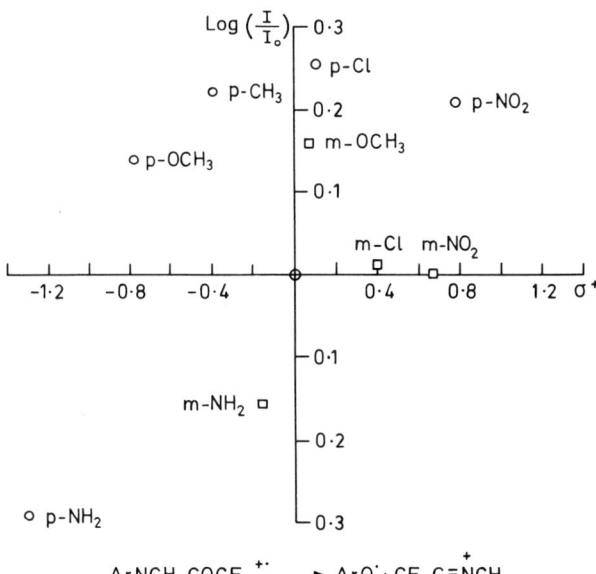

Fig. III-14. Substituent effect on decomposing trifluoramides. Log ([$CF_3C\equiv NCH_3^+$] / [$YC_6H_4N(CH_3)COCF_3^{+\cdot}$]) versus Brown's σ^+ constants [141]. (By courtesy of the Chemical Society, London.)

ecular ions influenced by the substituent.

b. *The energy distribution of the molecular ion, P(E)*. It has been shown by McLafferty et al. [4] that substituents may have a pronounced effect on P(E): Splitting of electronic states is increased by electron withdrawing substituents, additional states may arise from non-bonding electrons of the substituent and, more important, the relative population of the states may be changed considerably by various substituents. All of these factors may influence the fraction of molecular ions remaining undecomposed.

c. *Competing decompositions*. If the fragmentation reaction the substituent effect of which is being studied is not the decomposition of lowest activation energy, competing reactions which themselves are substituent dependent will influence the molecular ion intensity.

(2) *Factors influencing the fragment abundance.*

a. *The rate constant, k(E), of the process under study*. The shape of the k(E) curve is determined by the activation energy (discussed above) and the geometry of the activated complex. An example of a drastic influence of the substituent on the geometry of the activated complex has been reported by McLafferty et al. [4]. Whilst the formation of $C_7H_6Y^+$ from

122   III   Application of the QET

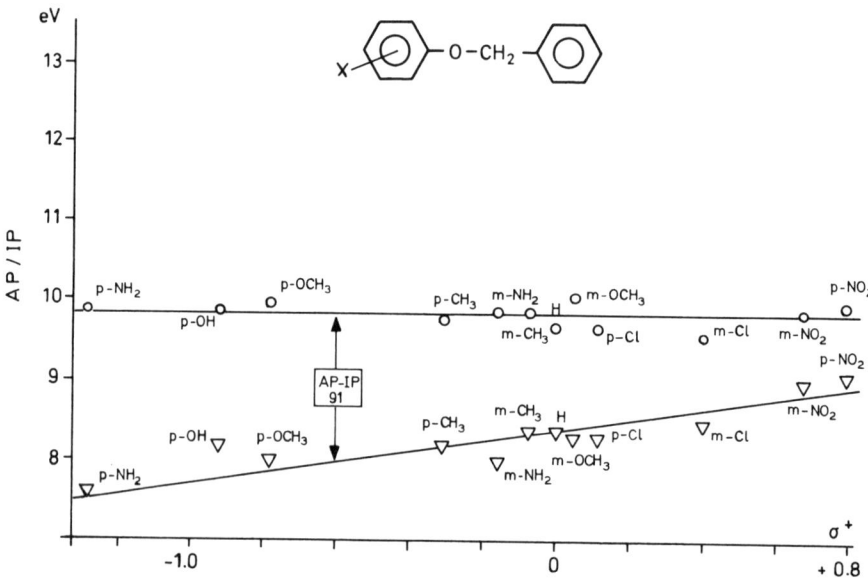

Fig. III-15. Substituent effect on the appearance potential for $C_7H_7^+$ formation (circles) and the ionization potential (triangles) of substituted phenyl benzyl ethers [138].

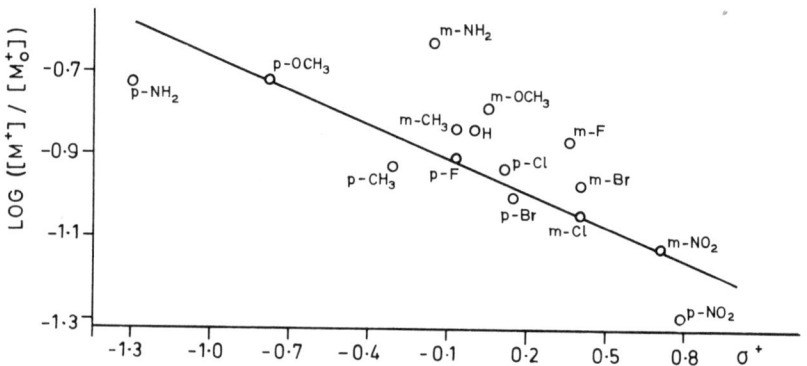

Fig. III-16. Substituent effect on the fraction of undecomposed molecular ions. $Log([M^+] / [M_o^+])$ of substituted acetophenones versus Brown's σ^+ constants [137] ($[M^+]$ = abundance of the molecular ion, $[M_o^+]$ = total ion abundance). (By courtesy of Heyden & Son Ltd.)

substituted 1,2-diphenylethanes usually proceeds via a loose transition state, resonance stabilization in p-aminodiphenylethane reduces the free rotation around the C-aryl-CH$_2$ bond as shown in Scheme III-4 and thus leads to a rather tight transition state.

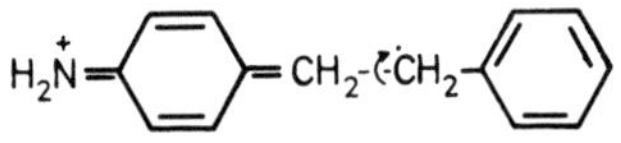

$$H_2\overset{+}{N}=\!\!\!\langle\rangle\!\!\!=CH_2\text{-}\overset{\cdot}{C}CH_2\text{-}\langle\rangle$$

Scheme III-4

b. *The rate constants of competing fragmentations.* Competing fragmentations may not only reduce the fraction of undecomposed molecular ions as discussed above, but also the relative abundance of the fragment ion, A$^+$. Substituents influence both the activation energy and the rate of these competing reactions. Moreover new reactions may be created by fragmentation of the substituent itself [1,155]. The influence of competing reactions is especially pronounced if decomposition processes with high activation energy are studied. In this case irregular substituent effects are frequently observed [1,155].

c. *Secondary decompositions.* The abundance of a primary fragment is also determined by secondary decompositions (as discussed in Section III-3.1) which themselves are substituent dependent. The influence of secondary decompositions can be reduced by using low electron energies which often leads to better correlations [152].

d. *Other factors.* There are a variety of less important factors which influence the relative fragment ion abundance e.g. the magnitude of the kinetic shift, the energy distribution in the primary fragment, the number of vibrational degrees of freedom of the neutral fragment [4].

A simple although only approximate kinetic treatment has been reported by Chin and Harrison [137] which allows the most important factors determining the relative fragment abundance to be discussed in a semi-quantitative fashion.

For a decomposition scheme of the general form

$$M^{+\cdot} \xrightarrow{k_1} A^+ \longrightarrow \text{secondary products}$$
$$\xrightarrow{k_i} \text{other products}$$

the ratio of fragment to molecular ion abundance is given by

$$\frac{[A^+]}{[M^{+\cdot}]} = f' \frac{k_1}{\sum k_i} (1/f - 1)$$

if f' is the fraction of primary fragments A$^+$ with insufficient energy to decompose, f the fraction of non-decomposing molecular ions. The rate constants used represent values averaged over the appropriate energy range of the reacting species. The equation demonstrates that the relative fragment intensity is determined by the fraction of undecomposed molecular and fragment ions and the extent of competing fragmentations.

QET (RRKM) calculations have been carried out by McLafferty et al. [4] for a series of substituted 1,2-diphenylethanes. The results demonstrate the strong influence of the substituent on k(E). Calculated metastable ion abundances were in fair agreement with experimental data.

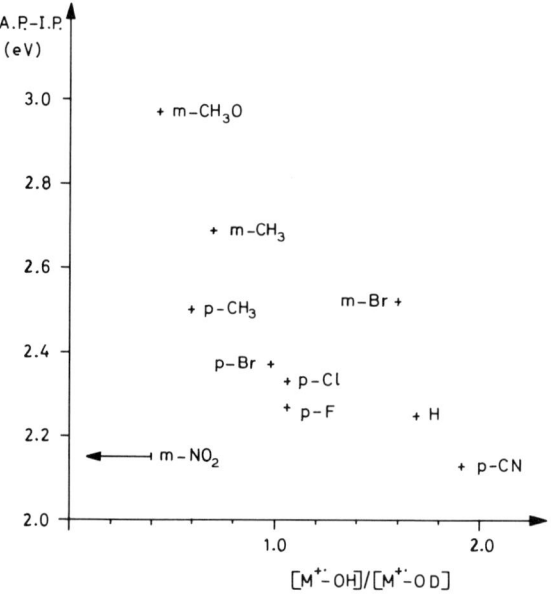

Fig. III-17. Substituent effect on the degree of H/D-scrambling in molecular ions of benzoic acid-O-d. [M$^{+\cdot}$ - OH] / [M$^{+\cdot}$ - OD] as a function of the difference between appearance and ionization potential [157].

Finally, a substituent effect on the extent of H/D-scrambling in the benzoic acid molecular ion has been observed [156,157]. It has been shown in Section III-2.3 that the degree of scrambling is a direct function of the internal energy of the decomposing ions which for metastable ions roughly corresponds to AP - IP. Thus, if the nitrocompound, for which the decomposition of the substituent interferes is excluded, an approximate correlation between AP - IP and the degree of scrambling in metastable ions has been observed [157] (Fig. III-17).

6.3 Competing Fragmentation of Disubstituted Aromatic and Aliphatic Compounds

The relative abundance of characteristic daughter ions A^+ and B^+ formed by fragmentation of the substituents X and Y in disubstituted benzenes has been discussed by Howe and Williams [158]. This abundance ratio depends on the k(E) curves for the two processes which are determined by the activation energy and the transition state geometry of the two reactions. Assuming that the introduction of a second substituent Y into a monosubstituted benzene, C_6H_5X, does not change the transition state configuration and the activation energy for fragmentation of the original substituent X significantly*), the appearance potential in the

Table III-5. Appearance Potentials for Fragmentation of Some C_6H_5X Compounds

X	Neutral fragment	AP (eV)
$COCH_3$	CH_3	9.99
$C(CH_3)_3$	CH_3	10.26
$CH(CH_3)_2$	CH_3	10.65
CO_2CH_3	OCH_3	10.80
$N(CH_3)_2$	H	10.80
CHO	H	10.99
C_2H_5	CH_3	11.25
OCH_3	CH_2O	11.30
I	I	11.46
OH	CO	11.67
CH_3	H	11.80
Br	Br	12.02
NO_2	NO_2	12.16
NH_2	HCN	12.50
Cl	Cl	13.20
CN	HCN	14.60
F	C_2H_2	14.73

monosubstituted compound can be used to predict the relative ease of fragmentation of the two groups X and Y. The process of lower activa-

---

* i.e. the substituent influences predominantly the ionization potential.

tion energy in the monosubstituted benzene will, in general, lead to the more abundant fragment in the disubstituted compound. Table III-5 shows the order of appearance potentials for the characteristic fragmentation of a variety of substituents in monosubstituted benzenes. In reality the above assumption that the activation energy for fragmentation of a given substituent does not change significantly on introduction of a second substituent does not hold strictly as discussed in Section III-6.1. This is also shown in Table III-6 for para-substituted bromobenzenes, demonstrating that the appearance potential for Br· loss varies from 10.8 to 13.2 eV. Nevertheless the ratio $[M^+ - Br]/[B^+]$*[)]

Table III-6. Variation of Relative Abundance of $M^+ - Br(A^+)$ to the Other Characteristic Fragment Ion B^+ in the 18-eV Spectra of a Series of para-Substituted Bromobenzenes, YC_6H_4Br

Substituent Y	B^+	$[M^+ - Br]/[B^+]$	AP $(M^+ - Br)$,[a] (eV)	AP (B^+),[a] (eV)
COMe	$M^+ - CH_3$	0.0042		10.58
NMe_2	$M^+ - H$	0.0090		11.15
C_2H_5	$M^+ - CH_3$	0.52	10.80	10.75
OMe	$M^+ - CH_2O$	0.83		
I	$M^+ - I$	0.022		12.04
OH	$M^+ - CO$	29	12.17	
Me	$M^+ - H$	7.2	11.30	12.48
NH_2	$M^+ - HCN$	61	11.95	
Cl	$M^+ - Cl$	63	12.70	
CN	$M^+ - HCN$	∼ 1000	13.21	

[a] The absence of an AP measurement means that the relative abundance of the ion was too low to permit an accurate measurement.

still approximates the order of appearance potentials in the monosubstituted compounds.

The influence of a substituent on the fragmentation of ionized bifunctional aliphatic compounds has been discussed by Remberg and Spiteller [159,160] using the model compound $C_2H_5-CH(X)-(CH_2)_4-CH(OCH_3)-C_2H_5$ with a variable substituent X. If the functional groups are arranged according to their increasing influence on the fragmenta-

---

* B^+ arises from fragmentation of the second substituent.

tion the following sequence is obtained: COOH < CH_2OH < OH < Cl < Br < $COOCH_3$ < SH < C=O < SCH_3 < OCH_3 < J < NH_2 < $N(CH_3)_2$. It is obvious that this sequence mainly reflects the "ionization potential of the functional group". Thus if the monofunctional compounds C_3H_7X are arranged in the order of decreasing ionization potentials one observes the following sequence [161] Cl > COOH, OH, Br, > $COOCH_3$ > $COCH_3$ > SH > J > SCH_3 > NH_2 in close correspondence to the order deduced by Remberg and Spiteller from ion abundances. This observation lends some support to the concept of charge or radical site localization discussed in Section IV-1.4.

6.4 Substituent Effects on the Kinetic Energy Release

Substituent effects on the kinetic energy release upon metastable decomposition have been reported in some instances [162-166]. A substituent effect is expected if the reverse activation energy contributes predominantly to the kinetic energy release. As result of resonance stabilization, electron-donating substituents should lead to more stable product ions than electron-withdrawing ones. Hence a larger reverse activation energy is expected for the former. If a constant fraction of this reverse activation energy is funneled into the translational channel, the fragmentation of compounds with electron-donating substituents should in general be accompanied by a larger kinetic energy release than those with electron-withdrawing groups. This is indeed observed experimentally as shown in Table III-7 for NO˙ loss from substituted nitrobenzenes [166].

Table III-7. Kinetic Energy Release for NO˙ Loss from p-Substituted Nitrobenzenes (in eV)[a]

Substituent	NH_2	OCH_3	OH	CH_3	H	Cl	F	CN
T	1.2	1.22	1.23	0.96	0.55	0.83	0.84	0.35
σ	-0.66	-0.27	-0.37	-0.17	0	+0.23	+0.06	+0.66

[a] NO˙ loss from nitrobenzene ions shows a composite metastable peak. The data refer to the broad component.

7 Metastable Ions

7.1 The Origin of Metastable Ions

Ions decomposing unimolecularly outside the ionization chamber of a mass spectrometer are termed "metastable ions". If the decompositions occur within the field free regions of a magnetic type instrument they give rise to a metastable peak. The origin of metastable peaks was first correctly interpreted by Hipple and Condon in 1945 [167]. Since then it has been shown that from these peaks a wealth of information can be extracted for fundamental studies, the elucidation of reaction mechanisms and ion structure assignments.

A considerable part of our knowledge of metastable ions is based on the work of Beynon, Cooks and their colleagues, who have reviewed this subject several times [104,168,169]. In view of the published literature the following chapter will be confined to a brief summary of the properties of metastable ions and some related fundamental studies while the use of metastable ions as a probe for reaction mechanisms and ion structures is discussed in the appropriate parts of this book (Chapter IV and V).

Depending on the geometry of the instrument metastable ions have lifetimes in the range of 10^{-6} to 10^{-5} s, corresponding to most probable rates of 10^6 to 10^5 s^{-1}. However, as discussed in Section III-1 the overall range of rate constants contributing to a measurable extent to the metastable decomposition extends from about 10^3 to 10^7 s^{-1}. As a result of the well defined lifetime window of metastable ions they have a rather narrow range of internal energies (about 0.1 - 1 eV above threshold [96,170]). The existence of metastable ions can be predicted directly from the QET: After primary excitation rapid radiationless transitions give rise to highly vibrationally excited ground state ions with a distribution of internal energies and thus a distribution of rate constants including those which predominantly lead to fragmentation in the metastable time range.

There are two additional sources for metastable decompositions [168]:
(1) Crossings of potential energy surfaces may lead to symmetry-forbidden predissociation which is slow enough to be detected within the metastable ion lifetime window. Forbidden predissociations via repulsive states are often observed in small di- and tri-atomic molecules. This is illustrated in Fig. III-18 for H_2 elimination from the H_2S molecular ion [171]. Predissociation has also been invoked to explain the exceptionally small kinetic energy released upon metastable decomposition of

halogenated acetophenones [172].

(2) Metastable decompositions may also be due to tunnelling through the dissociation barrier. Such tunnelling mechanisms have been assumed in order to rationalize the observation of hydrogen loss from metastable methane ions [173,174]. Tunnelling may also account for the abundant metastable ions associated with the charge separation fragmentation of many di- and triatomic ions [175].

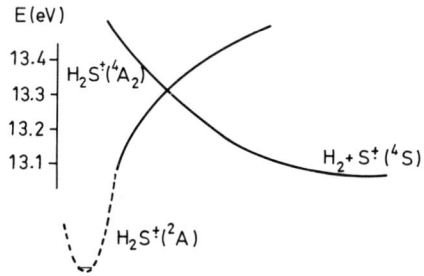

Fig. III-18. Crossing of potential energy surfaces in the $H_2S^{+ \cdot}$ system leading to predissociation of the ion [171].

7.2 Kinetic Energy Release

7.2.1 The Peak Shape of Metastable Ions

Upon decomposition part of the excess energy of an ion is released as kinetic energy, T, leading to a range of translational energies in the ion beam. The kinetic energy released upon *metastable decomposition* can be readily detected as peak broadening[*]. Small average T-values lead to narrow, gaussian-shaped peaks[**], whereas flat-topped or dish-shaped peaks are observed for larger T values [104] (Fig. III-19). The minimum observed in the center of a dish-shaped peak results from discri-

---

[*] This energy spread is not amplified for fragment ions formed in the source of the mass spectrometer as result of their low translational energy and is hence not detectable as peak broadening. However, a substantial amplification of this energy spread is observed for decomposing metastable ions which have a high translational energy (vide infra).

[**] In this case the non-fixed energy in the transition state is the main source for the kinetic energy release. This non-fixed energy is statistically partitioned upon fragmentation explaining the gaussian shape (vide infra).

130   III   Application of the QET

mination of ions with maximum velocity component in the collector slit direction by the finite slit length (see reference [104] for an exhaustive discussion) and is thus most pronounced for short slits and long ion paths from the point of decomposition to the collector.

Depending on the point of decomposition and the geometry of the mass spectrometer, metastable peaks can be recorded by scanning either the magnetic field strength, the electric sector potential or the acceleration voltage [104] or combinations of these. The experimental procedures employed for recording metastable peaks are briefly outlined in Section IV-2.3.1.

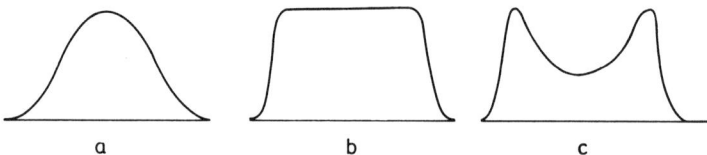

Fig. III-19. Typical peak shapes of metastable ions observed in magnetic sector instruments: (a) gaussian, (b) flat-topped, (c) dish-shaped.

7.2.2   Determination of Kinetic Energy Release

Decomposition of organic molecules leads to the release of a range of kinetic energies which may vary considerably (vide infra). The kinetic energy release can be determined using the deflection technique [176-181], by measuring the width of time of flight peaks [182-191], metastable peaks in conventional magnetic type instruments or by use of a Daly detector [192]. Only the analysis of metastable peak shapes in conventional instruments will be discussed in detail, as this technique is readily available for organic mass spectrometrists. It has been demonstrated that the average kinetic energy release is characterized by the peak width at 22% peak height for exactly gaussian peak shape, or by approximately the width at 50% peak height for a flat-topped peak*[)].

Formulae have been derived which allow the kinetic energy release, T, to be calculated from the peak width using mass, electric sector and acceleration potential scans [104]. These three equations can be reduced to one in which the kinetic energy is related to the fractional peak width $\Delta W/W$ as follows: ΔW is the width of the peaks in terms of

---

* It is obvious that for flat-topped peaks the position at which the peak width is measured is less critical.

mass units, electric sector voltage or acceleration voltage, W the mass, electric sector or acceleration voltage corresponding to the metastable peak center. For the reaction

$$m_1^{x+} \rightarrow m_2^{y+} + m_3$$

the relation between the fractional peak width and the kinetic energy release, T is then given by

$$\frac{\Delta W}{W} = 4 \left(\frac{m_3 T}{m_2 xeV} \right)^{1/2}$$

leading to

$$T = \frac{m_2 xeV}{16 m_3} \left(\frac{\Delta W}{W} \right)^2 \quad (III-5)$$

where V is the acceleration voltage [193]. If the metastable peak width is small one has to correct for the main beam width (i.e. by subtracting the square root of the differences of the squares [194]*). As the kinetic energy release values may be extremely small (vide infra) it is very important to note that the energy spread finally measured (e.g. on the electric sector energy scale) is considerably larger than T itself, i.e. the kinetic energy release is amplified [104]. The amplification factor is given as

$$A = 4 \left(\frac{xeV}{T} \right)^{1/2} \frac{(m_2 \cdot m_3)^{1/2}}{m_1} \quad (III-6)$$

It is obvious that the amplification factor is small for small neutral losses (e.g. hydrogen), but increases with decreasing kinetic energy release (for $T = 10^{-4}$ eV $A = 10^4$) [104]. This makes it possible to measure extremely small energy releases, the smallest value reported so far being 2×10^{-4} eV [133,172].

With instruments of high energy resolution (and collimating holes instead of slits) even vibrational fine structure has been resolved in metastable peak shapes (e.g. for the dissociation of HeH^+ [195,204], H_2^+ [202], H_3^+ [205], He_2^+ [203], NO^+ [200,201] and N_2^+ [196-199]. The structure can be explained to arise from rotational predissociation from various vibrational-rotational levels of the ground state ion.

---

*) $\Delta E = \sqrt{(\Delta E'')^2 - (\Delta E')^2}$ if $\Delta E'$ is the energy spread of the main beam and $\Delta E''$ that of the metastable peak.

It is obvious that this technique, termed "translational spectroscopy" gives valuable information on the energy spacing between vibrational-rotational states as highlighted in a recent study by Fournier et al. [204] of the process $HeH^+ \rightarrow He + H^+$. In general changes in ion source temperature [206,207] and electron energy [206] have little effect upon the kinetic energy release measured. A pronounced temperature dependence has so far only been observed for $H^\cdot$ loss from n-propanol [207] and $H^\cdot$ loss from methane [174] and has been explained by assuming tunnelling of the $H^\cdot$ atom through the centrifugal barrier.

Whilst the kinetic energy released upon metastable decomposition in conventional magnetic type mass spectrometers stems from ions with a range of internal energies (even if this range is rather narrow), the photoelectron-photoion-coincidence technique (PEPICO) can be used in conjunction with time-of-flight instruments [184-191] for studying the kinetic energy release as a function of the internal energy as already discussed in Section II-2.2.4. Such studies have revealed that the average kinetic energy release may be strongly dependent on the internal energy.

7.2.3 Kinetic Energy Release Distribution

In addition to the measurement of the average kinetic energy release the determination of the kinetic energy release distribution (KERD) has recently become possible by the analysis of peak shapes in magnetic type mass spectrometers [193,208-212] and time-of-flight instruments [184-190,213,214]. The latter method can be combined with the PEPICO technique and hence allows the determination of the kinetic energy release distribution as a function of the internal energy. The principle of the KERD determination has been outlined in Section II-2.2.4. As discussed in that chapter comparison of theoretical and experimental distributions constitute a rigorous test of the Quasi-Equilibrium Theory.

The data available so far reveal that the distribution of kinetic energies may vary considerably, as demonstrated in Fig. III-20 for the benzoyl ion formation from the benzaldehyde and α,α-dibromo acetophenone molecular ion [210]. Whilst the benzaldehyde system gives a sharply peaked distribution, a broad range of kinetic energies is released in the case of α,α-dibromo acetophenone, and this can be described as two dimensional Boltzmann distribution of the general form

$$f(T) \sim (T)^{1/2} e^{-cT} \qquad (III-7)$$

Figure III-20 shows that the peak shape directly reflects the energy distribution: Flat-topped peaks correspond to a narrow, gaussian-shaped peaks to a broad distribution*). Preliminary results seem to indicate that narrow distributions are observed for reactions with a large reverse activation energy [210].

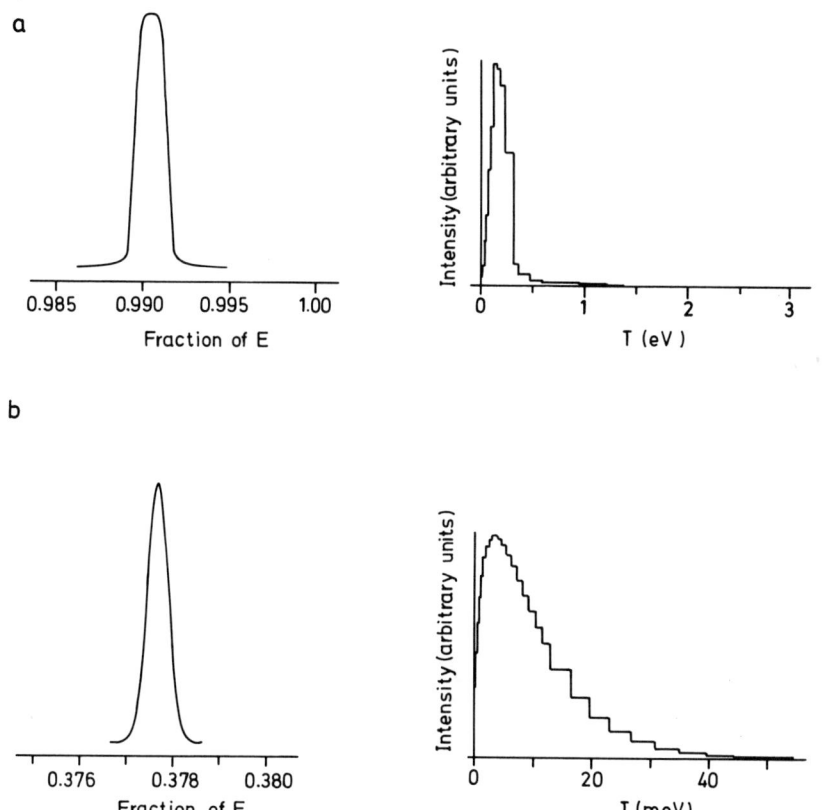

Fig. III-20. Peak shapes and distributions of kinetic energies released in benzoyl ion formation from metastable molecular ions of (a) benzaldehyde and (b) α,α-dibromo-acetophenone [210]. (By courtesy of Heyden & Son Ltd.)

---

* However even for a discrete energy release the metastable peak has a finite slope.

7.3 Energy Partitioning

7.3.1 Sources for Kinetic Energy Release

As illustrated in Fig. III-21 the kinetic energy released contains contributions from two main sources (neglecting tunnelling and predissociation):

(1) The energy above the potential barrier for decomposition, $\varepsilon^{\neq}$ (non-fixed energy), leading to the contribution $T^{\neq}$.
(2) The reverse activation energy, ε_o^r, leading to T^e. The total kinetic energy is given by $T = T^{\neq} + T^e$. There is considerable evidence that most simple bond cleavages have only small or no reverse activation energies, as the ion-free radical recombinations normally proceed with zero activation energy while rearrangement reactions may proceed with substantial reverse activation energy [170]*). $T^{\neq}$, but not T^e should depend on the internal energy and thus the ion lifetime of the decomposing metastable ion. Thus ion lifetime measurements may be employed to separate these two components [168].

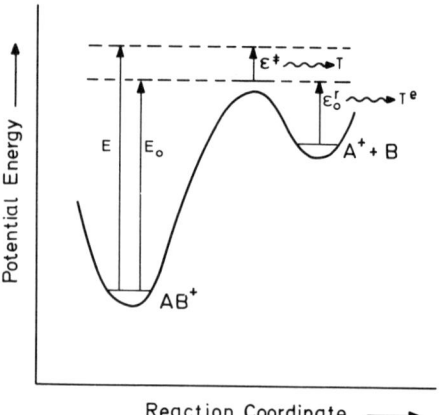

Fig. III-21. Contribution of the non-fixed energy of the activated complex and of the reverse activation energy to the observed kinetic energy release [104]. (By courtesy of Elsevier Scientific Publishing Company.)

---

*
   The contribution to the kinetic energy release from the reverse activation energy should be especially pronounced if one considers the decomposition of "hot" ions, i. e. precursor ions which have a higher heat of formation than the combined products. The molecular ion of hex-1-yne is one of the few hot ions identified so far. This ion decomposes inter alia to give the cyclopropenyl cation and a methyl radical, the combined heats of formation of which are about 110 kJ mol^{-1} below that of the precursor. Hence this decomposition process is accompanied by a kinetic energy release (0.165 eV) which is substantially larger than that of isomeric $C_6H_{10}^{+\cdot}$ ions [215].

7.3.2 Energy Partitioning

The total excess energy $\varepsilon_{excess} = \varepsilon^{\neq} + \varepsilon_0^r$ available upon decomposition is partitioned into translational and internal energy of the products. Only if both the ionic and neutral fragments are atoms is all the energy lost as kinetic energy. The energy partitioning quotient T/ε_{excess} represents the fraction of the excess energy, $\varepsilon_{excess} = \varepsilon^{\neq} + \varepsilon_0^r$, funnelled into translational energy of separation of the products. The individual partitioning quotients $T^{\neq}/\varepsilon^{\neq}$ and T^e/ε_0^r can be determined approximately for two limiting cases (Fig. III-22).

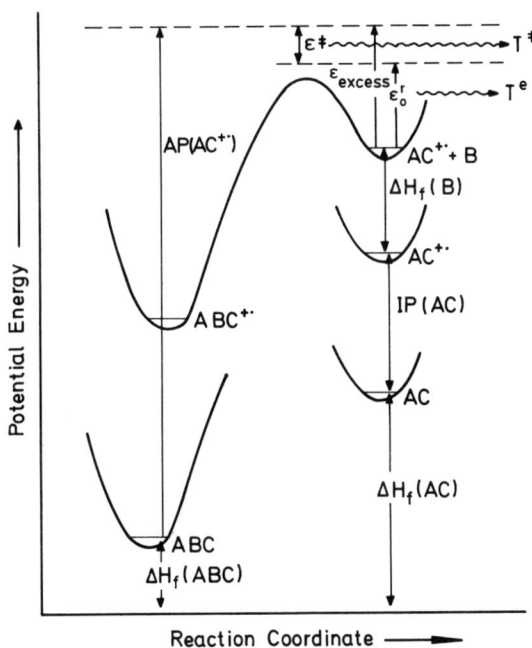

Fig. III-22. Experimental determination of the excess energy for the reaction $ABC^{+\cdot} \rightarrow AC^{+\cdot} + B$ [104]. (By courtesy of Elsevier Scientific Publishing Company.)

(1) *The reverse activation energy is zero* (as assumed for simple bond cleavages). A necessary, but not sufficient condition for this case is, that the average kinetic energy release is smaller than 0.1 eV [168]. In this instance $\varepsilon^{\neq} = \varepsilon_{excess}$ and $T^{\neq} = T$. The excess energy can be determined approximately by appearance potential measurements if the heats

of formation of the products are known*⁾. For the reaction

$$ABC^{+\cdot} \rightarrow AC^{+\cdot} + B$$

$$\varepsilon_{excess} = AP(AC^{+\cdot}) + \Delta H_f(ABC) - \Delta H_f(AC^{+\cdot}) - \Delta H_f(B) \quad (III-8)$$

Based on this approximation Haney and Franklin [183] found for a variety of decomposition processes the empirical correlation

$$\frac{T^{\neq}}{\varepsilon^{\neq}} = \frac{1}{\alpha N} \quad (III-9)$$

where N is the number of oscillators and α an arbitrary parameter with a mean value of 0.44.

If there is no reverse activation energy the energy partitioning coefficient can also be calculated using an appropriate form of the QET (Phase Space Theory) and assuming statistical partitioning of the excess energy (see Section II-1.6). Such calculations for $CH_3I^{+\cdot}$ [190] showed good agreement at low internal energies with values predicted by Haney and Franklin's correlation and the QET, as well as new data obtained by PEPICO experiments [190].

(2) *The decomposition process has a large reverse activation energy.* This is often but not always the case for rearrangement reactions. In this instance the excess energy will in general not be partitioned statistically between the translational and vibrational degrees of freedom. Thus the QET cannot be applied to calculate the energy partitioning coefficient. Rather, this information must be sought directly from experimental evidence [170].

If ε_o^r is large, the non-fixed energy, $\varepsilon^{\neq}$, can be neglected, i.e. $\varepsilon_{excess} = \varepsilon_o^r$ and $T = T^e$. (As a rule of thumb it has been suggested that $\varepsilon^{\neq}$ may be neglected if $\varepsilon_o^r > 0.5$ eV [168]). Using this approximation the reverse activation energy can be determined from appearance potential measurements and heats of formation data using equation III-8 with $\varepsilon_{excess} = \varepsilon_o^r$, whilst T^e is calculated from the metastable peak width. This approach has been used by Beynon, Cooks et al. to determine the energy partitioning quotient for the reverse activation energy, T^e/ε_o^r, for HCN loss from benzaldoxime methylesters [164,218], H_2CO loss from substituted anisoles [165], NO loss from aromatic nitrocompounds [166], and HX loss from alkyl halides [170,219]. If $\varepsilon^{\neq}$ is not negligible (as for instance for a variety of McLafferty rearrangements) $T^{\neq}$ can be cal-

---

*
 The heat of formation of the ionic product can be determined by ionization potential measurements, if ΔH_f of the neutral is known.

culated using the QET (RRKM Theory*)) and subtracted from the total kinetic energy released [170,211].

Recently a simple theoretical treatment of the energy partitioning has been reported by Christie et al.[220]. The potential energy loss of the activated complex whilst passing over the potential barrier is initially stored as kinetic energy in product modes. The extent of redistribution of this translational energy into vibrational energy of the products depends on the curvature of the minimum potential energy path from transition state to products. If the transition state reaction coordinate corresponds closely in its composition to a translational motion which separates the two fragments of the reaction, the reaction coordinate will be a straight line on the hypersurface. In this case the kinetic energy release will appear almost entirely as kinetic energy of separation. This situation most likely arises when the transition state is "late"; "late" transition states tend to occur with endoergic reactions. One therefore expects to find a relatively large proportion of the reverse activation energy appearing as kinetic energy of product separation for endoergic reactions. If, on the other hand, the reaction coordinate of the transition state is made up of motions that differ greatly from product separation, the potential path on the hypersurface will be curved. This will tend to channel energy into motions other than translation. Thus "early" transition states should favor partition of reverse activation energy into product vibrations and "early" transition states tend to occur with exoergic reactions.

Based on these considerations the authors developed a semiquantitative model to calculate partitioning quotients and observed good agreement with experimental data.

The experimental data available so far suggest that the energy partitioning quotient is quite similar for decomposition processes proceeding via identical mechanisms, demonstrating the great value of such data in elucidating reaction mechanisms, as discussed in Section IV-2.3.2. Preliminary results seem to indicate that the energy partitioning quotient is related to the geometry of the activated complex. *The tighter the activated complex the greater will be the fraction of reverse activation energy released as kinetic energy* [166,217].

The reliability of the energy partitioning data derived from metastable peak shapes depends on the quality of the AP measurements and the thermodynamic values used.

Very reliable data may be extracted from PEPICO measurements: Here

---

* Phase Space Theory calculations should give more accurate results (see Section II-2.2.4).

the internal energy of the decomposing ion is well defined and the adiabatic appearance potential can be determined with great accuracy although the measurement of the kinetic energy release is less accurate than for metastable ions. With this technique energy partitioning data have been determined for HCl loss from chloroalkanes as well as Cl˙ and Br˙ loss from propargyl chloride and bromide [191].

8 Collision Processes

8.1 Collision-Induced Dissociation

When ions having a high translational energy (a few hundred eV or more) collide inelastically with neutral atoms or molecules, part of their translational energy may be converted into excitation energy of the ions leading to subsequent decomposition [91,221-228]. A collision-induced dissociation of this kind can also be viewed as a dissociative ion-molecule reaction [231] according to equation

$$m_1^+ + N \rightarrow m_1^{+*} + N \rightarrow m_2^+ + m_3 + N$$

The whole group of collision-induced fragments of a given primary ion has been termed the collisional activation (CA) spectrum of this ion [91,227]. Since the place where collision-induced fragments are formed in the mass spectrometer is the same as for metastable ions (one of the field free regions of the instrument) they can be detected by the same methods [91,228].

The use of high translational energies (> 1 keV) and small target atoms (helium) leads primarily to vertical electronic excitation*) of the impacting ion [221]. It is reasonable to assume that the electronically excited ion returns to a vibrationally excited ground state ion by radiationless transitions with subsequent randomization of the internal energy over the entire ion prior to collision-induced decomposition in the same manner discussed in Section II-2.1.1 for electron impact excitation. Hence collision-induced decompositions may be described within the framework of the QET.

---

* It has, however, been pointed out by Franchetti et al. [230] that a high energy collision of an ion and a molecule is likely to result in products containing high angular momentum. Thus rotational energy may play an important role in collision-induced dissociations of organic ions.

In support of this assumption the electron impact and the collisional activation spectra of a molecular ion show the same types of fragments [91,225]. This does not necessarily imply that the internal energy distribution in a molecular ion must be the same after collision with electrons and atoms since this energy distribution is a function of the kinetic energy of the electrons or ions. However, the relative fragment intensities in 70 eV electron impact spectra are very similar to those obtained in collisional-activation spectra of the same molecular ions with 8-10 keV ions [91,225,229] and this makes it probable that the energy distributions are often comparable. In both cases a broad spectrum of excitation energies (0-10 eV) is produced*); in collisional activation the energy distribution after the collision is only slightly influenced by the energy distribution prior to the collision [91]. The similarity between the collisional activation spectra and the 70 eV electron impact spectra is due not only to a similar energy distribution but also to a comparable decomposition time, which for instrumental reasons is around 10^{-6} s in both cases.

As result of the high translational energy of the primary ion the law of conservation of momentum requires that the fraction of translational energy transferred to the target ion is negligible for collisions involving no or little change of direction of the primary ion**) (Only these ions are selected out of all colliding ions by the slit arrangement). Thus the excitation energy that the ion gains in the collision is of the same magnitude as the loss of its translational energy. The decrease of translational energy, Q, leads to a shift of the peak center to lower energies which can be measured directly and thus permits a simple determination of the excitation energy transferred. However, as result of the distribution of energies transferred to the ion upon collision, the observed peak will be composed of a large number of overlapping peaks of different widths and all displaced by different amounts from the position at which the peak due to unimolecular fragmentation would appear [168].

In direct correspondence with metastable decompositions part of

---

*
 Even higher energies may be transferred with low probability, as is apparent from the observation of doubly charged ions.

**
 If the mass of the ionic fragment is equal to that of the neutral target atom and the primary ion has an initial translational energy of 10^4 eV of which 10 eV are transferred, these 10 eV are partitioned between the primary ion and the target atom in a ratio of 4000 to 1 [168].

the non-fixed energy, $\varepsilon^{\neq}$, of the collisionally activated ion will be released as kinetic energy $T^{\neq}$. The total kinetic energy release, T, (including the fraction T^e from the reverse activation energy) can be deduced from the peak width as described for metastable ions. On average, the non-fixed energy of collisionally activated ions will be considerably larger than that of normal metastable ions, leading to larger T values and thus broader signals [231]. The relationship between the quantities Q and T is shown in Fig. III-23. Three types of information can thus be obtained from a collisional activation spectrum: (1) the relative intensities of the secondary ions; (2) the translational energy T released on dissociation; and (3) the excitation energy transferred, Q. Whilst the intensity ratios and the translational energy T re-

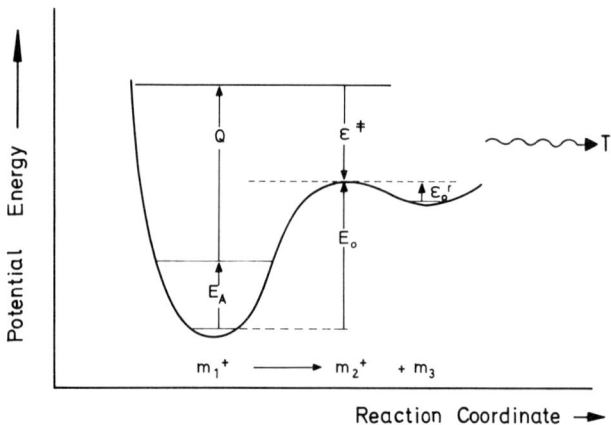

Fig. III-23. Energy transfer in collision-induced dissociation. The precursor ion M_1^+ may have a mean internal energy E_A before collision. On inelastic collision an amount Q of translational energy is converted into internal energy. During decomposition the non-fixed energy, $\varepsilon^{\neq}$, and the reverse activation energy, ε_o^r are released partly as translational energy, T.

leased in a dissociation (i.e. the peak half-width) are suitable for the characterization of an ion structure, determination of the excitation energy by means of the value Q (together with T) provides an insight into the thermochemical properties of an ion.

The intensity ratios of collision-induced fragments were studied by McLafferty et al. [91] as a function of various parameters. It was found that when the translational energy of the impacting ions increases there is not only an increase in the yield of the collision-induced secondary fragments but also a change in the intensity ratios between these fragments. As an immediate consequence of an increase in the mean transferred energies, the intensity of fragments having higher activation energies increases more than does that of fragments with lower activation energies. A similar increase in fragmentation processes with higher activation energy is found at high pressures of the collision gas; this is explained by the occurrence of multiple collisions. The nature of the collision gas, however, has no influence at all on the intensity ratios [91] (and also generally none on Q and T [231]), although it does influence the yield of collision-induced fragments: as the size of the atoms or molecules decreases, the yield increases, so that helium and H_2 are particularly suitable collision gases.

Especially important is the finding that the intensity ratio of the collision-induced secondary fragments from a given primary ion depends to only a very small extent on the excitation energy, i.e. on the energy distribution before the collision, as has been shown by varying the electron energy or by comparing the collisional activation spectra of field-ionized or electron-impact-ionized molecules [91,229]. Dependence on the excitation energy is found only, if at all, with processes involving the lowest activation energies [91] (see Section V-3.4.5.1).

The decrease in translational energy Q caused by energy transfer on collision, as also the energy T released on subsequent dissociation, have been studied in detail by Cooks et al. for the case of aliphatic alcohols [231] and by Wachs and McLafferty for the case of methane [232]. It was established, taking into account the internal energy present before the collision, that the decrease in translational energy Q in a specific collision-induced fragment is roughly equal to the excitation energy predicted from the breakdown diagram, at which the fragment in question is most probably formed, which again supports the conclusion that collision-induced decompositions are adequately described by the QET. Furthermore, the values of Q and T have been used to obtain information on the energy partitioning of the non-fixed energy. As $\varepsilon^{\neq}$ is generally large (vide supra) ε_o^r can often be neglected [231]. Using this approach it was demonstrated that within the limited accuracy of the measurements the energy partitioning quotient for H loss from the methanol molecular ion was consistent with statistical partitioning of the non-fixed energy [231]. Finally, for diatomic and triatomic ions it is possible to derive the electronic state in which the collision-induced fragments are formed from the quantity Q [233].

8.2 Charge Transfer Reactions

The collision of ions of high translational energy with neutral target atoms may not only lead to decomposition, but also to a variety of charge transfer reactions with simultaneous changes in the internal energy. The three main reaction types are charge exchange, charge stripping and charge inversion. Although these processes are not directly related to the QET they will be discussed here briefly.

8.2.1 Charge Exchange of Doubly Charged Ions

A small fraction of all ions formed under electron bombardment in the ion source of a mass spectrometer is doubly charged. These ions may undergo charge exchange reactions with a collision gas according to equation

$$m^{2+} + N \longrightarrow m^+ + N^+$$

If this charge exchange process occurs in front of the electric sector the originally doubly charged ions enter the sector with twice the translational energy but the same charge as the other ions. Hence they are only transmitted if the electric sector potential is increased by a factor of two. Complete spectra of these originally doubly charged ions (termed "2E spectra") can be obtained by scanning the magnetic sector [234]. Thus the charge exchange process allows the separation of the weak doubly charged ions from singly charged ones. The doubly charged ion spectra of a variety of compounds [234-238] have been reported with special emphasis on unsaturated hydrocarbons. These hydrocarbon ions show characteristic fragment peaks of the general type $C_nH_2^{++}$ (n = 2-5) and $C_nH_6^{++}$ (n > 6), having a proposed structure

$$\overset{+}{HC}=(C)_n=\overset{+}{CH} \text{ and } H_3C-\overset{+}{C}=(C)_n=\overset{+}{C}-CH_3$$

Independent evidence for this linear structure is available from the kinetic energy released upon charge separation fragmentation. In such reactions the total coulombic energy gained on separating the charges is released as kinetic energy. This kinetic energy release is so large (T = 1-4 eV) that contributions from the non-fixed energy and from the non-coulombic portion of the reverse activation energy may be neglected. Thus

$$T = \frac{e^2}{R} \tag{III-10}$$

where R is the intercharge distance in the transition state. Using this approach an interchange distance of 5.5 Å was calculated for the reaction $C_6H_6^{2+} \rightarrow C_5H_3^+ + CH_3^+$, which is consistent with an acyclic doubly charged benzene ion [239].

Charge separation fragmentation of doubly charged ions in the spectrum of deuterium labelled toluene is accompanied by nearly complete H/D randomization [236]; the degree of randomization depends on the decomposition process. While the loss of H· in the ion source (fast direct cleavage) occurs preferentially from the side chain, all hydrogens are essentially randomized prior to H_2 loss (slow rearrangement) [240].

8.2.2 Charge Stripping and Charge Inversion

Collision of a singly charged ion of high translational energy with a neutral target gas may lead to further ionization

$$m^+ \rightarrow m^{2+} + e^-$$

The neutral target gas may become excited or even ionized. Such charge stripping reactions may be conveniently observed by reducing the electric sector potential by a factor two ("E/2 spectra") [241,242]. Finally, collision may lead to charge inversion

$$m^+ + N \rightarrow m^- + N^{2+}$$

Such charge inversion reactions may be studied by reversing the polarity of the electric sector potential ("-E spectra") [243]. Alternatively charge inversion may also be observed if negative ions are fired into a neutral target gas [244-246]

$$m^- + N \rightarrow m^+ + N^{2-}$$

More information on the energetics of the various charge transfer reactions can be obtained if the electric sector potential is scanned over the peak of interest. The peak fine structure due to energy lost or gained under collision gives information on the electronic states of the ions [247,248] or, more important, allows the determination of the second or third ionization potentials [249].

Finally a modification of a double focussing mass spectrometer for angular resolution has been reported. This allows the measurement of energy loss spectra at different scattering angles [250].

9 References

1. R.G. Cooks, I. Howe and D.H. Williams, Org. Mass Spectrom., **2**, 137 (1969).
2. M.L. Vestal, in "Fundamental Processes in Radiation Chemistry" (P. Ausloos, Ed.), Wiley, New York, 1968.
3. D.H. Williams and I. Howe, "Principles of Organic Mass Spectrometry", McGraw-Hill, London, 1972.
4. F.W. McLafferty, T. Wachs, C. Lifshitz, G. Innorta and P. Irving, J. Am. Chem. Soc., **92**, 6867 (1970).
5. F.W. McLafferty and W.T. Pike, J. Amer. Chem. Soc., **89**, 5951 (1967).
6. W.T. Pike and F.W. McLafferty, J. Amer. Chem. Soc., **89**, 5953 (1967).
7. M.L. Gross and F.W. McLafferty, Chem. Commun., 254 (1968).
8. J.L. Occolowitz, J. Amer. Chem. Soc., **91**, 5202 (1969).
9. A.N.H. Yeo and D.H. Williams, J.Amer. Chem. Soc., **93**, 395 (1971).
10. C.W. Tsang and A.G. Harrison, Org. Mass Spectrom., **7**, 1377 (1973).
11. C.W. Tsang and A.G. Harrison, Org. Mass Spectrom., **3**, 647 (1970).
12. B.S. Rabinovitch and D.W. Setser, in "Advances in Photochemistry", Vol. 3, Wiley, New York, 1964, p. 1.
13. D.H. Williams and R.G. Cooks, Chem. Commun., 663 (1968).
14. R.H. Shapiro, Org. Mass Spectrom., **1**, 907 (1968).
15. R.H. Shapiro and J.W. Serum, Org. Mass Spectrom., **2**, 533 (1969).
16. R.H. Shapiro and K.B. Tomer, Org. Mass Spectrom., **2**, 579 (1969).
17. R.H. Shapiro and T.F. Jenkins, Org. Mass Spectrom., **2**, 771 (1969).
18. K.B. Tomer, J. Turk and R.H. Shapiro, Org. Mass Spectrom., **6**, 235 (1972).
19. I. Howe, N.A. Uccella and D.H. Williams, J.C.S. Perkin II, 76 (1973).
20. J.C. Tou, J. Phys. Chem., **75**, 1903 (1971).
21. K. Levsen and H.D. Beckey, Int. J. Mass Spectrom. Ion Phys., **7**, 341 (1971).
22. P. Brown, Org. Mass Spectrom., **3**, 1175 (1970).
23. J.R. Dias and C. Djerassi, Org. Mass Spectrom., **6**, 385 (1972).
24. F.W. McLafferty and R.B. Fairweather, J. Am. Chem. Soc., **90**, 5915 (1968).
25. J.F. Elder, J.H. Beynon and R.G. Cooks, Org. Mass Spectrom., **10**, 273 (1975).
26. F.W. McLafferty, D.J. McAdoo and J.S. Smith, J.Am. Chem. Soc., **91**, 5400 (1969).
27. H.D. Beckey, "Principles of Field Ionization and Field Desorption Mass Spectrometry", Pergamon, Oxford, 1977.
28. P.J. Derrick in "Mass Spectrometry", Int. Rev. Science, Phys. Chem. Series II, Vol. 5, (A. Maccoll, Ed.) Butterworths, 1975.
29. H.D. Beckey, H. Hey, K. Levsen and G. Tenschert, Int. J. Mass Spectrom. Ion Phys., **2**, 101 (1969).
30. F. Borchers, K. Levsen, H. Schwarz, C. Wesdemiotis and R. Wolfschütz, J. Am. Chem. Soc., **99**, 1716 (1977).
31. K. Levsen and H.D. Beckey, Int. J. Mass Spectrom. Ion Phys., **9**, 63 (1972).
32. H.D. Beckey and K. Levsen in "Recent Topics in Mass Spectrometry" (R.I. Reed, Ed.) Gordon and Breach, New York, 1971.
33. P. Schulze and W.J. Richter, Int. J. Mass Spectrom. Ion Phys., **6**, 131 (1971).
34. K. Levsen and H.D. Beckey, Int. J. Mass Spectrom. Ion Phys., **15**, 333 (1974).

35. P. J. Derrick, A.M. Falick and A.L. Burlingame, J. Chem. Soc., Faraday Trans. I, 71, 1503 (1975).
36. J. van der Greef, C.B. Theissling and N.M.M.Nibbering, "Advances in Mass Spectrometry", Vol. 7, Heyden, London, 1977.
37. D.G. Patterson, R.B. Scott and P. Brown, Org. Mass Spectrom., 12, 395 (1977).
38. A.N.H. Yeo and D.H. Williams, J. Am. Chem. Soc., 91, 3582 (1969).
39. I. Howe and F.W. McLafferty, J. Am. Chem. Soc., 92, 3797 (1970).
40. I. Howe and F.W. McLafferty, J. Am. Chem. Soc., 93, 99 (1971).
41. K. Levsen, F.W. McLafferty and D.M. Jerina, J. Am. Chem. Soc., 95, 6332 (1973).
42. J.S. Smith and F.W. McLafferty, Org. Mass Spectrom., 5, 483 (1971).
43. J.H. Beynon, R.A. Saunders and A.E. Williams, Z. Naturforsch., 20a, 180 (1965).
44. R. Neeter and N.M.M. Nibbering, Org. Mass Spectrom., 5, 735 (1971).
45. T.A. Molenaar-Langeveld, N.M.M. Nibbering and Th.J. de Boer, Org. Mass Spectrom., 5, 725 (1971).
46. A. Venema, N.M.M. Nibbering and Th.J. de Boer, Org. Mass Spectrom., 3, 1589 (1970).
47. F. de Jong, H.J.M. Sinnige and M.J. Janssen, Org. Mass Spectrom., 3, 1539 (1970).
48. R.G. Cooks, I. Howe, S.W. Tam and D.H. Williams, J. Am. Chem. Soc., 90, 4064 (1968).
49. K. Levsen and H.D. Beckey, Int. J. Mass Spectrom. Ion Phys., 14, 45 (1974).
50. P.J. Derrick, A.M. Falick, A.L. Burlingame, Adv. Mass Spectrom., 6, 877 (1974).
51. P.J. Derrick, A.M. Falick and A.L. Burlingame, J. Amer. Chem. Soc., 94, 6794 (1972).
52. P.J. Derrick, A.M. Falick, S. Lewis and A.L. Burlingame, Org. Mass Spectrom., 7, 887 (1973).
53. P.J. Derrick, A.M. Falick, A.L. Burlingame and C. Djerassi, J. Amer. Chem. Soc., 96, 1054 (1974).
54. P.J. Derrick and A.L. Burlingame, Acc. Chem. Res., 7, 328 (1974).
55. P. Brown and C. Fenselau, Org. Mass Spectrom., 7, 305 (1973).
56. P.J. Derrick and A.L. Burlingame, J. Am. Chem. Soc., 96, 4909 (1974).
57. J. van der Greef and N.M.M. Nibbering, in preparation.
58. R.P. Morgan and P.J. Derrick, in "Advances in Mass Spectrometry", Vol. 7, Heyden, London, 1977.
59. P. Wolkoff, J. van der Greef and N.N.M. Nibbering, J. Am. Chem. Soc., 100, 541 (1978).
60. F. Borchers, K. Levsen, H. Schwarz and C. Wesdemiotis, Int. J. Mass Spectrom. Ion Phys., in press.
61. S. Meyerson and J.L. Corbin, J. Am. Chem. Soc., 87, 3045 (1965).
62. R.H. Shapiro, K.B. Tomer, R.M. Caprioli and J.H. Beynon, Org. Mass Spectrom., 3, 1333 (1970).
63. R.H. Shapiro, K.B. Tomer, J.H. Beynon and R.M. Caprioli, Org. Mass Spectrom., 3, 1593 (1970).
64. F. Benoit, Org. Mass Spectrom., 7, 295 (1973).
65. C.E. Parker, M.M. Bursey and L.G. Pedersen, Org. Mass Spectrom., 9, 204 (1974).
66. J.L. Holmes and F. Benoit, Org. Mass Spectrom., 4, 97 (1970).
67. F.M. Benoit and A.G. Harrison, Org. Mass Spectrom., 11, 1056 (1976).
68. G.V. Pfeiffer and J.G. Jewett, J. Am. Chem. Soc., 90, 2143 (1970).
69. P.K. Bischof and M.J.S. Dewar, J. Am. Chem. Soc., 97, 2278 (1975).

70. A.J. Lorquet and J.C. Lorquet, J. Chem. Phys., 49, 4955 (1968).
71. I.H. Suzuki and K. Maeda, Int. J. Mass Spectrom. Ion Phys., 13, 293 (1974).
72. I.H. Suzuki and K. Maeda, Int. J. Mass Spectrom. Ion Phys., 15, 281 (1974).
73. R. Stockbauer and M.G. Inghram, J. Chem. Phys., 62, 4862 (1975).
74. G.A. Gallup, D. Steinheider and M.L. Gross, Int. J. Mass Spectrom. Ion Phys., 22, 185 (1976).
75. P.J. Derrick, A.M. Falick and A.L. Burlingame, J. Chem. Soc. Perkin Trans. II, 98 (1975).
76. D.J. McAdoo, F.W. McLafferty and T.E. Parks, J. Am. Chem. Soc., 94, 1601 (1972).
77. G.G. Meisels, J.Y. Park and B.G. Giessner, J. Am. Chem. Soc., 91, 1555 (1969).
78. K.R. Jennings, Z. Naturforsch., 22a, 454 (1967).
79. W.O. Perry, J.H. Beynon, W.E. Baitinger, J.W. Amy, R.M. Caprioli, R.N. Renauld, L.C. Leitch and S. Meyerson, J. Am. Chem. Soc., 92, 7236 (1970).
80. E. Wentrup-Byrne, F.O. Gülaçar, P. Müller and A. Buchs, Org. Mass Spectrom., 12, 636 (1977).
81. H. Budzikiewicz and R. Stolze, Monath., 108, 869 (1977).
82. D.H. Williams, R.G. Cooks, J. Ronayne and S.W. Tam, Tetrahedron Lett., 1777 (1968).
83. D.H. Williams and J. Ronayne, Chem. Commun., 1129 (1967).
84. R.P. Morgan, P.J. Derrick and A.G. Harrison, J. Am. Chem. Soc., 94, 4189 (1977).
85. D.J. McAdoo, P.F. Bente, M.L. Gross and F.W. McLafferty, Org. Mass Spectrom., 9, 525 (1974).
86. M.B. Wallenstein and M. Krauss, J. Chem. Phys., 34, 929 (1961).
87. H.M. Rosenstock and M. Krauss, in "Advances in Mass Spectrometry", Vol. 2 (R.M. Elliot, Ed.), Pergamon, Oxford, 1963, p. 270.
88. R.G. Cooks and D.H. Williams, Chem. Commun., 627 (1968).
89. P.F. Bente, F.W. McLafferty, D.J. McAdoo and C. Lifshitz, J. Phys. Chem., 79, 713 (1975).
90. Y.N. Lin and B.S. Rabinovitch, J. Phys. Chem., 74, 1769 (1970).
91. F.W. McLafferty, P.F. Bente, R. Kornfeld, S.-C. Tsai and I. Howe, J. Am. Chem. Soc., 95, 2120 (1973).
92. H.M. Rosenstock, Int. J. Mass Spectrom. Ion Phys., 20, 139 (1976).
93. R.W. Kiser, "Introduction to Mass Spectrometry and its Applications", Pentice-Hall, Englewood Cliffs, 1965.
94. W.A. Chupka, J. Chem. Phys., 30, 191 (1959).
95. J.R. Gilbert and A.J. Stace, Org. Mass Spectrom., 10, 1032 (1975).
96. I. Hertel and Ch. Ottinger, Z. Naturforsch., 22a, 40 (1967).
97. R.D. Hickling and K.R. Jennings, Org. Mass Spectrom., 3, 1499 (1970).
98. J.H. Beynon, J.A. Hopkinson and G.R. Lester, Int. J. Mass Spectrom. Ion Phys., 2, 291 (1969).
99. R.E. Fox and A. Langer, J. Chem. Phys., 18, 460 (1950).
100. B. Steiner, C.F. Giese and M.G. Inghram, J. Chem. Phys., 34, 189 (1961).
101. W.A. Chupka and J. Berkowitz, J. Chem. Phys., 47, 2921 (1967).
102. K.M.A. Refaey and W.A. Chupka, J. Chem. Phys., 48, 5205 (1968).
103. T.W. Bentley, R.A.W. Johnstone and B.N. McMaster, Chem. Commun., 510 (1973).
104. R.G. Cooks, J.H. Beynon, R.M. Caprioli and G.R. Lester, "Metastable Ions", Else-

vier, 1973, p. 93.

105. J.H. Beynon, R.G. Cooks, K.R. Jennings and A.J. Ferrer-Correia, Int. J. Mass Spectrom. Ion Phys., $\underline{18}$, 87 (1975).

106. D.H. Williams, in "Advances in Mass Spectrometry", Vol. 7, Heyden, London, 1977.

107. M.L. Gross, Org. Mass Spectrom., $\underline{6}$, 827 (1972).

108. R.C. Dunbar, J. Am. Chem. Soc., $\underline{95}$, 472 (1973).

109. R.C. Dunbar and E.W. Fu, J. Am. Chem. Soc., $\underline{95}$, 2716 (1973).

110. C. Lifshitz, A.M. Peers, M. Weiss and M.J. Weiss, in "Advances in Mass Spectrometry", Vol. 6 (A.R. West, Ed.), Applied Science Publisher, Barking, 1974, S. 871.

111. S.M. Gordon and N.W. Reid, Int. J. Mass Spectrom. Ion Phys., $\underline{18}$, 379 (1975).

112. C. Lifshitz, M. Weiss and S. Landau-Gefen, Proceedings of the 25th Conference of the American Society for Mass Spectrometry, Washington, 1977, p. 512.

113. R. Stockbauer and H.M. Rosenstock, Int. J. Mass Spectrom. Ion Phys., in press.

114. C. Lifshitz and F.A. Long, J. Chem. Phys., $\underline{41}$, 2468 (1964).

115. A.G. Harrison in "Topics in Organic Mass Spectrometry" (A.L. Burlingame, Ed.), Wiley-Interscience, New York, 1970.

116. I. Howe, in "Mass Spectrometry", Vol. 2, (D.H. Williams, Ed.), The Chemical Society, London, 1973, p. 36.

117. W.A. Chupka, J. Chem. Phys., $\underline{54}$, 1936 (1971).

118. M.L. Vestal and G. Lerner, "Fundamental Studies Related to the Radiation Chemistry of Small Organic Molecules", Aerospace Research Laboratory Report 67-0114 (1967).

119. M. Corval, Bull. Soc. Chim. France, 2878 (1970).

120. M. Corval and P. Masclet, Org. Mass Spectrom., $\underline{6}$, 511 (1972).

121. S.M. Gordon, G.J. Krige and N.W. Reid, Int. J. Mass Spectrom. Ion Phys., $\underline{14}$, 109 (1974).

122. P.C. Haarhoff, Mol. Phys., $\underline{7}$, 101 (1963).

123. U. Löhle and Ch. Ottinger, J. Chem. Phys., $\underline{51}$, 3097 (1969).

124. M. Bertrand, J.H. Beynon and R.G. Cooks, Org. Mass Spectrom., $\underline{7}$, 193 (1973).

125. F.W. McLafferty, D.J. McAdoo, J.S. Smith and R. Kornfeld, J. Am. Chem. Soc., $\underline{93}$, 3720 (1971).

126. R. Neeter and N.M.M. Nibbering, Org. Mass Spectrom., $\underline{7}$, 1091 (1973).

127. R.D. Smith and J.H. Futrell, Org. Mass Spectrom., $\underline{11}$, 309 (1976).

128. G. Eadon and R. Zawalski, Org. Mass Spectrom., $\underline{12}$, 599 (1977).

129. N.M.M. Nibbering, to be published.

130. J.H. Beynon, A.E. Fontaine and G.R. Lester, Int. J. Mass Spectrom. Ion Phys., $\underline{1}$, 1 (1968),

131. C. Lifshitz and R. Sternberg, Int. J. Mass Spectrom. Ion Phys., $\underline{2}$, 303 (1969).

132. P.R. Briggs, W.L. Parker and T.W. Shannon, Chem. Commun., 727 (1968).

133. J.H. Beynon, R.M. Caprioli and T. Ast, Org. Mass Spectrom., $\underline{5}$, 229 (1971).

134. M. Bertrand, J.H. Beynon and R.G. Cooks, Int. J. Mass Spectrom. Ion Phys., $\underline{9}$, 346 (1972).

135. L.P. Hammett, "Physical Organic Chemistry", McGraw-Hill, New York, 1940, Chapter VII.

136. H.C. Brown and Y. Okamoto, J. Am. Chem. Soc., $\underline{80}$, 4979 (1958).

137. M.S. Chin and A.G. Harrison, Org. Mass Spectrom., $\underline{2}$, 1073 (1969).

138. P. Brown, Org. Mass Spectrom., 4, 519 (1970).
139. P. Brown, Org. Mass Spectrom., 4, 533 (1970).
140. A. Buchs, G.P. Rosetti and B.P. Susz, Helv. Chim. Acta, 47, 1563 (1964).
141. R.A.W. Johnstone and D.W. Payling, Chem. Commun., 601 (1968).
142. J.M.S. Tait, T.W. Shannon and A.G. Harrison, J. Am. Chem. Soc., 84, 4 (1962).
143. M.M. Bursey and F.W. McLafferty, J. Am. Chem. Soc., 88, 529 (1966).
144. J. Michnowicz and B. Munson, Org. Mass Spectrom., 4, 481 (1970).
145. G.O. Phillips, W.G. Filby and W.L. Mead, Chem. Commun., 1269 (1970).
146. H.D. Beckey and M.D. Migahed, Org. Mass Spectrom., 6, 923 (1972).
147. J.H. Bowie and B. Nussey, Org. Mass Spectrom., 6, 429 (1972).
148. M.M. Bursey, Org. Mass Spectrom., 1, 31 (1968).
149. I. Howe, in "Mass Spectrometry" Vol. 1 (D.H. Williams, Ed.), The Chemical Society, London, 1971, Chapter 2.
150. I. Howe, in "Mass Spectrometry" Vol. 2 (D.H. Williams, Ed.), The Chemical Society, London, 1973, Chapter 2.
151. M.M. Bursey in "Advances in Linear Free Energy Relationships", (J. Shorter, Ed.), Plenum, 1972.
152. T.W. Bentley and R.W.A. Johnstone, in "Advances in Physical Organic Chemistry", Vol. 8 (V. Gold, Ed.), Academic Press, 1970, p. 229.
153. F.W. McLafferty, Chem. Commun., 956 (1968).
154. T.W. Bentley, R.A.W. Johnstone and D.W. Payling, J. Am. Chem. Soc., 91, 3978 (1969).
155. D.H. Williams and I. Howe, "Principles of Organic Mass Spectrometry", McGraw-Hill, London (1972), p. 62.
156. F. Benoit, Org. Mass Spectrom., 7, 295 (1973).
157. K. Levsen, unpublished results.
158. I. Howe and D.H. Williams, J. Am. Chem. Soc., 91, 7137 (1969).
159. G. Remberg and G. Spiteller, Chem. Ber., 103, 3640 (1970).
160. G. Remberg, E. Remberg, M. Spiteller-Friedmann and G. Spiteller, Org. Mass Spectrom., 1, 87 (1968).
161. H.M. Rosenstock, K. Draxl, B.W. Steiner and J.T. Herron, J. Phys. Chem. Ref. Data, Vol. 6 (1977).
162. M.M. Bursey and F.W. McLafferty, J. Am. Chem. Soc., 88, 5023 (1966).
163. D. Davis and D.H. Williams, Chem. Commun., 412 (1970).
164. R.G. Cooks, D.W. Setser, K. Jennings and S. Jones, Int. J. Mass Spectrom. Ion Phys., 7, 493 (1971).
165. R.G. Cooks, M. Bertrand, J.H. Beynon, M.E. Rennekamp and D.W. Setser, J. Am. Chem. Soc., 95, 1732 (1973).
166. J.H. Beynon, M. Bertrand and R.G. Cooks, J. Am. Chem. Soc., 95, 1739 (1973).
167. J.A. Hipple and E.U. Condon, Phys. Rev., 68, 54 (1945).
168. R.G. Cooks and J.H. Beynon, in "Mass Spectrometry", International Review of Science, Phys. Chem. Series II, Vol. 5, Butterworths, London (1975), Chapter 5.
169. J.H. Beynon and R.G. Cooks, Int. J. Mass Spectrom. Ion Phys., 19, 107 (1976).
170. K.C. Kim, J.H. Beynon and R.G. Cooks, J. Chem. Phys., 61, 1305 (1974).
171. D.T. Terwilliger, R.G. Cooks and J.H. Beynon, Int. J. Mass Spectrom. Ion Phys., 18, 43 (1975).
172. R.G. Cooks, K.C. Kim and J.H. Beynon, Chem. Phys. Lett., 26, 131 (1974).

173. H.M. Rosenstock, in "Advances in Mass Spectrometry", Vol. 4, The Institute for Petroleum, London, 1968, p. 523.
174. B.H. Solka, J.H. Beynon and R.G. Cooks, J. Phys. Chem., $\underline{79}$, 859 (1975).
175. R.G. Cooks, D.T. Terwilliger and J.H. Beynon, J. Chem. Phys., $\underline{61}$, 1208 (1974).
176. C.E. Berry, Phys. Rev., 78 597 (1950).
177. J.D. Morrison and H.E. Stanton, J. Chem. Phys., $\underline{28}$, 9 (1958).
178. R.Taubert, Z. Naturforsch., $\underline{19a}$, 911 (1964).
179. C.G. Rowland, J.H.D. Eland and C.J. Danby, Int. J. Mass Spectrom. Ion Phys., $\underline{2}$. 457 (1969).
180. C.G. Rowland, Int. J. Mass Spectrom. Ion Phys., $\underline{7}$, 79 (1971).
181. D.K. Sen Sharma and J.L. Franklin, Int. J. Mass Spectrom. Ion Phys., $\underline{13}$, 139 (1974).
182. J.L. Franklin, P.M. Hierl and D.A. Whan, J. Chem. Phys., $\underline{47}$, 3148 (1967).
183. M.A. Haney and J.L. Franklin, J. Chem. Phys., $\underline{48}$, 4093 (1968).
184. J.H.D. Eland, Int. J. Mass Spectrom. Ion Phys., $\underline{9}$, 397 (1972).
185. I.G. Simm and C.J. Danby, J.C.S. Farad. Trans. II, $\underline{72}$, 860 (1976).
186. I.G. Simm, C.J. Danby, J.H.D. Eland and P.I. Mansell, J.C.S. Farad. Trans. II, $\underline{72}$, 426 (1976).
187. R. Stockbauer, Int. J. Mass Spectrom. Ion Phys., $\underline{25}$, 89 (1977).
188. B. Brehm, J.H.D. Eland, R. Frey and A. Küstler, Int. J. Mass Spectrom. Ion Phys., $\underline{12}$, 197 (1973).
189. B.Brehm, J.H.D. Eland, R.Frey and A. Küstler, Int. J. Mass Spectrom. Ion Phys., $\underline{12}$, 213 (1973).
190. D.M. Mintz and T. Baer, J. Chem. Phys., $\underline{65}$, 2407 (1976).
191. B.P. Tsai, A.S. Werner and T. Baer, J. Chem. Phys., $\underline{63}$, 4384 (1975).
192. N.R. Daly, A. McCormick, R.E. Powell and R. Hayes, Int. J. Mass Spectrom. Ion Phys., $\underline{11}$, 255 (1973).
193. D.T. Terwilliger, J.H. Beynon and R.G. Cooks, Proc. R. Soc. Lond. A $\underline{341}$, 135 (1974).
194. Ch. Ottinger, Phys. Lett., $\underline{17}$, 269 (1965).
195. J. Schopman and J. Los, Physica, $\underline{48}$, 190 (1970).
196. P.G. Fournier, J.B. Ozenne and J. Durup, J. Chem. Phys., $\underline{53}$, 4095 (1970).
197. P.G. Fournier, C.A. van de Runstraat, T.R. Govers, J. Schopman, F.J. de Heer and J. Los., Chem. Phys. Letters, $\underline{9}$, 426 (1971).
198. P.G. Fournier, T.R. Govers, C.A. van de Runstraat, J. Schopman and J. Los, J. Phys. (Paris), $\underline{33}$, 755 (1972).
199. R. Locht, J. Schopman, H. Wankenne and J. Momigny, Chem. Phys., $\underline{7}$, 393 (1975).
200. T.R. Govers and J. Schopman, Chem. Phys. Letters, $\underline{12}$, 414 (1971).
201. D. Pham, M. Bizot, J. Durup, B. Fourmann and J.B. Ozenne, in: Electronic and atomic collisions, VII ICPEAC, Abstracts of papers, Vol. 1, eds. L.M. Branscomb, H. Ehrhardt, R. Geballe, F.J. de Heer, N.V. Fedorenko, J. Kistemaker, M. Barat, E.E. Nikitin and A.C.H. Smith (North-Holland, Amsterdam, 1971) p. 427.
202. J.G. Maas, N.P.F.B. van Asselt and J. Los, Chem. Phys., $\underline{8}$, 37 (1975).
203. J.G. Maas, N.P.F.B. van Asselt, P.J.C.M. Nowak, J. Los and S.D. Peyerimhoff, Chem. Phys., $\underline{17}$, 217 (1976).
204. P.G. Fournier, G. Comtet, R.W. Odom, R. Locht, J.G. Maas, N.P.F.B. van Asselt and J. Los, Chem. Phys. Lett., $\underline{40}$, 170 (1976).

205. R.G. Cooks and J.H. Beynon, Chem. Commun., 1282 (1971).
206. E.G. Jones, L.E. Bauman, J.H. Beynon and R.G. Cooks, Org. Mass Spectrom., 7, 185 (1973).
207. M. Medved, R.G. Cooks and J.H. Beynon, Int. J. Mass Spectrom. Ion Phys., 19, 179 (1976).
208. J.L. Holmes and A.D. Osborne, Int. J. Mass Spectrom. Ion Phys., in press.
209. D.T. Terwilliger, R.G. Cooks and J.H. Beynon, Int. J. Mass Spectrom. Ion Phys., 18, 43 (1975).
210. J.F. Elder, R.G. Cooks and J.H. Beynon, Org. Mass Spectrom., 11, 423 (1976).
211. J.F. Elder, J.H. Beynon and R.G. Cooks, Org. Mass Spectrom., 10, 273 (1973).
212. J.L. Holmes, A.D. Osborne and G.M. Weese, Int. J. Mass Spectrom. Ion Phys., 19, 207 (1976).
213. D.M. Mintz and T. Baer, Int. J. Mass Spectrom. Ion Phys., 25, 39 (1977).
214. C.E. Klots, D. Mintz and T. Baer, J. Chem. Phys., 66, 5100 (1977).
215. J.L. Holmes, presented at the Euchem Conference on "The Chemistry of Ion Beams", Nordwijk (Holland), (1977).
216. R. Stockbauer, Int. J. Mass Spectrom. Ion Phys., 25, 401 (1977).
217. R.K. Boyd and J.H. Beynon, Int. J. Mass Spectrom. Ion Phys., 23, 163 (1977).
218. J.H. Beynon, M. Bertrand and R.G. Cooks, Org. Mass Spectrom., 7, 785 (1973).
219. R.G. Cooks, K.C. Kim and J.H. Beynon, Int. J. Mass Spectrom. Ion Phys., 15, 245 (1974).
220. J.R. Christie, P.J. Derrick and G.J. Rickard, J. Chem. Soc. Faraday Trans. II, 74, 304 (1978).
221. H. Yamaoka, P. Dông and J. Durup, J. Chem. Phys., 51, 3465 (1969).
222. J. Durup, P. Fournier and P. Dông, Int. J. Mass Spectrom. Ion Phys., 2, 311 (1969).
223. J. Los, Ber. Bunsenges. Phys. Chem., 77, 640 (1973).
224. W.F. Haddon and F.W. McLafferty, J. Am. Chem. Soc., 90, 4745 (1968).
225. K.R. Jennings, Int. J. Mass Spectrom. Ion Phys., 1, 227 (1968).
226. J.H. Beynon, R.M. Caprioli and T. Ast, Int. J. Mass Spectrom. Ion Phys., 7, 88 (1971).
227. F.W. McLafferty, R. Kornfeld, W.F. Haddon, K. Levsen, I. Sakai, P.F. Bente, S.C. Tsai and H.D.R. Schuddemage, J. Am. Chem. Soc., 95, 3886 (1973).
228. K. Levsen and H. Schwarz, Angew. Chem. (Int. Ed.), 15, 509 (1976).
229. K. Levsen and H.D. Beckey, Org. Mass Spectrom., 9, 570 (1974).
230. V. Franchetti, B.S. Freiser and R.G. Cooks, Org. Mass Spectrom., 13, 106 (1978)
231. R.G. Cooks, L. Hendricks and J.H. Beynon, Org. Mass Spectrom., 10, 625 (1975).
232. T. Wachs and F.W. McLafferty, Int. J. Mass Spectrom. Ion Phys., 23, 243 (1977).
233. K.C. Kim, M. Uckotter, J.H. Beynon and R.G. Cooks, Int. J. Mass Spectrom. Ion Phys., 15, 23 (1974).
234. J.H. Beynon, A. Mathias and A.E. Williams, Org. Mass Spectrom., 5, 303 (1971).
235. T. Ast, J.H. Beynon and R.G. Cooks, Org. Mass Spectrom., 6, 749 (1972).
236. T. Ast, J.H. Beynon and R.G. Cooks, Org. Mass Spectrom., 6, 741 (1972).
237. H. Sakurai, A. Tatematsu and H. Nakata, Bull. Chem. Soc. Japan, 47, 2731 (1974).
238. H. Sakurai, A. Tatematsu and H. Nakata, Bull. Chem. Soc. Japan, 49, 2800 (1976).
239. J.H. Beynon and A.E. Fontaine, Chem. Commun., 717 (1966).

240. T. Ast, J.H. Beynon and R.G. Cooks, J. Am. Chem. Soc., 94, 1834 (1972).
241. R.G. Cooks, J.H. Beynon and T. Ast, J. Am. Chem. Soc., 94, 1004 (1972).
242. T. Ast, J.H. Beynon and R.G. Cooks, J. Am. Chem. Soc., 94, 6611 (1972).
243. T. Keough, J.H. Beynon and R.G. Cooks, J. Am. Chem. Soc., 95, 1695 (1973).
244. J.H. Bowie and T. Blumenthal, J. Am. Chem. Soc., 97, 2959 (1975).
245. J.H. Bowie and T. Blumenthal, Aust. J. Chem., 29, 115 (1976).
246. J.H. Bowie, P.Y. White, J.C. Wilson, F.C.V. Larsson, S.-O. Lawesson, J.Ø. Madsen, C. Nolde and G. Schroll, Org. Mass Spectrom., 12, 191 (1977).
247. T. Keough, J.H. Beynon and R.G. Cooks, Int. J. Mass Spectrom. Ion Phys., 16, 417 (1975).
248. T. Ast, R.G. Cooks and J.H. Beynon, Adv. Mass Spectrom., 6, 815 (1974).
249. R.G. Cooks, T. Ast and J.H. Beynon, Int. J. Mass Spectrom. Ion Phys., 11, 490 (1973).
250. V. Franchetti, J.J. Carmody, D.A. Krause and R.G. Cooks, Int. J. Mass Spectrom. Ion Phys., 26, 353 (1978).

Chapter IV. Reaction Mechanisms

1 The Mechanistic Approach

Although the QET was already developed whilst organic mass spectrometry was still in its infancy, during the first decade after its introduction this kinetic theory was only accepted reluctantly by organic mass spectrometrists who preferred to rationalize the mass spectral behavior of organic molecules by using generalizations of mechanistic organic chemistry such as those employing inductive, resonance and similar effects, i.e. the fragmentation behavior was discussed predominantly in energetic terms extrapolating from solution chemistry to the gas-phase. It will be shown that this mechanistic approach to organic mass spectrometry which is still used with great success in rationalizing or even predicting mass spectrometric fragments and their abundance is in general not in contradiction with the QET.

1.1 Product Stabilities and Bond Strengths

According to this mechanistic approach [1] the relative fragment ion abundance is determined by

(1) *The stability of the reaction products*. The relative abundance of a given ion will increase with an increase in the stability of either the ionic or the neutral product.

(2) *The lability of the bond cleaved*. The decreasing strength of a bond is reflected in an increasing abundance of the product formed by cleavage of this bond.

The validity and usefulness of this concept has been discussed in detail by Johnstone [2]. Neglecting excess energy terms, it is the activation energy of a given fragmentation which directly reflects the heats of formation of the products and thus their stability i.e. the validity of Hammond's postulate is assumed [3]. For two competing reactions we have:

$$M^{+\cdot} \begin{array}{c} \nearrow A_1^+ + B_1^{\cdot} \\ \searrow A_2^+ + B_2^{\cdot} \end{array}$$

Scheme IV-1

$$\Delta E_o = [\Delta H_f(A_1^+) + \Delta H_f(B_1^\cdot)] - [\Delta H_f(A_2^+) + \Delta H_f(B_2^\cdot)] \qquad (IV-1)$$

Moreover the lability of a bond is also characterized by the activation energy for cleavage of this bond.

According to the QET fragment ion abundances are determined predominantly by the activation energies if ions of low internal energies (e.g. metastable ions) are sampled or if the geometries of the activated complexes are similar. Only in these instances do the criteria of product stability and bond lability allow a safe prediction of the relative fragment abundances (provided the reaction does not involve a reverse activation energy). However, even for ions of high internal energy (70 eV) the product stability rule should in general allow the prediction at least of the occurrence of a given fragment.

A direct consequence of the criterion of product stability is the *even-electron rule* [4]. According to this rule *odd-electron ions* (such as molecular ions or fragments formed by rearrangement) *decompose by loss of radicals or even-electron molecules, whereas even-electron ions* (such as protonated molecules or fragments formed by direct cleavage) *fragment by loss of even-electron molecules*. The reason for this fragmentation behavior lies in the higher thermodynamic stability of even-electron species compared with odd-electron ones. Thus, if possible, those reactions in which the electrons are kept paired will be preferred.

This is demonstrated in Table IV-1 for the metastable decomposition of decyl ions, $C_{10}H_{21}^+$, and decene ions, $C_{10}H_{20}^{+\cdot}$, generated by ethyl or ethane loss from the n-dodecane molecular ion [5]. Although these two ions differ only by one hydrogen atom they show completely different metastable decompositions. The even-electron decyl ions decompose exclusively by loss of alkene molecules (propene to hexene) forming again even-electron secondary fragments, $C_nH_{2n+1}^+$. Thus, both the ionic and neutral fragments are stable even-electron species. However, the loss of an alkyl radical from the even-electron precursor would lead to both odd-electron ionic and neutral fragments and is hence not observed for metastable decompositions [6]. Loss of alkyl radicals from even electron fragments is, however, observed at high internal energies (collision-induced decomposition) as here not only the activation energies (product stabilities), but also the geometries of the activated complexes determine the relative fragment abundances.

In contrast to the simple metastable decomposition pattern of the even-electron decyl ion, the odd-electron decene ion loses both neutral alkane and alkene molecules, but also alkyl radicals. These losses lead to either a stable even-electron ionic fragment and an unstable odd-electron neutral fragment or vice versa. Thus in both cases the sum of

Table IV-1. Metastable Decomposition of Decene ($C_{10}H_{20}^{+\cdot}$) and Decyl ($C_{10}H_{21}^{+}$) Fragment Ions Generated from n-Dodecane

m/e	$C_{10}H_{20}^{+\cdot}$ Loss of	Rel. Int. (%)	$C_{10}H_{21}^{+}$ Loss of	Rel. Int. (%)
125	$CH_3^{\cdot}$	1.4		
112	C_2H_4	10		
111	$C_2H_5^{\cdot}$	23		
110	C_2H_6	5.5		
99			C_3H_6	26
98	C_3H_6	9.3		
97	$C_3H_7^{\cdot}$	16		
96	C_3H_8	6.8		
85			C_4H_8	42
84	C_4H_8	10		
83	$C_4H_9^{\cdot}$	4.3		
82	C_4H_{10}	5.7		
71			C_5H_{10}	28
70	C_5H_{10}	5.0		
57			C_6H_{12}	3.9
56	C_6H_{12}	0.5		
55	$C_6H_{13}^{\cdot}$	1.1		

the product stabilities will be comparable. A quantitative comparison of the heats of formation of the products supports this view [5,7].

Whilst the even electron rule seems to hold for aliphatic hydrocarbon ions, a variety of exceptions are known for ions of other compound classes [8-11]. Thus the most abundant metastable transition from the even electron $C_6H_5-C{\equiv}C-\overset{+}{C}(CH_3)_2$ ion is due to the loss of a methyl radical [10].

The importance of the product stabilities in directing mass spectrometric reactions has been reemphasized by Williams and Bowen [12] in a recent study on the metastable elimination of HX and ˙X˙ from substituted benzenes*).

---

* As metastable ions are under consideration here entropic factors, i.e. the transition state geometries, are less important than energetic factors, i.e. the heats of formation of the products.

IV-1 The Mechanistic Approach

$$C_6H_5X^{+\cdot} \begin{cases} \longrightarrow C_6H_4^{+\cdot} + HX \\ \longrightarrow C_6H_5^+ + X^{\cdot} \\ \searrow \text{competing fragmentations} \end{cases}$$

Scheme IV-2

The difference in heats of formation of $C_6H_4^{+\cdot}$ and $C_6H_5^+$ is ~ 281 kJ mol^{-1}. Therefore $\Delta H_f(HX)$ must be less than $\Delta H_f(X^{\cdot})$ by at least 281 kJ mol^{-1} before metastable peaks will be observed for HX loss

Table IV-2. Thermochemical Data for some Hypothetical and Observed Reactions of Ionized Monosubstituted Benzenes $(C_6H_5X)^a$

X	Ion	(ΔH_f^I)	Neutral	(ΔH_f^N)	$\Delta H_f^I + \Delta H_f^N$	Obs.
F	$C_6H_4^{+\cdot}$	(1474)	HF	(−272)	1202	+
	$C_6H_5^+$	(1193)	F$^{\cdot}$	(80)	1273	−
CN	$C_6H_4^{+\cdot}$	(1474)	HCN	(134)	1608	+
	$C_6H_5^+$	(1193)	CN$^{\cdot}$	(419)	1612	−
Cl	$C_6H_4^{+\cdot}$	(1474)	HCl	(−92)	1382	−
	$C_6H_5^+$	(1193)	Cl$^{\cdot}$	(121)	1314	+
Br	$C_6H_4^{+\cdot}$	(1474)	HBr	(−38)	1436	−
	$C_6H_5^+$	(1193)	Br$^{\cdot}$	(113)	1306	+
I	$C_6H_4^{+\cdot}$	(1474)	HJ	(25)	1499	−
	$C_6H_5^+$	(1193)	J$^{\cdot}$	(109)	1302	+
OH	$C_6H_4^{+\cdot}$	(1474)	H$_2$O	(−243)	1231	−
	$C_6H_5^+$	(1193)	$^{\cdot}$OH	(38)	1231	−
	$C_5H_6^{+\cdot}$	(1001)	CO	(−109)	892	+
NH$_2$	$C_6H_4^{+\cdot}$	(1474)	NH$_3$	(−46)	1428	−
	$C_6H_5^+$	(1193)	$^{\cdot}$NH$_2$	(172)	1365	−
	$C_5H_6^{+\cdot}$	(1001)	HNC	(197)	1198	+
CH$_3$	$C_6H_4^{+\cdot}$	(1474)	CH$_4$	(−75)	1398	−
	$C_6H_5^+$	(1193)	$^{\cdot}$CH$_3$	(138)	1331	−
	$C_7H_7^+$	(875)	H$^{\cdot}$	(218)	1093	+

[a] All data in kJ mol^{-1}

in preference to X˙ loss, provided there are no competing reactions and the reverse activation energy is zero. In Table IV-2 the thermochemical data for HX and X˙ loss from a variety of substituted benzene ions are listed. The last column indicates whether the reaction is accompanied by a metastable loss or not. It is obvious that the condition $\Delta H_f(HX) + 281$ kJ $\leq \Delta H_f(X˙)$ is satisfied when X = F and CN but not satisfied if X = Cl, Br and I. Accordingly, a metastable transition is observed for loss of HF and HCN from ionized fluorobenzene and benzonitrile, but a metastable transition is found for Cl˙, Br˙ and I˙ loss from ionized chloro-, bromo-, and iodobenzene. In the case of phenol, aniline, and toluene competing fragmentations (loss of CO, HCN, and H) lead to ionic and neutral fragments of even lower total heats of formation (second last column). Hence only these reactions are accompanied by a metastable peak.

Similar enthalpy considerations have been used by Bowen and Williams to predict or rationalize the metastable decomposition of $C_nH_{2n+1}^+$ carbenium ions [6] and $C_nH_{2n}^{+\cdot}$ radical cations [7]. However, the relative heats of formation of all hypothetical product combinations can only be used to predict the occurrence of metastable ions if the reverse activation energy is negligible or zero as expected for direct bond cleavages and symmetry allowed 1,1-eliminations with σ-bond formation (e.g. H_2 loss and CH_4 loss). For all other reactions a transition state energy significantly greater than the sum of the heats of formation of the products is expected.

Consider as example the metastable decomposition of butyl ions [6]. Although the formation of a 1-methylallyl cation and a hydrogen molecule ($\Delta H_f = 854$ kJ mol^{-1}) is the most favorable product combination, formation of an allyl ion and a methane molecule ($\Delta H_f = 871$ kJ mol^{-1}) is observed experimentally, demonstrating that hydrogen loss from butyl ions (e.g. by 1,2-elimination) must involve a substantial reverse activation energy (estimated to range between 6 and 77 kJ mol^{-1}). This example demonstrates that the presence or absence of a metastable transition can be used to obtain at least semiquantitative information about the reverse activation energy of a unimolecular decomposition process, i.e. to deduce lower and upper limits for this quantity.

1.2 Stevenson's Rule

Stevenson was the first one to explore the possibility of determining the dissociation energy (D) from appearance potential measurements [13]. He found the general rule that for the reaction $A-B - e^- \rightarrow A^+ + B˙$ the appearance potential, $AP(A^+)$, is only given by the equality

$$AP(A^+) = IP(A^\cdot) + D(A-B)$$

if $IP(A^\cdot) < IP(B^\cdot)$. Later this rule was extended and reformulated by Audier [14]:

In the dissociation of an ion the positive charge will remain on the fragment of lower ionization potential*). For the reaction

$$AB^{+\cdot} \diagup\!\!\!\!\!\!\searrow \begin{matrix} A^+ + B^\cdot \\ B^+ + A^\cdot \end{matrix}$$

one can deduce from Figure IV-1 (neglecting a reverse activation energy)

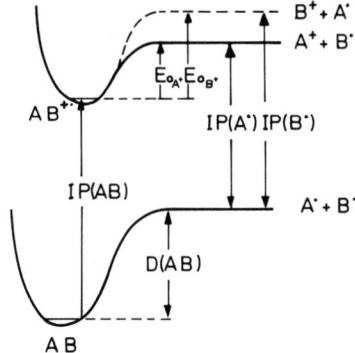

Fig.IV-1. Potential energy diagram illustrating Stevenson's rule.

that

$$E_{oA^+} = IP(A^\cdot) + D(A-B) - IP(AB) \qquad (IV-2)$$

$$E_{oB^+} = IP(B^\cdot) + D(A-B) - IP(AB) \qquad (IV-3)$$

Thus

$$\Delta E_o = IP(A^\cdot) - IP(B^\cdot) \qquad (IV-4)$$

The differences in activation energies are determined by the differences in ionization potentials. As the geometry of the activated complex is identical for the formation of both fragments their relative rates and thus their relative abundances are exclusively determined by the

---

* In the meantime Audier's formulation has been used as Stevenson's rule in several textbooks [4,16,17]. Being aware that this is not Stevenson's original rule we continue to use this term to avoid further confusion.

Table IV-3. Correlation of R_1^+ and R_2^+ Relative Abundances with Radical Ionization Potentials

R_1 - R_2	$IP(R_1)$[a]	Abundance (R_1^+)	$IP(R_2)$[a]	Abundance (R_2^+)	Base Peak
$(CH_3)_2CH$ - CH_2OH	7.55	100	7.6	66.8	R_1^+
$(CH_3)_2CH$ - $CH(OH)CH_3$	7.55	14.5	6.9	100	R_2^+
$(CH_3)_2CH$ - CH_2OCH_3	7.55	8.2	6.95	100	R_2^+
$(CH_3)_2CH$ - CH_2SH	7.55	75.9	8.0	48	$C_3H_5^+$
$(CH_3)_2CH$ - $CH_2C_6H_5$	7.55	14.7	7.27	100	R_2^+
$(CH_3)_3C$ - CH_2OH	6.93	100	7.6	7.4	R_1^+
$(CH_3)_3C$ - CH_2SH	6.93	100	8.0	16.5	R_1^+
$(CH_3)_3C$ - $CH_2C_6H_5$	6.93	100	7.27	39.8	R_1^+
$(CH_3)_3C$ - CH_2NH_2	6.93	7.7	6.2	100	R_2^+
$(CH_3)_3C$ - $CH(OH)CH_3$	6.93	100	6.9	79.2	R_1^+
CH_3OCH_2 - CH_2OH	6.95	100	7.6	25.7	R_1^+
CH_3OCH_2 - $CH_2C_6H_5$	6.95	100	7.27	38.8	R_1^+
H_2NCH_2 - $CH(OH)CH_3$	≈ 6.2	100	6.9	9.5	R_1^+
$C_6H_5CH_2$ - CH_2CHCH_2	7.27	100	8.05	46	R_1^+
$C_6H_5CH_2$ - $CH(OH)CH_3$	7.27	55.6	6.9	31.1	$C_7H_8^{+\cdot}$
		5.1		14.2	$C_7H_8^{+\cdot}$ 10 eV
CH_2CHCH_2 - CH_2OH	8.07	31.6	7.6	53.3	$C_3H_6^+$
CH_2CHCH_2 - $CH(OH)CH_3$	8.07	10.0	6.9	100	R_2^+

[a] eV

differences in activation energies. Thus Stevenson's rule is in agreement with the QET.

Harrison et al. [15] applied this rule to a variety of compounds listed in Table IV-3. It is obvious that in all cases where the radical ionization potentials differ by at least 0.3 eV the species of lower ionization potential predominates in the mass spectrum. Only in one case, $C_6H_5CH_2$-CH(OH)CH_3, the group of highest ionization potential, C_7H_7, is more abundant in the 70 eV mass spectrum; however this ion is formed to a significant extent by fragmentation of the $C_7H_8^{+\cdot}$ base peak. If this secondary formation route is suppressed at low electron energies the abundance ratio reflects the ionization potentials correctly.

Stevenson's rule is not restricted to direct bond cleavages, but can also be used for rearrangement reactions to predict which fragments will carry the charge [4].

1.3 Proton Affinities

In Stevenson's rule the decomposition of an organic ion is rationalized as the competition of the two fragments for the charge. This reasoning could be carried one step further arguing that in certain hydrogen rearrangement reactions the two moieties of the dissociating ion are competing for a proton.

Williams [18] has used this argument to explain the ethylene elimination from $CH_2=OCH_2CH_3^+$. This reaction would be symmetry forbidden if it occurred as a concerted process via a four-membered transition state as shown in Scheme IV-3a. Alternatively one could envision a transition state in which formaldehyde and ethylene are loosely bound

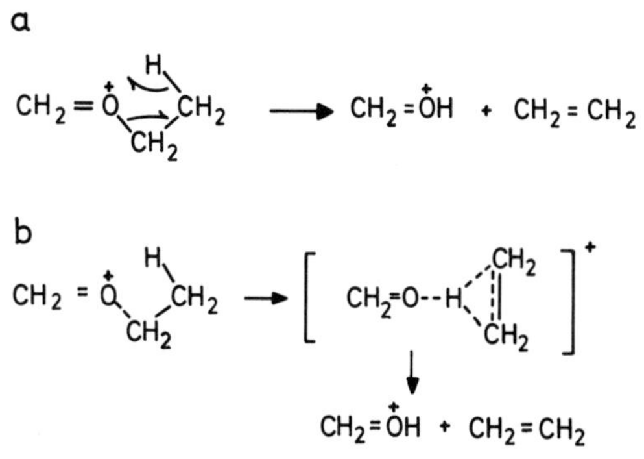

a

$CH_2=\overset{+}{O}\underset{CH_2}{\overset{H}{\diagdown}}CH_2 \longrightarrow CH_2=\overset{+}{O}H + CH_2=CH_2$

b

$CH_2=\overset{+}{O}\underset{CH_2}{\overset{H}{\diagdown}}CH_2 \longrightarrow \left[CH_2=O--H\underset{CH_2}{\overset{CH_2}{\diagdown}} \right]^+$

$\downarrow$

$CH_2=\overset{+}{O}H + CH_2=CH_2$

Scheme IV-3

to a central hydrogen (Scheme IV-3b). Both molecules are competing for the proton. As the proton affinity of formaldehyde is larger than that of ethylene, not $C_2H_5^+$ but $CH_2=OH^+$ is formed upon fragmentation.

1.4 Charge Localization — Radical Site Localization

Mass spectrometric fragmentation reactions have been successfully rationalized or even predicted assuming that the reaction is either initiated by the radical site (α-cleavage) or by the charge site (i-cleavage) [1,19]. According to this concept reaction initiation at the radical site arises from the strong tendency for electron pairing: The odd-electron is donated to form a new bond to an adjacent atom accompanied by cleavage of the α-bond, as depicted in Scheme IV-4 for the ionized diethyl ether*)

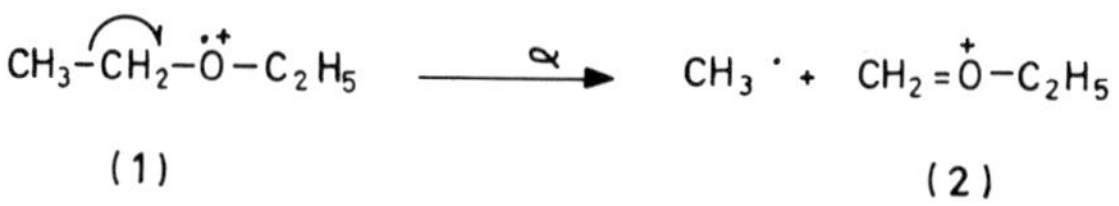

$$CH_3-CH_2-\overset{\cdot+}{O}-C_2H_5 \quad \xrightarrow{\alpha} \quad CH_3\cdot \;+\; CH_2=\overset{+}{O}-C_2H_5$$

$$(1) \hspace{4cm} (2)$$

Scheme IV-4

Alternatively initiation of a reaction by the positive charge involves attraction of an electron pair leading to cleavage of the bond adjacent to the heteroatom accompanied by charge migration as shown again for diethylether in Scheme IV-5

$$C_2H_5\overset{\cdot+}{\frown O}-C_2H_5 \quad \longrightarrow \quad C_2H_5^+ \;+\; \cdot OC_2H_5$$

$$(1) \hspace{4cm} (3)$$

Scheme IV-5

The radical site initiation (α-cleavage) is the more important mechanism and has often been referred to as "charge localization concept" [8].

The validity and usefulness of this qualitative approach has been repeatedly questioned [20-23]. Thus charge distribution calculations

---

*) A half arrow indicates transfer of a single electron (homolytic cleavage), a full arrow transfer of an electron pair (heterolytic cleavage) [8].

IV-1 The Mechanistic Approach

not only for hydrocarbon ions [24] but also for a variety of heteroatom-containing ions (including the molecular ions of ethylamine [22] and estrone [25] as well as the fragment ions $C_6H_5CO^+$ [26], $C_2H_3O^+$ [27], and $C_2H_6N^+$ [28] have shown that the positive charge is spread over the whole ion and often is preferentially localized at the hydrogen atoms [28] which sharply contrasts with the formalism used in the mechanistic approach outlined above.

Attempts to find experimental evidence against the concepts of charge or radical site localization were less successful. Thus Bentley et al. [20] studied the electron-impact-induced fragmentation of methionine and selenomethionine and concluded that the charge localization concept would predict the following reaction sequence.

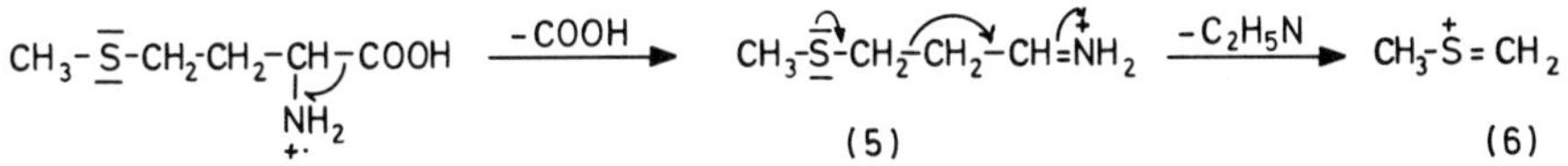

$$CH_3-\underline{\bar{S}}-CH_2-CH_2-CH-COOH \xrightarrow{-COOH} CH_3-\underline{\bar{S}}-CH_2-CH_2-CH=NH_2 \xrightarrow{-C_2H_5N} CH_3-\overset{+}{S}=CH_2$$

(4)   (5)   (6)

Scheme IV-6

In contrast to this "expected" mechanism the authors deduce a complicated multistep mechanism for the formation of the $C_2H_5Se^+$ and $C_2H_5S^+$ ion (although $C_2H_5S^+$ can also be formed via a second route according to Scheme IV-6). They conclude that a localized charge may not be necessary at the fragmentation site and offer an alternative explanation, resonance of the charge through sigma bonds to the active site.

However, Budzikiewicz and Pesch [29] demonstrated for a series of alkyl amines ω-substituted with a second functional group (O, S, Se) that charge migration below the ionization potential of primary amines from the second heteroatom to the nitrogen does not occur via the σ-bonds, but through space (i.e. via a cyclic transition state). Moreover, van den Heuvel and Nibbering [30] reinvestigated the methionine system using deuterium labelling and high resolution. They report that 90% of the $C_2H_5S^+$ ion is generated according to Scheme IV-6 as predicted by the charge localization concept.

Charge localization has also been invoked to explain the drastic dependence of the rate for the McLafferty rearrangement on the substituent in acetates of the general formula $CH_3COO(CH_2)_nX$ [31].

The concept and role of charge localization has recently been reinvestigated by Williams and Beynon [32]. The authors reemphasize that the important factor is not the charge but the localized radical site. They provide evidence from electron spin resonance, photoelectron spectroscopy, free radical reactions, ultraviolet spectroscopy and the fragmentation of doubly charged ions for the preferential (although not exclusive) localization of the unpaired electron density in certain molecular orbitals, although upon ionization the electron may be removed from various orbitals, i.e. radical site localization does not mean that the lone electron is localized in a single molecular orbital. If an unpaired electron is indeed largely localized at a heteroatom (e.g. at the oxygen in diethylether in Scheme IV-4) the α,β-C-C bond will be weakened leading to the observed α-cleavage. As far as data are available they support the hypothesis that the α,β-C-C bond strength in heteroatom-containing compounds decreases upon ionization (see Table IV-4).

Table IV-4. Comparison of the Activation Energies for α-Cleavage in some Neutral and Ionic Systems (in kJ mol^{-1}).

Reaction (Heats of Formation)	Activation Energy
$(CH_3)_2C=O \longrightarrow CH_3\dot{C}=O + CH_3\cdot$ (-218) (-17) (138)	339
$(CH_3)_2C=O^{+\cdot} \longrightarrow CH_3\overset{+}{C}=O + CH_3\cdot$ (716) (636) (138)	58
$CH_3COOCH_3 \longrightarrow CH_3\dot{C}=O + \dot{O}CH_3$ (-414) (-17) (-4)	393
$CH_3\overset{+\cdot}{COO}CH_3 \longrightarrow CH_3\overset{+}{C}=O + \dot{O}CH_3$ (578) (636) (-4)	54

Thus the bond between CO and CH_3 in ionized acetone is much weaker than in acetone and the bond between CO and OCH_3 in ionized methylacetate much weaker than in methylacetate.

Hence the partial localization of the unpaired electron density in certain orbitals does not cause the dissociation, but reduces the activation energy for the cleavage of certain bonds and thus increases the rate constant for this reaction. Thus although the QET does not take any charge or radical site localization into account this concept is not in contradiction with it.

Concluding, there is some experimental evidence for the concept of "charge localization" (radical site localization) and, keeping its limitations in mind, it is certainly a useful approach to the rationalization of many fragmentation pathways of excited organic ions.

2 Methods for the Elucidation of Ionic Reaction Mechanisms

During the early development of organic mass spectrometry decomposition mechanisms were mainly deduced by the use of chemical intuition, but today a whole panoply of methods is available to substantiate this intuition.

2.1 Isotopic Labelling

The specific incorporation of heavy isotopes (e.g. ^2H, ^{13}C, ^{18}O, ^{15}N) is certainly the most frequently and most successfully used technique for the investigation of ionic fragmentation mechanisms. The analysis of the data may be hampered by the occurrence of isotope effects and atom "scrambling". Both phenomena have been discussed extensively in Section III-5 and III-2.3. The determination of deuterium isotope effects may be complicated by the occurrence of atom scrambling. In this case at least two differently labelled compounds have to be used. For instance both the degree of scrambling, α, prior to H· loss and the deuterium isotope effect, i, can be determined for the toluene molecular ion if the two differently labelled toluenes (7) and (8) are studied

(7)   (8)

For toluene-α-d_3 (7) the intensity ratio for H and D loss is given as

$$\frac{[M^{+\cdot} - H]}{[M^{+\cdot} - D]} = \frac{5i\alpha/(3 + 5i)}{3\alpha/(3 + 5i) + (1 - \alpha)} \qquad (IV-5)$$

whilst the same ratio for the ring deuterated compound (8) is given as [33]

$$\frac{[M^{+\cdot} - H]}{[M^{+\cdot} - D]} = \frac{3i\alpha/(5 + 3i) + (1 - \alpha)}{5\alpha(5 + 3i)} \qquad (IV-6)$$

From equation IV-5 and IV-6 i and α can readily be determined.

Deuterium isotope effects do not necessarily compromise the interpretation of isotopic labelling data, but may give valuable mechanistic information as shown in Section IV-2.4. Atom scrambling may complicate the elucidation of non-specific rearrangements, i.e. hydrogen rearrangements involving several transition states. It will be shown below how energy dependence and ion lifetime measurements can be used to differentiate between these two processes. Atom scrambling can be reduced by sampling ions of high internal energy or short ion lifetimes which may be very useful in elucidating reaction mechanisms as discussed in Section IV-2.6.3.

2.2 Steric Blocking

Deuterium labelling is the most powerful tool for elucidating the mechanism of hydrogen rearrangement reactions. However, if only the primary fragmentation is studied such labelling only gives information on the origin of the transferred hydrogen (provided that there is no hydrogen randomization), but no indication of the site to which the hydrogen is transferred. Such information may be obtained by studying secondary decompositions using metastable ions or collision-induced dissociation. A rather simple alternative method uses introduction of blocking groups. For instance the hydrogen rearrangement in n-butylbenzene may involve either a six membered transition state in which the hydrogen is transferred to the ortho position of the ring to form the methylene-cyclohexadiene ion (11) or a four membered transition state leading to the toluene molecular ion (10) (see Scheme IV-7). Substitution of the two ortho positions by methyl groups suppresses the hydrogen rearrangement

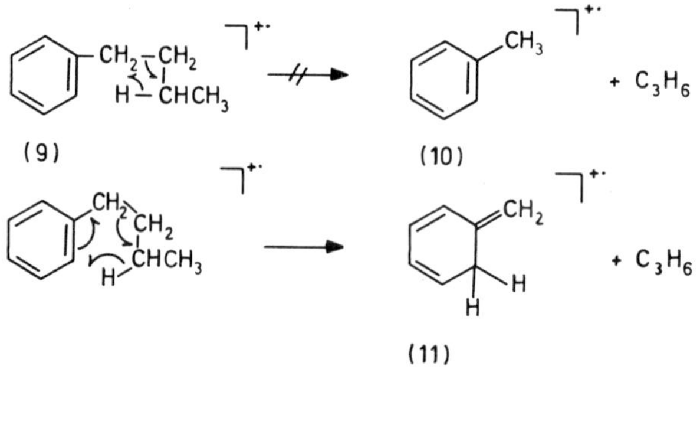

Scheme IV-7

completely indicating that the reaction indeed involves a six-membered transition state [34]. This conclusion has been supported by ion cyclotron resonance [35,36] and isotopic labelling studies [37] demonstrating that $C_7H_8^{+\cdot}$ ions from toluene and n-butylbenzene differ in their initial structure. Steric blocking has also been used to elucidate the mechanisms of hydrogen rearrangements in acetanilides [38] and phenyl acetates [34] where the hydrogen is transferred to the heteroatom via a four membered transition state (Scheme IV-8).

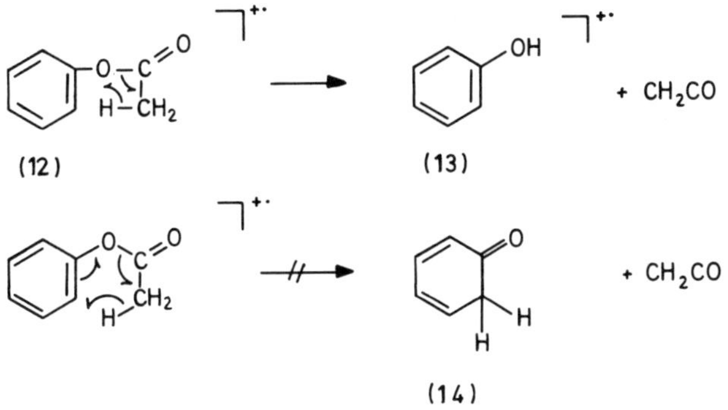

Scheme IV-8

In the case of acetanilide the same conclusion was reached in the isotope effect study [39] discussed in Section IV-2.4, whilst the mechanism for ketene elimination from phenylacetate was corroborated by an ICR study [40].

Steric blocking has also been used to elucidate the mechanism of HCN loss from benzaldoxime [41] (see Section IV-2.3.2), $CH_3CO^{\cdot}$ loss from α,α,α',α'-tetramethyl-N,N'-tetramethylene diacetamide [42] and propene loss from phenyl-n-propyl ether [43].

In using this method one has to ensure that the reduction of a given fragment intensity on substitution by blocking groups is not due to competing reactions involving the blocking group.

2.3 Metastable Ions

2.3.1 Fragmentation Pathways

Metastable ions allow precursor and daughter ions to be linked together and are thus extremely valuable for deducing the decomposition pathways of a given ion. Thus metastable ions are now routinely used to establish the complete fragmentation pattern of a given molecular ion. New methods in recording metastable ion spectra have removed earlier ambiguities in the correct assignment of precursor and daughter ions and make it possible to detect metastable ions with high sensitivity [16].

Metastable ions recorded by an acceleration voltage scan link all precursor ions to a given daughter ion, whilst the electric sector scan, if preceeded by mass analysis in an instrument of reversed geometry, links all daughter ions to a given precursor (DADI [44,45] or MIKE techniques [46])*). However, as result of the kinetic energy released upon metastable decomposition there may still be a substantial overlapping of adjacent peaks. Metastable ion spectra of high mass resolution may be obtained by scanning two fields of a double focussing mass spectrometer simultaneously, thus discriminating against all decomposition processes with non-zero kinetic energy. A variety of such "linked scans" was proposed recently [47-50] the most useful being that in which the magnetic (B) and electrostatic (E) fields are scanned so that the ratio B/E is constant. With such linked scan again all daughter ions of a given precursor will be transmitted.

When using metastable ions to derive the fragmentation pathways of a given molecular ion one should be aware of two facts:
(1) Only processes of lowest activation energy give rise to abundant metastable ions. Thus not every decomposition process needs to be accompanied by a metastable peak of measurable intensity.
(2) The observation of a metastable ion does not prove unambiguously that the reaction is a one step process. If a slow reaction is followed by a fast one, a metastable peak will be observed for the combined reaction as well as for the first step. In most cases a metastable peak will also be detected for the second step [51-55].

2.3.2 Kinetic Energy Release and Energy Partitioning Data

The analysis of metastable peak shapes is one of the most powerful tools for the investigation of reaction mechanisms, as shown by the pioneering work of Beynon, Cooks et al. [56-58]. In order to resolve fine structure in such peaks high energy resolution is required (which can be achieved, for example, using a double focussing instrument with a narrow energy resolving slit) [16]. Metastable peak shape analysis at high energy resolution led to the detection of composite metastable

---

*
  DADI = direct analysis of daughter ions
  MIKE = mass analysed ion kinetic energy

peaks in a large variety of reactions [41,56-68]. Fig. IV-2 shows such a composite metastable peak for HCN loss from benzaldoxime [41], where

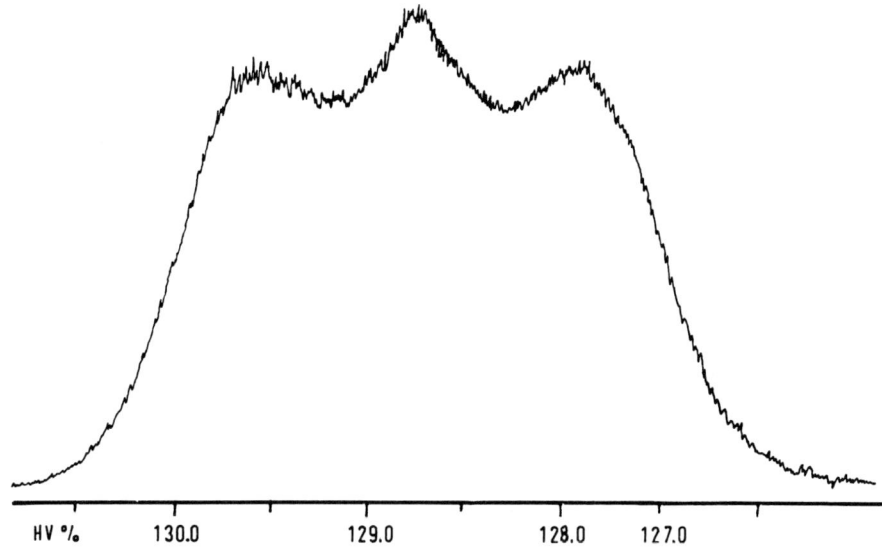

Fig. IV-2. Composite metastable peak obtained for loss of HCN from benzaldoxime [41]. (By courtesy of Heyden & Son Ltd.)

a gaussian-shaped peak with an average kinetic energy release of T = 0.05 eV is superimposed onto a dish-shaped peak with T = 0.67 eV.

If a decomposition reaction of the *molecular ion* leads to a composite metastable peak this may result (1) from isolated electronic states of the molecular ion or (2) from competing fragmentations via different transition states. As isolated electronic states in gaseous organic ions are rarely encountered a composite metastable peak will in general indicate the presence of two competing fragmentation mechanisms.

The situation is more complex if a composite metastable peak is observed upon decomposition of a *fragment ion*. In this case either the formation of the fragment occurred via two reaction channels leading to two non-interconverting structures of identical composition or the primary fragment has a unique structure, but subsequent decomposition involves two competing mechanisms.

For the above mentioned HCN loss from benzaldoxime both a four- and five-membered transition state according to Scheme IV-9[*] has been

---

[*] For the sake of simplicity proposed intermediate steps in this reaction sequence have been omitted.

suggested.

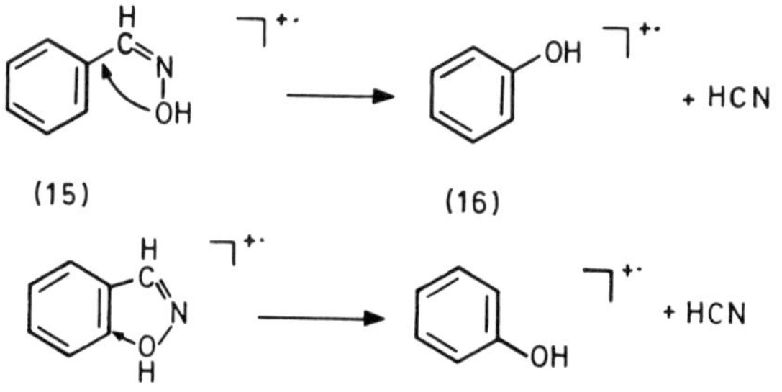

Scheme IV-9

It should, however, be pointed out that the mere observation of a composite metastable peak does not give any information on the detailed mechanism, in particular it is not known a priori which component of the composite peak corresponds to which mechanism. In the case of the benzaldoxime, blocking of the two ortho positions by substitution with chlorine atoms (see Section IV-2.2) led to the disappearance of the narrow component of the composite peak, suggesting that the broad component can be assigned to the four-centered transition state, the narrow component to the five-centered one [41].

Correspondingly the composite metastable peak observed for H_2CO loss from substituted anisoles [56] (Scheme IV-10) has been interpreted in terms of two competing decompositions via four- and five-membered transition states.

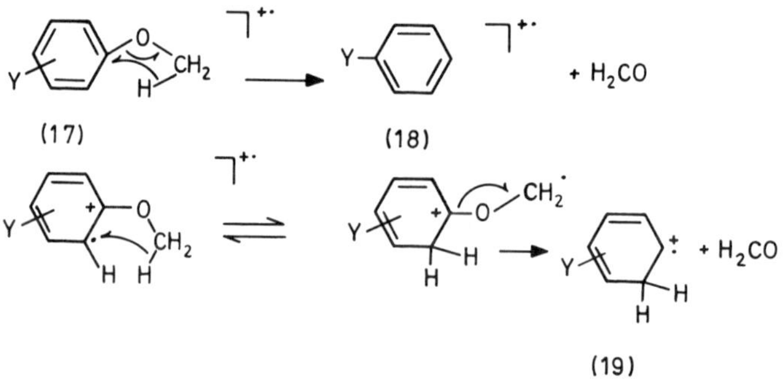

Scheme IV-10

Here two different product ions are formed, at least initially*). It was concluded that the formation of the more stable product (the substituted benzene ion) must be accompanied by a larger reverse activation energy and thus a larger kinetic energy release. Hence the large T value corresponds to the four-center, the small T value to the five-center reaction.

Similarly composite metastable peaks led to the detection of competing four- and five-centered transition states for HCN loss from p-methoxy benzaldoxime-O-methylester [58] and HCl loss from 2,3-dichlorobutane [62], while NO loss from substituted nitrobenzenes seems to involve three- and four-centered transition states [57]. In all these reactions the larger T values were associated with the activated complex of smaller ring size. Thus the data available so far show that *the fraction of reverse activation energy released as kinetic energy is the larger the tighter the transition state***). The following quantitative values for the partitioning quotient characteristic for a transition state of given ring size have temptatively been suggested [69]:

3-membered ring $0.6 < (T^e/\varepsilon_o^r) < 1.0$
4-membered ring $0.1 < (T^e/\varepsilon_o^r) < 0.75$
5- or 6-membered rings $0 < (T^e/\varepsilon_o^r) < 0.1$

If further experimental values support this conclusion energy partitioning data should be of great diagnostic value in the assignment of transition state geometries.

It should however be emphasized that competing mechanisms for a given transition state need not necessarily lead to a composite metastable ion. Thus it has been shown by ion cyclotron resonance [70] and collisional activation [71] in conjunction with deuterium labelling that phenoxyethyl halide ions form $C_6H_6O^{+\cdot}$ fragments via three distinct mechanisms, although all three reactions are accompanied by the same kinetic energy release which points to a stepwise mechanism [72]. Finally it has been pointed out that comparison of experimental and calculated data for the kinetic energy release, $T^{\neq}$, originating from the non-fixed energy may give information on the activated complex configuration provided it is possible to determine T^e, the fraction of kinetic energy release due to the reverse activation energy [73].

---

*
  One should, however, keep in mind that the kinetic energy release values give no information about the final structure of the ions.

**
  In a theoretical study of energy partitioning it has been pointed out by Christie et al. [74] that the ring size of the transition state can affect the energy partitioning only in an indirect way.

2.3.3 Kinetic Energy Release as Potential Energy Surface Probe

Williams and Bowen [75] proposed using kinetic energy release data in conjunction with transition state energies to infer the transition state structure. The authors discussed the ethylene elimination from $CH_3CH_2CH=XH^+$ (X = O, NH) and proposed potential energy surfaces for these reactions as shown in Fig. IV-3. In this diagram it is assumed

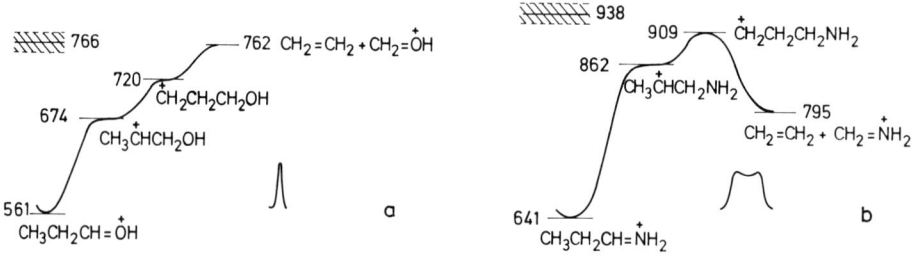

Fig. IV-3. Potential energy surface for ethylene loss from $CH_3CH_2CH=OH^+$ (a) and $CH_3CH_2CH=NH_2^+$ (b); the hatched areas represent the experimentally determined energy required for dissociation, and the kinetic energy release profile is shown schematically [75]. (By courtesy of the American Chemical Society.)

that dissociation into $CH_2=XH^+$ and $CH_2=CH_2$ occurs via direct cleavage of the intermediate ion $^+CH_2CH_2CH_2XH$ which is readily reached by two successive 1,2-hydrogen shifts. Heat of formation data suggest that this dissociation is endothermic for oxonium ions, but exothermic for ammonium ions. The potential energy diagram is directly supported by the observation that decomposition of the oxonium ion is accompanied by a small kinetic energy release and that of the ammonium ion by a large kinetic energy release (see peak shapes in Fig. IV-3). Appearance potential measurements for ethylene loss (hatched area in Fig. IV-3) further corroborate that the transition state geometry has been correctly chosen.

2.3.4 Kinetic Energy Release and Molecular Orbital Symmetry Considerations

The concept of symmetry conservation of molecular orbitals ("Woodward-Hoffmann rules") has been shown to be extremely useful in understanding or predicting thermal or photochemical reactions. The significance of this concept for unimolecular reactions as observed in a mass spectrometer has been explored by Williams and Hvistendahl [76-79] who studied the metastable H_2 elimination for the following reactions

IV-2 *Methods for the Elucidation of Mechanisms*   171

$$CH_3CH_3^{+\cdot} \longrightarrow CH_2CH_2^{+\cdot} + H_2$$
$$CH_2=\overset{+}{O}H \longrightarrow HC\equiv O^+ + H_2$$
$$CH_2=\overset{+}{N}H_2 \longrightarrow HC\equiv \overset{+}{N}H + H_2$$
$$CH_2=\overset{+}{S}H \longrightarrow HC\equiv \overset{+}{S} + H_2$$
$$CH_3-\overset{+\cdot}{N}H_2 \longrightarrow CH_2=\overset{+\cdot}{N}H + H_2$$

Deuterium labelling revealed that in all five reactions H_2 elimination proceeds via a specific 1,2-elimination. The observation of an abundant metastable peak for H_2 loss suggests (but does not prove) that this is a concerted (one step) reaction, and this is supported by thermochemical data. Thus if these 1,2-eliminations occur via a planar transition state they should be symmetry forbidden as demonstrated in the correlation diagram for H_2 loss from ethane (Scheme IV-11). As bond stretching

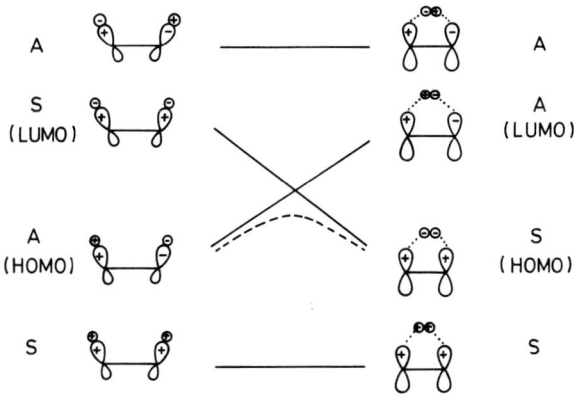

Scheme IV-11

proceeds, the energy of the anti-symmetric highest occupied orbital increases rapidly. As the highest occupied, A(HOMO), and lowest unoccupied, S(LUMO), orbital cross the electrons transfer into the symmetric S molecular orbital. As the transition state is passed the electronic reorganization results in a $H_2-C_2H_4^{+\cdot}$ complex in which there is an occupied symmetric molecular orbital, S(HOMO), characterized by mutual repulsion between the carbon and hydrogen nuclei. Dissociation should therefore be accompanied by the release of a relatively large amount of kinetic energy. A significant kinetic energy release has indeed been found for all reactions leading to the observation of dish-shaped peaks in each case, thus confirming the author's conclusion*).

Using this criterion it is now possible to confirm or reject the possibility of 1,2-eliminations in those cases where hydrogen randomization thwarts the interpretation of deuterium labelling experiments, which is generally the case in even electron hydrocarbon ions. For instance, $C_6H_7^+$ ions lose H_2 in a concerted reaction preceded by complete randomization of all hydrogens. This reaction leads to a narrow gaussian-shaped metastable peak, i.e. only little kinetic energy is released upon decomposition, and a 1,2-elimination can be discounted. It is suggested that H_2 loss from $C_6H_7^+$ (which is assumed to have the protonated benzene structure) occurs via a symmetry-allowed 1,1-elimination leading to the stable phenyl ion [77]. In contrast, H_2 loss from $C_7H_9^+$ is accompanied by a substantial kinetic energy release of ~ 0.9 eV, indicating a symmetry-forbidden 1,2- or 1,3-elimination leading to the tropylium ion [77,80].

Dewar and Rzepa [81] recently deduced the transition states for H_2 elimination from various small organic ions using MINDO/3 calculations. Their calculations demonstrate that the energetically most favorable H_2 elimination from $C_2H_6^{+\cdot}$, $CH_2=OH^+$ and $CH_2=SH^+$ indeed proceeds via a concerted 1,2-elimination as predicted by Williams and Hvistendahl. This elimination is, however, not synchronous: The transition state for H_2 elimination from $C_2H_6^{+\cdot}$, for instance, is very unsymmetrical, one C-H bond being almost broken (1.55 Å) and the other much less (1.24 Å), while the H-H bond is incompletely formed. Furthermore, their calculations demonstrate that the most favorable H_2 elimination from $C_2H_5^+$ is indeed the symmetry-allowed 1,1-elimination predicted by Williams and Hvistendahl, but again the transition state has an unexpected geometry, one C-H bond being completely broken and the other almost so.

2.4 Isotope Effects

Although the occurrence of deuterium isotope effects may complicate the interpretation of isotopic labelling data it has been shown mainly by Williams et al. [39,82-84], that such isotope effects may yield important mechanistic information. *The observation of a substantial deuterium isotope effect in a reaction demonstrates that the transfer of a hydrogen occurs in the rate determining step.* Consider $C_2H_4O_2^{+\cdot}$ ions formed by McLafferty re-

---

*
Note, however, that the theoretical treatment developed by Christie et al. [74] for the calculation of partitioning quotients predicts a large kinetic energy release for H_2 loss from $C_2H_6^{+\cdot}$ without invoking orbital symmetry arguments.

arrangement from aliphatic acids which according to collisional activation and metastable ion spectra have the enolic structure (Scheme IV-12) [85]. The metastable $C_2H_4O_2^{+\cdot}$ loses a hydroxyl radical. If this hydroxyl group is eliminated by direct cleavage and no isotope effect is present then the deuterium-labelled ion $CH_2=C(OH)OD^{+\cdot}$ (20) and $CD_2=C(OH)OD^{+\cdot}$ (21) should lose OH and OD in equal proportions, whilst complete H/D randomization amongst all hydrogen atoms should lead to a ratio $[C_2(H,D)_4O_2^{+\cdot}-OH]:[C_2(H,D)_4O_2^{+\cdot}-OD]$ = 3:1 for (20) and 1:3 for (21). The actually observed ratio is 1:2.6 for both (20) and (21), which clearly rules out complete or partial randomization of the hydrogens, but suggests the presence of a primary isotope effect, demonstrating that the rate determining step involves transfer of a hydrogen. The isotope effect can be explained assuming a rearrangement of the enolic ion (20) to the acetic acid molecular ion (22) during decomposition in which the transfer of a hydrogen with subsequent loss of OD is

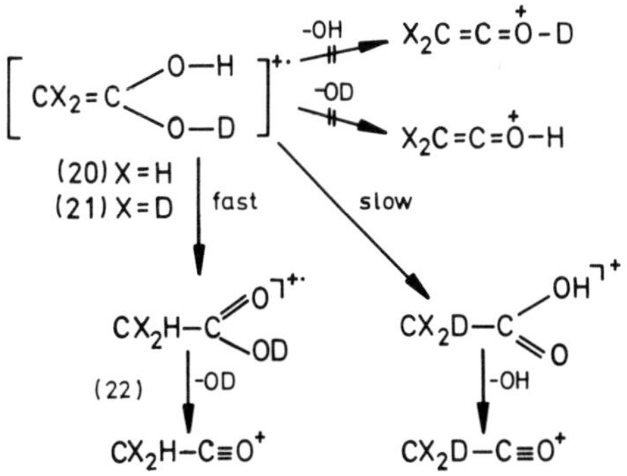

Scheme IV-12

favored over the transfer of a deuterium with subsequent loss of OH (Scheme IV-12).

Howe and Williams [82] studied the ethylene elimination from p-bromophenetole. Two mechanisms are conceivable: The hydrogen is either transferred via a four-membered transition state to the oxygen (23 → 24) or via a six-membered transition state to the ortho position of the aromatic ring (23 → 25) as outlined in Scheme IV-13. The authors determined the metastable ratio for further loss of Br$^\cdot$ and CO from both the $C_6H_5BrO^{+\cdot}$ fragment ion generated from p-Br-phenetole as well as the $C_6H_4DBrO^{+\cdot}$ ion generated from the p-Br-phenetole deuterated in the side chain and observed a substantial isotope effect on CO loss. No primary isotope effect should be observed on CO loss from (25) while hydrogen

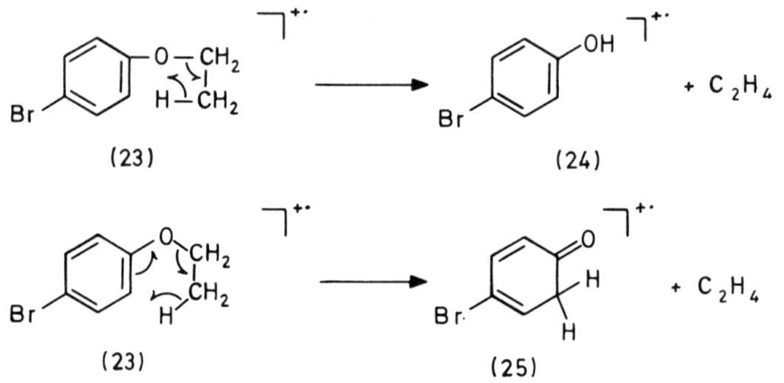

Scheme IV-13

transfer must precede CO loss from (24), which explains the isotope effect and clearly demonstrates that the ethylene elimination proceeds via a four-membered transition state to give the p-bromophenol ion.

It is noteworthy that in this method a competing metastable transition for which no isotope effect would be expected (Br$^{\cdot}$ loss) is used as "reference" to detect the isotope effect.

The same approach has been used by Uccella et al. to differentiate between a four-centered and six-centered transition state for ketene elimination from p-chloroacetanilide (Scheme IV-14) [39].

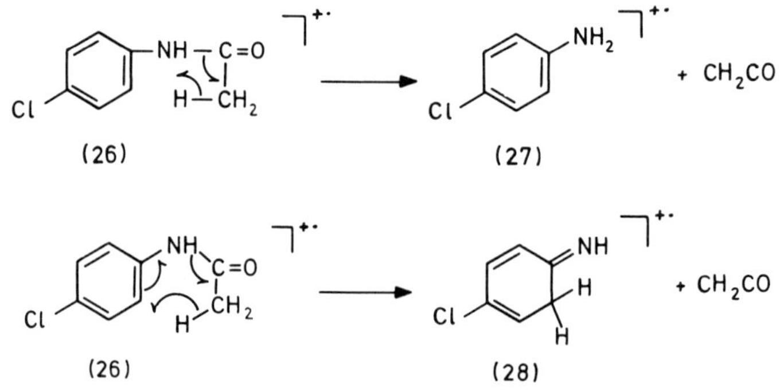

Scheme IV-14

Here an isotope effect on HCN loss was observed for competing metastable Cl$^{\cdot}$ and HCN loss from the $C_6H_6NCl^{+\cdot}$ fragment ion generated from p-chloroacetanilide and the side-chain deuterated analogue which again reveals that HCN loss must be preceded by hydrogen transfer as expected if the p-chloroaniline molecular ion is generated through a four-centered transition state. This conclusion has been supported by a later study of various deuterated acetanilides [86]. Moreover high resolution measurements demonstrated that the ions decomposing in the ion source also have the aniline structure [87]. The same conclusions have been achieved by blocking the two ortho positions by methyl groups as discussed in Section IV-2.2 [38].

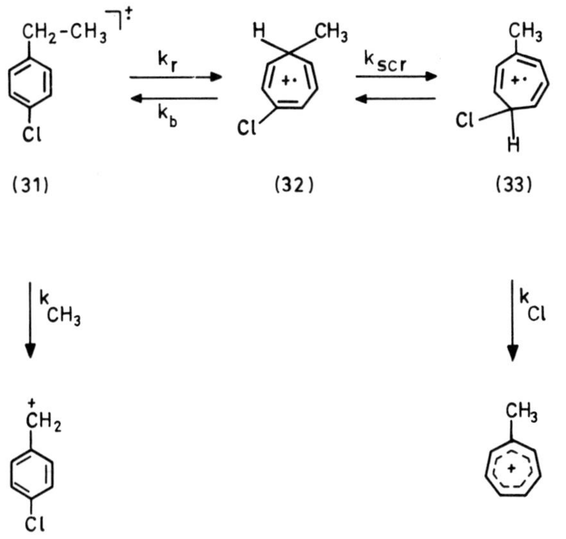

Scheme IV-15

Whilst no isotope effect on halogen loss has been assumed in the preceding discussion a significant isotope effect on Cl˙ loss has been observed in p-chloroethylbenzene [83]. If the α-hydrogens of the side chain are substituted by deuterium the rate of Cl˙ loss is suppressed relative to $CH_3^˙$ loss by almost a factor of three. This result has been explained by assuming that ring expansion to a seven-membered ring occurs prior to Cl˙ elimination whilst methyl is lost from the unrearranged ion according to Scheme IV-15. The ring expansion is followed by scrambling of hydrogen and deuterium atoms as shown in Scheme IV-15 until the structure (33) which is favorable for chlorine elimination is reached.

Deuterium isotope effects have been employed by Howe [88] to unravel the decomposition mechanism of the 2-methylhexane ion. This molecular ion only shows abundant metastable transitions for loss of $CH_3^˙$ and $C_3H_7^˙$, demonstrating that these two processes must be in competition and hence should have similar activation energies. Deuterium labelling reveals that an n-propyl (and not an iso-propyl radical) is eliminated. If $C_3H_7^{\cdot}$ is lost by a simple direct cleavage (reaction b in Scheme IV-16) the heat of formation data (in kJ) shown in parenthesis in Scheme IV-16 indicate that this process yields product ions having a total ground

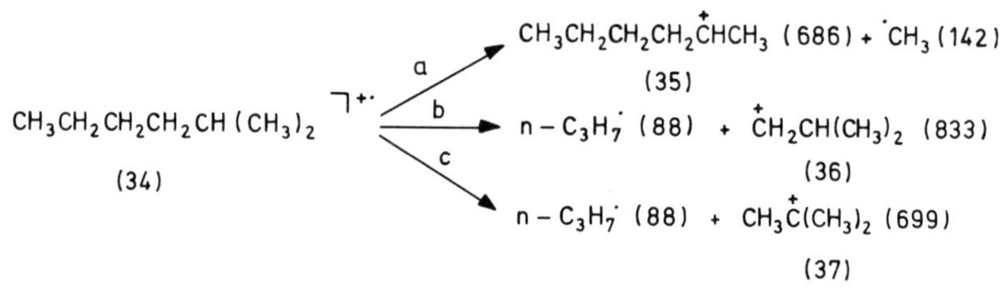

$$CH_3CH_2CH_2CH_2CH(CH_3)_2 \rceil^{+\cdot} \xrightarrow{a}_{c}^{b} \begin{array}{l} CH_3CH_2CH_2CH_2\overset{+}{C}HCH_3 \ (686) + {}^{\cdot}CH_3 \ (142) \\ (35) \\ n-C_3H_7^{\cdot} \ (88) + \overset{+}{C}H_2CH(CH_3)_2 \ (833) \\ (36) \\ n-C_3H_7^{\cdot} \ (88) + CH_3\overset{+}{C}(CH_3)_2 \ (699) \\ (37) \end{array}$$

(34)

Scheme IV-16

state energy of ∼ 1.0 eV higher than products formed by methyl loss (a). This energy difference is too large to allow competition within the framework of the QET. However, if the t-butyl ion is formed by elimination of n-$C_3H_7^{\cdot}$ (reaction c) the products of reaction (a) and (c) differ only by ∼ 0.4 eV in their heats of formation and (if a reverse activation energy is considered for (c)) should have similar activation energies. Hence reaction (c) is allowed within the framework of the QET.

This conclusion is supported by the observation of a large isotope effect favoring $C_3H_7^{\cdot}$ elimination from the unlabelled 2-methylhexane (34) as compared to $C_3H_7^{\cdot}$ elimination from 2-methylhexane-2-^{2}H, $CH_3CH_2CH_2CH_2CD(CH_3)_2$, (38). This isotope effect has been detected using the methyl elimination as "reference" metastable transition. The ratio for metastable $C_3H_7^{\cdot}/CH_3^{\cdot}$ loss from (34) is more than four times larger than from (38), demonstrating that a hydrogen transfer must accompany $C_3H_7^{\cdot}$ elimination in support of mechanism (c).

Finally, a quite different approach for the elucidation of reaction mechanisms has been used by Hvistendahl and Williams [84] who demonstrated that the loss of molecular hydrogen from ethylene occurs through transition states in which two C-H bonds are stretched synchronously. It is known that for such concerted H_2 eliminations in partially labelled compounds equation IV-7 applies, whilst for non-concerted reactions equation IV-8 is valid:

$$[k_{HD}/k_{D_2}]^2 = k_{H_2}/k_{D_2} \qquad (IV-7)$$

$$k_{HD} = [k_{H_2} + k_{D_2}]/2 \qquad (IV-8)$$

For trans-$(1,2-{}^2H_2)$- and $(1,1-{}^2H_2)$- ethylenes the observed k_{H_2}/k_{D_2} ratio of 700-740 requires a theoretical value of k_{HD}/k_{D_2} 26-27 for a perfectly concerted reaction which is in good agreement with the observed ratio of 23.

2.5 The Detection of Functional Group Interaction in Apparent Direct Bond Cleavages

Many fragments which according to their mass and isotopic labelling data seem to result from a simple direct bond split in reality involve functional group interaction (for instance, anchimeric assistance) in their formation [29,89-121]*). Functional group interaction is frequently encountered in bi- and polyfunctional alkanes, in aromatic compounds in which the aromatic ring contains two substituents in ortho positions, and in other compound classes. Functional group interaction (the term "neighboring group interaction" is also used) in bifunctional alkanes has been reviewed recently [42]. The occurrence of unusual decomposition processes gives the first indication of functional group interaction: Thus bifunctional alkanes differ in their fragmentation from monofunctional alkanes, ortho-substituted aromatic compounds from meta- and para-substituted ones.

Functional group interaction in apparent direct bond cleavages leads to a tight transition state. Thus this interaction can be detected by methods which reflect the tightness of the activated complex configuration as discussed below.

2.5.1 Energy Dependence

It has been shown in Section III-2 that the energy dependence of competing decomposition processes reflects the transition state geometry: In most cases the abundance of rearrangement reactions (tight transition state) increases relative to that of direct bond cleavages if the electron energy is lowered. Thus the energy dependence can be used to differentiate between rearrangement reactions and direct bond cleavages. This approach is especially useful for detecting a functional group interaction (e.g. anchimeric assistance) which always involves a relatively tight transition state. Thus Shapiro and Jenkins [89] postulated a phenyl participation in the expulsion of Br˙ from β-phenylethylbro-

---

* Functional group interaction is of course not confined to apparent direct bond cleavages but is also encountered in various types of rearrangement reactions.

mide according to Scheme IV-17. Their observation that the loss of Br·

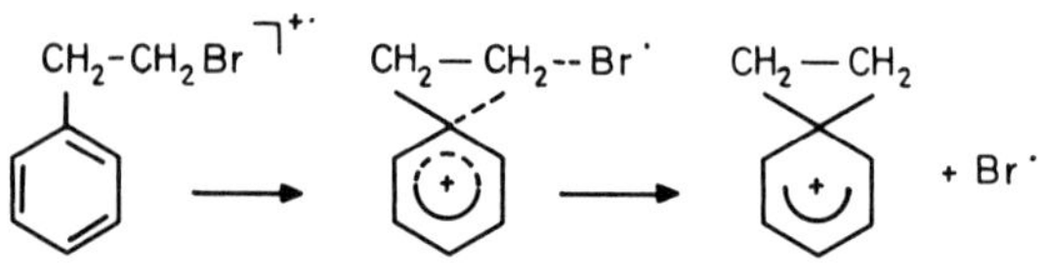

Scheme IV-17

relative to loss of ·CH_2Br increases with decreasing electron energy supports this assumption. Similar conclusions were reached for a series of ring-substituted 5-bromo-2-phenyl pent-2-enes [90]. The energy dependence was also used as the criterion for postulating a functional group interaction for CH_3COCH_2· loss from ketocarbonic esters of the type $CH_3CO(CH_2)_5COOC_2H_5$ [91].

It should, however, be noted that the energy dependence supports the involvement of a functional group interaction but does not prove it unambiguously. In particular, the energy dependence does not give any direct evidence as to the specific mechanism involved and ion structure formed. Hence additional experimental information has to be sought to confirm the mechanism. Thus the mechanism for Br· expulsion from the β-phenylethylbromide outlined in Scheme IV-17 has been corroborated by Nibbering et al. [92] and Köppel and McLafferty [93] who demonstrated by deuterium labelling that the α- and β-methylene groups in $C_8H_9^+$ ions from β-phenylethylbromide are equivalent prior to collision induced decomposition, i.e. ions generated by Br· loss from β-phenylethylbromide deuterated either in the α- or in the β-position lose equal amounts of CH_2 and CD_2, suggesting the formation of an ethylenebenzenium ion by phenyl participation at least at low internal energies*).

The energy dependence is useful not only for detecting functional group interactions but is helpful in unravelling other rearrangement reactions which cannot be recognized as such by isotopic labelling. Thus in p-chloroethylbenzene the ratio of $[M^+ - Cl]/[M^+ - CH_3]$ increases with decreasing energy, demonstrating that a rearrangement (ring ex-

---

* In contrast, phenyl participation in the β-phenylethyl chloride molecular ion leads to the formation of protonated benzocyclobutene ions at low internal energies whilst direct cleavage with apparent subsequent isomerization to ethylenebenzenium and/or the protonated benzocyclobutene ion prevails at high internal energies [93].

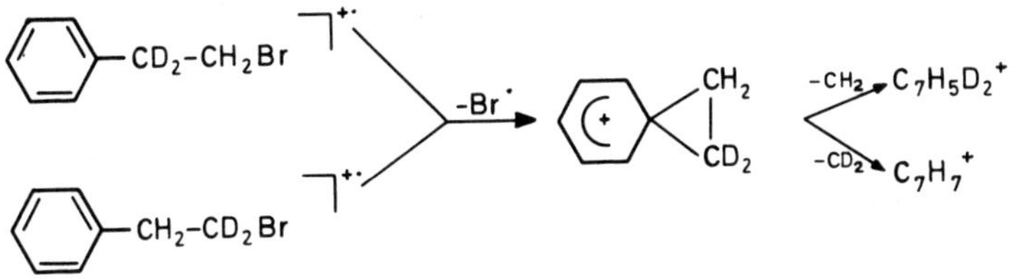

Scheme IV-18

pansion) must be involved prior to Cl˙ loss [83].

Here the unexpected isotope effect observed for Cl˙ loss from the α-deuterated compound has been used to confirm that ring expansion to a seven-membered ring precedes Cl˙ loss as discussed in Section IV-2.4. Finally the energy dependence can also be used to differentiate between hydrogen randomization and non-specific hydrogen rearrangements as outlined in Section IV-2.6.3.

2.5.2 Abundant Metastable Ions

As reactions with tight transition states usually dominate at low internal energies they lead to abundant metastable ions. Thus intense metastable ions accompanying formal direct bond cleavages may point to a more complex mechanism involving functional group interaction. Using this criterion McLafferty et al. [94] concluded that the alkyl loss by formal direct cleavage from a variety of compounds leads to a five membered ring according to Scheme IV-19. For example the 1-chlorohexane ion shows an abundant metastable peak for ethyl loss, but hardly any for methyl or propyl loss, consistent with the formation of a five-

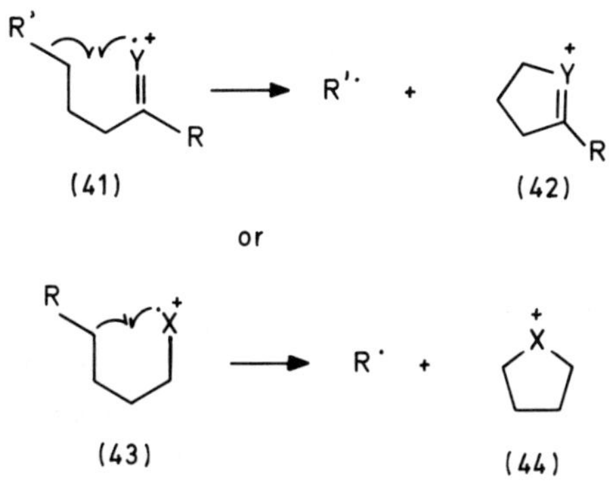

Scheme IV-19

membered ring.

It should, however, be kept in mind that the detection of a tight transition state from the observation of an abundant metastable ion may be compromised if there are competing rearrangements with even lower activation energies. Hence it is obvious that further experimental evidence is necessary to substantiate the conclusion. In the case of chloroalkanes, additional support for the formation of a five-membered ring chloronium ion has again been provided by analysis of the collision-induced fragments of suitable deuterium-labelled precursors as outlined in Scheme IV-20 [95].

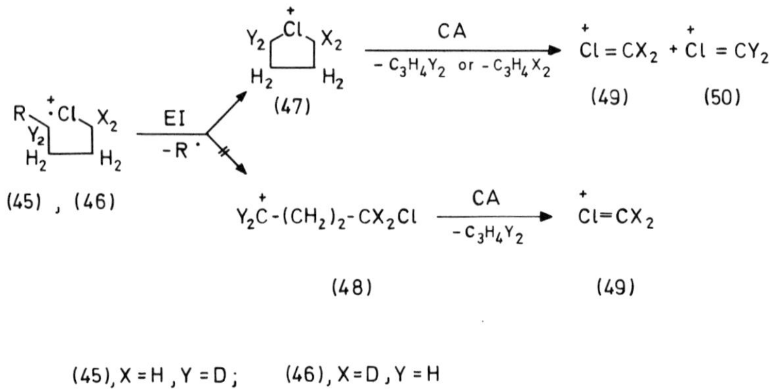

$$\begin{array}{c} R\overset{+}{-}Cl\diagup X_2 \\ Y_2\square H_2 \\ H_2 \end{array} \xrightarrow{EI}_{-R^{\cdot}} \begin{array}{c} Y_2\overset{+}{-}Cl\diagup X_2 \\ H_2\square H_2 \\ (47) \end{array} \xrightarrow[-C_3H_4Y_2\text{ or }-C_3H_4X_2]{CA} \overset{+}{Cl}=CX_2 + \overset{+}{Cl}=CY_2 \\ (49) \qquad (50)$$

$$\overset{+}{Y_2C}-(CH_2)_2-CX_2Cl \xrightarrow[-C_3H_4Y_2]{CA} \overset{+}{Cl}=CX_2$$

(45), (46)

(48) (49)

(45), X = H, Y = D; (46), X = D, Y = H

Scheme IV-20

The observation that (49) and (50) are formed in the same ratio from both precursors (45) and (46) shows the formation of the cyclic structure (47) unequivocally.

Thus abundant metastable ions accompanying formal direct bond cleavages give a first indication of a possible tight transition state, but no direct information on the mechanism. This information must be gained from additional experiments.

2.5.3 Relative Fragment Abundances

The relative abundance of a given fragment as function of the chain length in a series of homologous bifunctional alkanes has been repeatedly used to identify functional group interaction. If loss of one of the two substituents were due to simple direct bond split the relative abundance of the fragment formed by this cleavage should vary smoothly with increasing chain length. However, pronounced maxima in the relative fragment abundance at a certain chain length indicate that the

IV-2 *Methods for the Elucidation of Mechanisms* 181

fragment formation is favored by functional group interaction which may lead to a cyclic fragment. Such cyclization via functional group interaction has for instance been postulated for the elimination of $CH_3OCOCH_2\cdot$ from α,ω-dicarbonicacid dimethylesters [42,96],

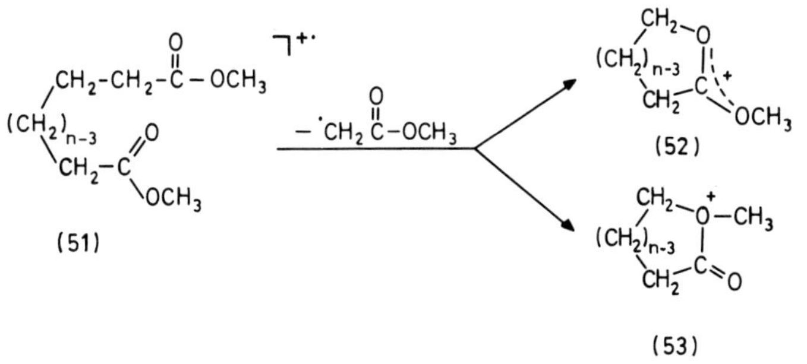

$$\begin{array}{c}
\underset{(51)}{\left[\begin{array}{c}CH_2-CH_2-\overset{O}{\overset{\|}{C}}-OCH_3\\(CH_2)_{n-3}\\CH_2-C\underset{OCH_3}{\overset{O}{\diagup}}\end{array}\right]^{+\cdot}} \xrightarrow{-\,\cdot CH_2\overset{O}{\overset{\|}{C}}-OCH_3}
\begin{array}{c}
\underset{(52)}{\begin{array}{c}CH_2-O\\(CH_2)_{n-3}\quad\|\\CH_2-C\underset{OCH_3}{\overset{+}{\diagup}}\end{array}}\\[2em]
\underset{(53)}{\begin{array}{c}CH_2-\overset{+}{O}-CH_3\\(CH_2)_{n-3}\\CH_2-C\approx O\end{array}}
\end{array}
\end{array}$$

Scheme IV-21

$CH_3OCO(CH_2)_nCOOCH_3$, as depicted in Scheme IV-21. In support of this assumption the $(M - CH_2COOCH_3)^+$ fragment reaches a maximum relative abundance for n = 4 and 5 (see Fig. IV-4) where functional group interaction can lead to a relatively stable five- or six-membered cyclic ion. The same approach has been used to identify functional group interaction for acetyl loss from N,N'-tetramethylene diacetamide and its homologues [97], $CH_3CONH(CH_2)_nNHCOCH_3$, for C_2H_5 + piperidine loss from N(γ-acetylaminopropyl)-2-ethylpiperidine [98], for $CH_3COCH_2\cdot$ loss from 7-oxooctanoic acid ethylester and its homologues [91], $CH_3CO(CH_2)_n$-$COOC_2H_5$, for $CH_3COCH_2\cdot$ loss from O-acetyl derivatives of ω-hydroxy methylketones [91] and HNO_2 loss from γ-nitrobutyric acid methylester [108]. Here too must be emphasized that the variation of the relative fragment abundance with increasing hydrocarbon chain length reveals the presence of a functional group interaction, but does not give any unambiguous information on the specific mechanism or the ion structure (although certain mechanisms may be excluded). Thus two alternative mechanisms have for instance been proposed for the above discussed $CH_3OCOCH_2\cdot$ elimination from α,ω-dicarbonicacid dimethylesters [96] as shown in Scheme IV-21 and it is open to question whether in this case a cyclic structure is formed at all.

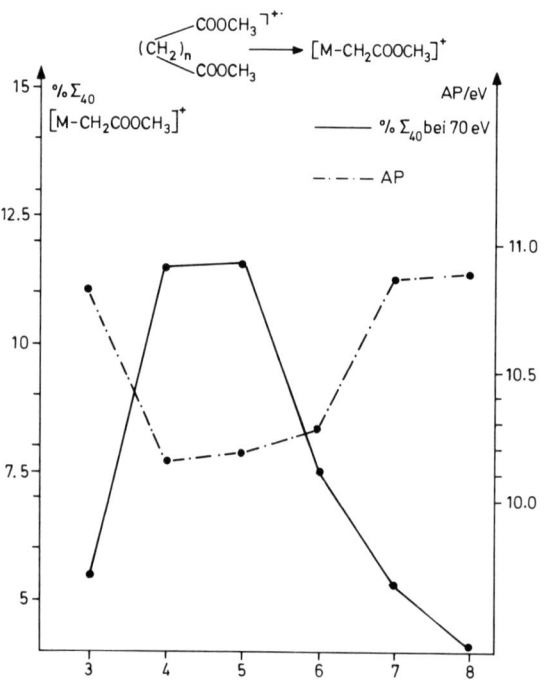

Fig. IV-4. Relative abundance and appearance potentials of the $(M-CH_2COOCH_3)^+$ fragment from α,ω-dicarbonic acid dimethylesters as a function of the chain length [96].

2.5.4 Appearance Potentials

The comparison of appearance potentials of a given fragment as function of the chain length constitutes an alternative approach for unraveling functional group interactions in bifunctional alkane molecular ions. If the decomposition of such a bifunctional alkane ion occurs by simple direct bond split the appearance potential should be approximately constant as the transition state geometry is independent of the chain length. Assuming that the excess energy is constant, which should be the case for such a homologous series, the appearance potential difference for the reaction

$$[A-(CH_2)_n-A]^{+\cdot} \xrightarrow{-A^\cdot} [M-A]^+$$

Scheme IV-22

is given as

$$AP_n - AP_{n-1} = [\Delta H_f(M_n-A)^+ - \Delta H_f(M_{n-1}-A)^+] - [\Delta H_f(M_n) - \Delta H_f(M_{n-1})] \quad (IV-9)$$

Since the difference between the heats of formation of the fragments $(M-A)^+$ from two homologous compounds should be of the same magnitude as the difference of the heats of formation of the two homologous molecules the net difference between the APs should be zero or close to zero.

If however functional group interaction is involved in the elimination of A leading e.g. to a cyclic ion $\overline{A(CH_2)_n}^+$ the appearance potential should reflect both the ring size of the transition state and of the ion formed in the reaction. This concept has been used by Budzikiewicz and Pesch [29] who demonstrated that the APs for the formation of $CH_2=\overset{+}{N}H_2$ from the molecular ion of the homologous series $CH_3X(CH_2)_nNH_2$ (X = S, Se) reach a minimum value for n = 4 pointing at a functional group interaction via a six-membered cyclic transition state. The same criterion has frequently been used by Schwarz et al. [96,99-101]. The authors demonstrated that functional group interaction need not be invoked to explain the methyl loss from the molecular ion of α,ω-bis-(trimethyl silyl)-ethers, $(CH_3)_3SiO(CH_2)_nOSi(CH_3)_3$ and $\cdot S-C_6H_4CH_3$ loss from S,S^L-di-p-tolyl-α,ω-bis-thioates, $CH_3C_6H_4SCO(CH_2)_nCOSC_6H_4CH_3$ [101] as the APs do not depend on the chain length. On the other hand, the APs for $CH_3OCOCH_2\cdot$ loss from α,ω-dicarbonicacid dimethylesters show a pronounced minimum for n = 4-6 (see Fig. IV-4) suggesting that in this case functional group interaction leads to a cyclic transition state or even cyclic ion which is also evident from the dependence of the fragment ion abundance on the alkane chain length as discussed in Section IV-2.5.3. The same reasoning has been used to reveal that loss of a phenoxy radical from compounds of the general type $C_6H_5O(CH_2)_nX$ (X = OC_6H_5, OCH_3, SC_6H_5, Br and n = 2-6) is the result of a functional group interaction [100].

2.5.5 Symmetry Arguments

As emphasized in the foregoing, the methods discussed above give a first indication that functional group interaction may be involved. The detailed mechanism must be confirmed by more direct techniques, of which isotopic labelling is one of the most important. If only the isotopic labelling data for the primary fragmentation which involves functional group interaction are analyzed, in many instances no conclusion can be reached about the mechanism as many fragmentations involving functional group interactions appear to be direct cleavage processes.

However, label analysis of the secondary decomposition using either metastable or collision-induced fragmentation often gives the desired information. Collision-induced decompositions are in general more valuable for such mechanistic interpretation, as they include structurally significant direct bond cleavages whilst rearrangement reactions prevail with metastable decompositions. As it is frequently assumed that functional group interaction leads to cyclic ions or at least cyclic transition states the elucidation of the mechanism is often based on symmetry arguments.

One of the first cases where this approach was used was reported by Shapiro and Tomer [122]. The authors concluded that loss of Br' from β-bromophenylacetate leads to a cyclic ion (see Scheme IV-23). In sup-

Scheme IV-23

port of this conclusion the authors showed by oxygen labelling that both oxygens become equivalent before or during the secondary decomposition of this ion. Several cases involving this approach have recently been reported by Van de Sande [102-107]. For instance, it has been suggested that the electron impact induced loss of a phenoxy radical from the molecular ions of ω-phenoxyalkyl methylethers, $C_6H_5O(CH_2)_nOCH_3$ (n = 2-6), does not result from a simple direct cleavage, but is due to a functional group interaction (SN_i type reaction) as depicted in Scheme IV-24 [102]. According to this scheme a cyclic O-methyl oxonium ion

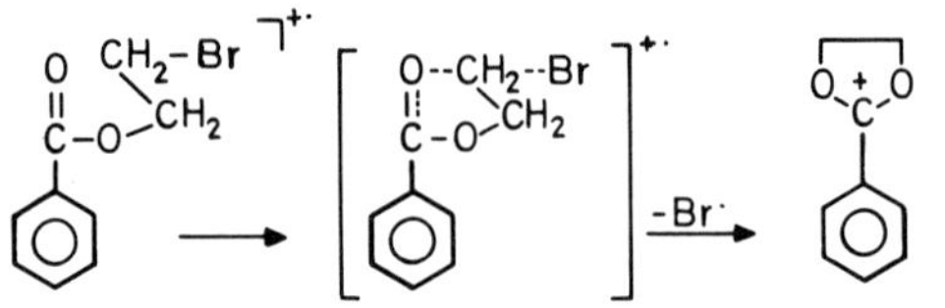

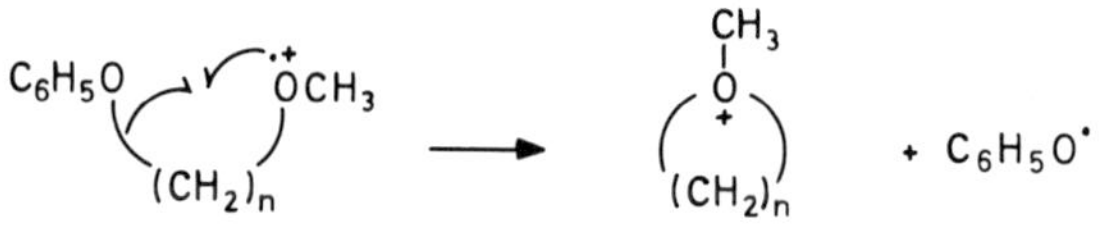

(54)   (55)

Scheme IV-24

is formed (at least in the transition state) which in the case of n = 4 should have the structure of the O-methyltetrahydrofuranium ion (59).

Ion (59) loses C_3H_6 to give a $\overset{+}{CH_2}=O-CH_3$ fragment which, if a cyclic structure is formed, must contain one of the two methylene groups vicinal to the oxygen with equal probability as result of the symmetry of the tetrahydrofuranium ions (Scheme IV-25). The deuterium-labelled

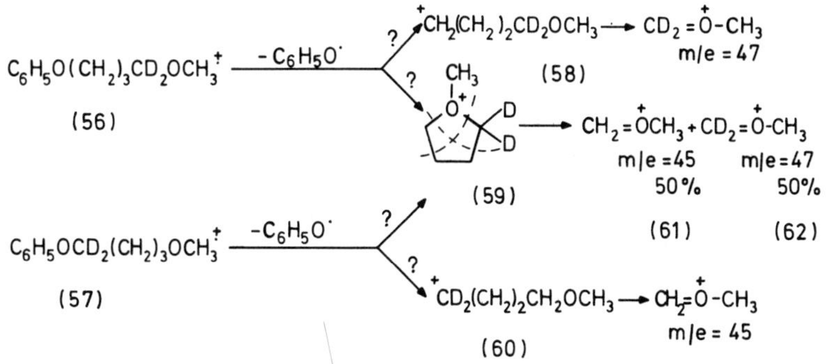

Scheme IV-25

phenoxybutyl methylesters (56) and (57) have been used to prove that the symmetry requirements for a cyclic structure (59) are fulfilled. A cyclic structure (59) should lead to equal amounts of $CH_3-\overset{+}{O}=CH_2$ (m/e 45) and $CH_3-\overset{+}{O}=CD_2$ (m/e 47) whereas an open structure can only yield $CH_3-\overset{+}{O}=CD_2$ in the case of (56), but $CH_3-\overset{+}{O}=CH_2$ in the case of (57) (see Scheme IV-25). The experimental results demonstrate unequivocally that the symmetry requirements for a cyclic structure or at least a cyclic transition state are fulfilled in support of mechanism IV-24[*].

Similar symmetry arguments have been used to demonstrate that the loss of a phenoxy radical from α,ω-bis-aryloxyalkanes, $C_6H_5O(CH_2)_nOC_6H_5$ (n = 2-7) [103], from ω-phenylthioalkylphenylethers, $C_6H_5O(CH_2)_nSC_6H_5$ (n = 2-6) [104], ω-phenoxyalkylbromides, $C_6H_5O(CH_2)_nBr$ (n = 2-6) [105] and from 4-(ω-phenoxymethyl) tetrahydropyranes [106] is the result of a functional group interaction in which cyclic ions are formed. If suitable metastable transitions are not available to elucidate the symmetry of the primary fragment formed via such a functional group interaction

---

[*] The interpretation of the actual results is complicated by the occurrence of partial hydrogen scrambling prior to decomposition of (59). However C_3H_6, C_3H_5D and $C_3H_4D_2$ are lost in the same ratio from (59) independent of whether this ion is formed from the labelled precursor (56) or (57) and this ratio differs considerably from that calculated for total scrambling. This result is only compatible with a cyclic structure.

collision induced decomposition can be used, as discussed for alkyl loss from chloroalkanes [95] in Section IV-2.5.2 and Br loss from β-phenylethylbromide in Section IV-2.5.1 [92,93].

Using such symmetry arguments in conjunction with metastable and collision-induced decomposition, it has been shown that loss of a phenoxy radical from the molecular ion of $C_6H_5OCH_2CH_2XR^{+\cdot}$ (X = O, S and R = CH_3, C_6H_5) involves functional group interaction in each case [78]. However, this interaction leads to a three-membered cyclic structure only if X = S whilst a reciprocal hydrogen shift mechanism seems to operate if X = O (see Scheme IV-26).

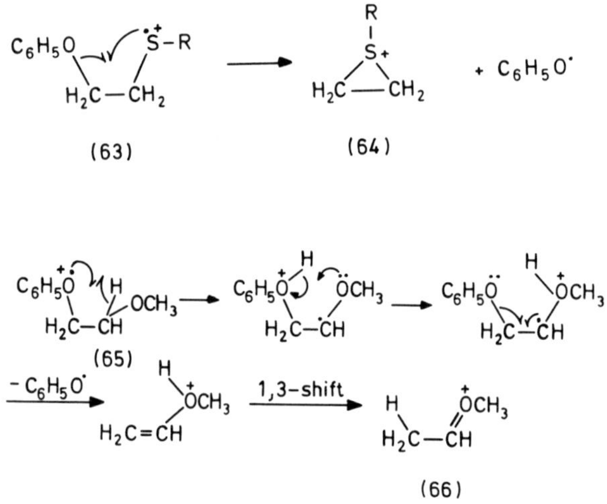

Scheme IV-26

2.6 Field Ionization Kinetics

The strong dependence of ionic decomposition processes on time has already been stressed in Section III-2 while discussing the ion lifetime dependence of competing rearrangement reactions and direct bond cleavages as well as atom scrambling. Hence it is obvious that the study of fragmentation processes as function of the ion lifetime (especially if combined with isotopic labelling) may shed additional light on the detailed reaction mechanisms. It will be shown below that the mechanistically important events often occur within the first nanosecond after ionization. Thus among the experimental methods available for ion lifetime measurements discussed in Section II-2.1.3 only the field ionization technique is suited for the elucidation of reaction mechanisms as it gives a time-resolved view of the decomposition processes from 10^{-11} to 10^{-5} s after ionization. The principle of ion lifetime measurements using the field ionization technique has been briefly outlined in Section II-2.1.3. For detailed information the reader is referred to the original publications [123-131]. It is noteworthy that mechanistic studies using the field ionization kinetic (FIK) technique are not confined to volatile compounds which can be introduced into the ion source via an indirect or direct inlet system, but may include thermally labile, involatile compounds whose gas phase ion chemistry cannot be studied by any other method. This can be achieved using the FIK technique in conjunction with the field desorption method [132].

2.6.1 Mechanisms for Hydrogen Randomization

Hydrogen exchange processes prior to decomposition ("hydrogen scrambling" or "hydrogen randomization") constitute a specific type of rearrangement or isomerization reaction which, as result of its tight transition state, will be largely suppressed at short decomposition times (t < 10^{-11} s) as shown in Section III-2.3.

It has been demonstrated by Derrick et al. [133-137] that the mechanisms of such hydrogen exchange processes can often be unravelled using the field ionization kinetic technique. Consider the cyclohexene molecular ion which loses ethylene via a retro-Diels-Alder reaction (Scheme IV-27). Fig. IV-5 shows the ion current for cyclohexene-3,3,6,

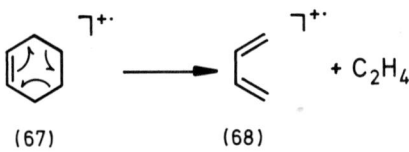

(67)   (68)

Scheme IV-27

6-d_4 as function of the molecular ion lifetime [133] from 1×10^{-11} - 1×10^{-9} s after ionization. From the time sequence of the ion intensity maxima, $(M^{+\cdot} - C_2H_4) < (M^{+\cdot} - C_2H_2D_2) < (M^{+\cdot} - C_2HD_3) < (M^{+\cdot} - C_2H_3D)$ a mechanism for hydrogen - deuterium exchange as shown in Scheme IV-28 can be deduced.

Hence hydrogen - deuterium exchange occurs via specific 1,3-allylic hydrogen shifts leading to a complete randomization of all hydrogens within 10^{-9} s after ionization. Moreover, it was shown that the hydrogen randomization in the 2-methylpropene and 1-butene molecular ion prior to methyl loss also proceeds via 1,3-allylic hydrogen shifts [134,135].

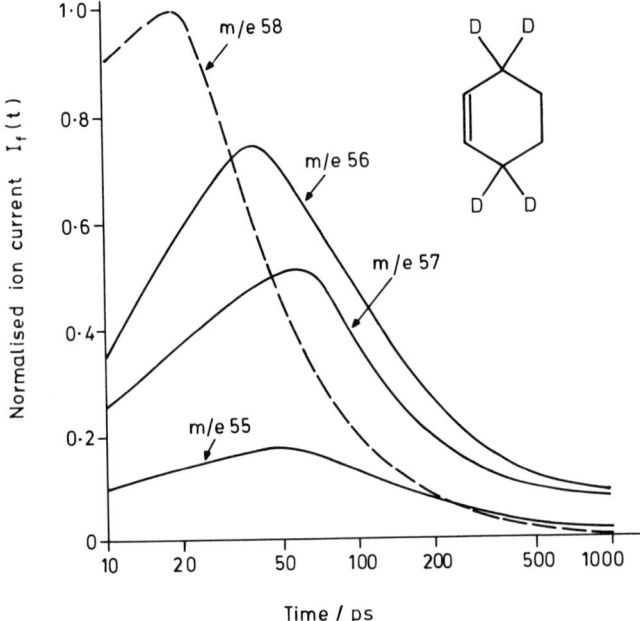

Fig. IV-5. Fragment ion currents due to loss of ethylene as a function of time following field ionization of cyclohexene-3,3,6,6-d_4 [133]. (By courtesy of the American Chemical Society.)

A study of the hydrogen randomization processes in straight chain aliphatic ketones as well as α-branched ketones on the one hand [136] and γ-branched ketones on the other hand [137] revealed substantial differences. Whilst hydrogen randomization in the straight chain and α-branched ketones is not observed within 7×10^{-10} s after ionization, the hydrogen-deuterium exchange in 7-methyl-4-octanone-7-d_1 already started at 2×10^{-11} s. The authors argue that the tertiary radical initially formed by γ-hydrogen transfer in a γ-branched molecular ion is significantly more stable than the secondary radical formed with a straight chain or α-branched molecular ion. Thus in the former case H/D exchange in the hydrocarbon chain will compete more effectively with the McLafferty rearrangement than in the latter case. It is evident that this rationalization implies that the McLafferty rearrangement occurs stepwise after at least 2×10^{-11} s (vide infra).

Finally it has been shown by FIK measurements that the formation of $C_7H_8^{+\cdot}$ ions from the molecular ion of n-pentylbenzene is preceded

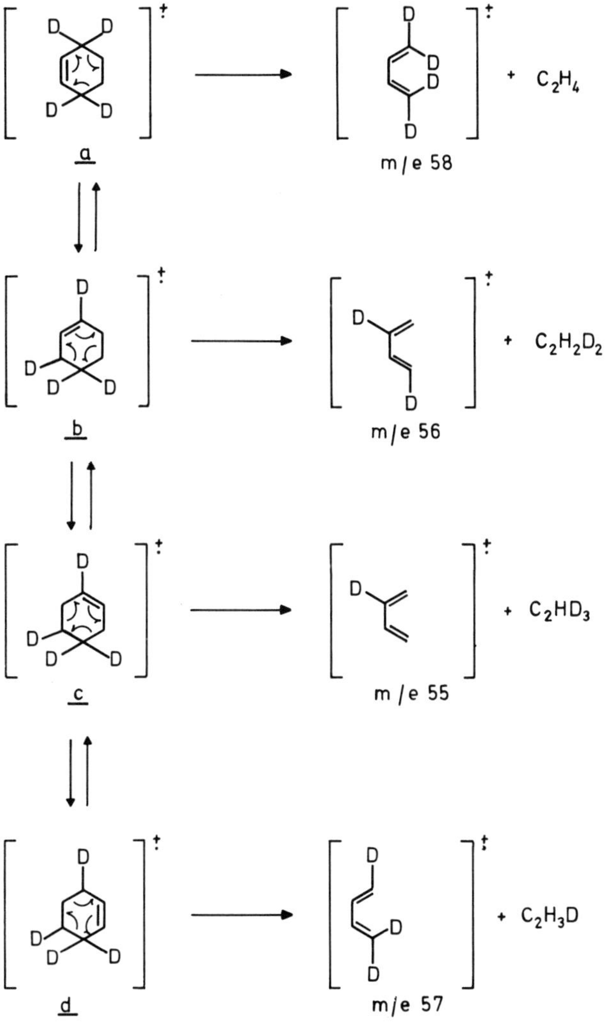

Scheme IV-28

by a remarkably slow hydrogen exchange between the ortho-positions of the ring and the 3- and 4- positions of the side chain involving combinations of six- and seven-membered transition states. This exchange process only becomes significant at $\sim 10^{-6}$ s after ionization and does not lead to a statistical equilibration of the hydrogens involved within the mass spectrometric time scale (10^{-5} s) [138].

2.6.2 Suppression of Atom Randomization

It is obvious that complete H/D randomization in a molecular ion or part of this ion obscures the elucidation of the mechanism for any subsequent hydrogen rearrangement. Here the FIK technique is invaluable: By studying the rearrangement process at the shortest resolvable time it is possible to reduce interference by hydrogen-deuterium exchange considerably, allowing the mechanism to be determined unambiguously. Nibbering et al. [139-140] have used this approach repeatedly to separate the mechanism for hydrogen rearrangement from the competing hydrogen randomization process. Thus all hydrogens in the molecular ion of 3-phenyl-propanal, $C_6H_5CH_2CH_2CHO$, are extensively scrambled prior to electron-impact-induced loss of C_2H_2O and C_3H_4O which prevents the achievement of any definite conclusion on the mechanism [141]. In an earlier EI study by Venema et al. [141] it was suggested that both hydrogen rearrangements probably start with a transfer of a hydrogen from the 2-position. In a recent FIK study of this system Wolkoff et al. [139] demonstrated that at 2×10^{-10} s after ionization loss of C_2H_2O and C_3H_4O from 3-phenylpropanal occurs predominantly via transfer of the aldehyde hydrogen, presumably via a six-membered transition state to the ortho-positions of the ring (see Scheme IV-29 and Fig. IV-6 for C_3H_4O loss).

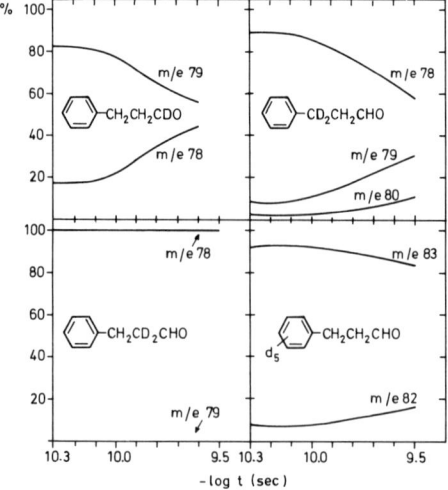

Fig. IV-6. Relative abundances for loss of $C_3(H,D)_4O$ from differently deuterated 3-phenylpropanal molecular ions as a function of the ion lifetime [139].

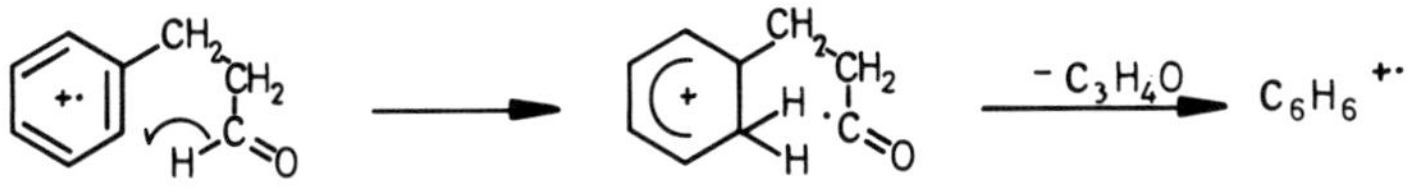

Scheme IV-29

The authors conclude that the subsequent hydrogen randomization occurs via two separated processes: (a) H-exchange between the aldehyde and benzylic hydrogen atoms, (b) between the aldehyde and ortho hydrogen atoms.

Similarly, the benzyl cyanide molecular ion, $C_6H_5CH_2CN^{+\cdot}$, is completely scrambled prior to electron-impact-induced HCN loss [142]. In this case also the elimination mechanism could be established by reducing the hydrogen randomization at short ion lifetimes: The FIK data demonstrated that HCN loss from the benzyl cyanide molecular ion occurs via a formal 1,1-elimination [140], although the molecular ion of benzyl cyanide may have rearranged to (72) prior to HCN loss (Scheme IV-30) as has been proposed for the propionitrile ion [143]. The same

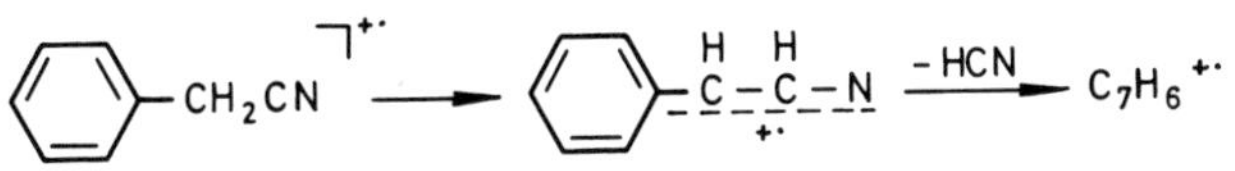

Scheme IV-30

approach can be used to reduce the extent of carbon scrambling prior to fragmentation. Thus the carbon atoms in the 2-methylpropene-2-^{13}C molecular ion, $(CH_3)_2{}^{13}C=CH_2{}^{+\cdot}$, are completely randomized prior to ethylene elimination at 2×10^{-10} s after ionization while at 2×10^{-11} s the central carbon atom is retained with high specificity [144]. To rationalize this observation a decomposition mechanism via a methylcyclopropane intermediate has been assumed. The FIK technique has also been employed successfully to unravel the fragmentation mechanisms in the molecular ions of 1-pentene [145], 1-heptene [146] and 1-octene [147].

2.6.3 Differentiation between Hydrogen Randomization and Non-specific Hydrogen Rearrangements

Many hydrogen rearrangements are to a greater or lesser extent non-specific, i.e. the transferred hydrogen originates from various sites of the molecular ion. In this instance a differentiation between hydrogen randomization (i.e. hydrogen exchange prior to decomposition) and non-specific hydrogen rearrangement (i.e. hydrogen rearrangement via competing transition state geometries) is not possible using 70 eV EI spectra alone. It was shown that the dependence of the relative fragment abundance on both the electron energy and the ion lifetime can be used as a criterion for differentiating between these two processes [43,148-152]. Hydrogen randomization processes increase with increasing ion lifetime (decreasing electron energy) leading to an apparent increase in the unspecificity of the hydrogen rearrangement process. The opposite observation, i.e. an increase in specificity at longer ion lifetimes (lower electron energy) clearly demonstrates that a non-specific hydrogen rearrangement rather than a hydrogen randomization process is operating: The transition state with lowest activation energy is favored at long ion lifetimes (low electron energies). Using this criterion it was shown both by energy dependence and ion lifetime measurements that the alkene loss from phenyl-n-alkyl ethers results from an unspecific hydrogen rearrangement involving all the hydrogens of the side chain [43,148,150]. Table IV-5 shows the percentage of hydrogen trans-

Table IV-5. Percentage of Hydrogen Transfer from the Various Positions of the Side Chain of Phenyl-n-Propyl Ether as a Function of the Internal Energy (EI)

Position of the transferred hydrogen	Source 70 eV	12 eV	1st field-free region, 70 eV	2nd field-free region, 70 eV	12 eV	Random distribution
1	28	27	27	26	26	28.6
2	25	23	20	18	16	28.6
3	47	50	53	57	58	42.8

fer from various positions of the side chain of the molecular ion of phenyl-n-propyl ether as function of the energy, Table IV-6 as function of the ion lifetime. Whilst at the shortest decomposition time the per-

Table IV-6. Percentage of Hydrogen Transfer from the Various Positions of the Side Chain of Phenyl-n-Propyl Ether as a Function of the Decomposition Time (FI)

Position of the transferred hydrogen	t (s)					Random distrib.
	10^{-11}	4×10^{-10}	7×10^{-9}	5×10^{-8} -5×10^{-7}	6×10^{-6} -10^{-5}	
1	28	28	26	27	24	28.6
2	27	25	25	23	21	28.6
3	45	47	49	50	55	42.8

centage of hydrogen transfer from the various positions is close to the ratio expected for complete randomization, the transfer of a γ-hydrogen becomes more important at longer ion lifetimes, which clearly demonstrates that a non-specific hydrogen transfer via transition states of varying ring sizes and not hydrogen randomization is operating. In addition it has been shown both by steric blocking [43] (Section IV-2.2), using 2,6-dimethylphenyl-n-propylether, and by collision-induced dissociation [150] of the $C_6H_6O^{+\cdot}$ ion formed by this hydrogen rearrangement that the hydrogens are exclusively transferred to the oxygen, leading to the formation of a phenol ion (Scheme IV-31).

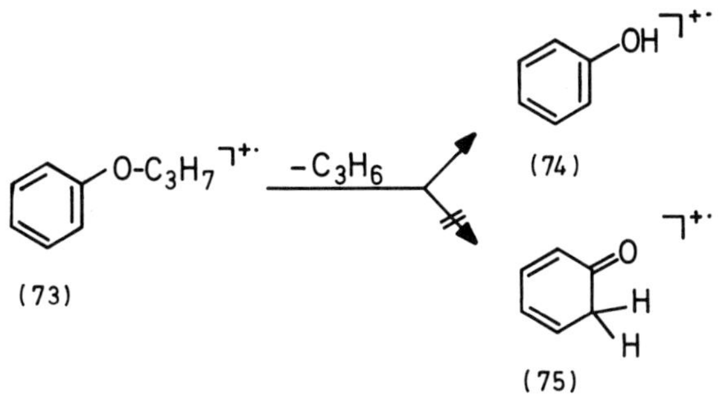

Scheme IV-31

Both energy dependence [149] and ion lifetime measurements [151] have been used to show that the in general minor unspecificity of the hydrogen rearrangement in aliphatic and aromatic esters does not result from a hydrogen randomization process. Similarly, it has been demonstrated by FIK measurements that the unspecificity for the loss of water from n-hexanol is not due to H/D-exchange reactions prior to decomposition, but to hydrogen rearrangements via competing five- and six-membered transition states, where the five-membered transition state dominates at short times (high internal energies) [152].

2.6.4 Other Systems

Considerable efforts have been made to deduce from FIK data whether the McLafferty rearrangement with and without charge migration in the molecular ions of aliphatic aldehydes is a concerted or a stepwise process [153-155]. Charge retention at the carbonyl group leads to a $C_2H_4O^{+\cdot}$ fragment (77), charge migration to an ionized alkene ion (78), e.g. $C_4H_8^{+\cdot}$ in the case of hexanal (Scheme IV-32).

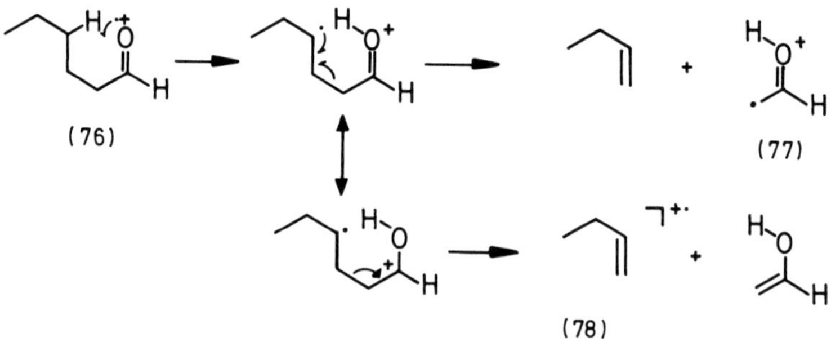

Scheme IV-32

Fig. IV-7 represents the ratio for the rates of formation of $C_2H_4O^{+\cdot}$ and $C_4H_8^{+\cdot}$ from hexanal and 3-methyl-pentanal from 2×10^{-11} to 10^{-9} s after ionization. The fact that the McLafferty rearrangement with charge retention dominates over that with charge migration in hexanal at the shortest time was originally interpreted as evidence that the former reaction is concerted (and thus faster) while the latter occurs stepwise [153]. If this were the case a similar behavior should be observed for the McLafferty rearrangement in 3-methylpentanal [154]. Fig. IV-7, however, reveals that in this system the rearrangement with charge retention is suppressed relative to that with charge migration even at the shortest times [154]. Hence the FIK results give no information about the degree to which the McLafferty rearrangement is concerted. Rather they reflect the charge competition between C_2H_4O and C_4H_8. The hexanal results have been rationalized assuming that C_2H_4O has rearranged to acetaldehyde, C_4H_8 to 2-butene. The results discus-

IV-2 *Methods for the Elucidation of Mechanisms* 195

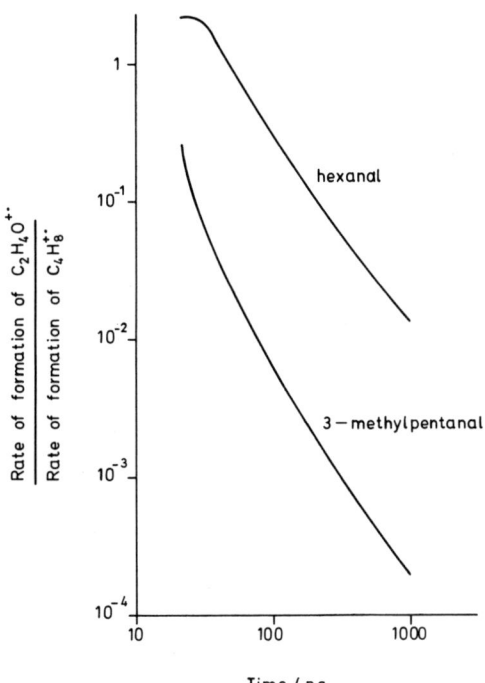

Fig. IV-7. Relative rates of formation of $C_2H_4O^{+\cdot}$ and $C_4H_8^{+\cdot}$ from 3-methylpentanal and from hexanal as a function of the ion lifetime [154]. (By courtesy of the Chemical Society (London).)

sed above demonstrate that an unequivocal determination of a reaction mechanism based on FIK data is difficult if the mechanistic step of interest, e.g. the hydrogen transfer in the McLafferty rearrangement prior to β-cleavage does not show up as mass change.

The complex process for water elimination from the cyclohexanol molecular ion, which has been the subject of extensive EI studies [156-158], has also been investigated using the FIK method [156]. One of the intriguing results of this study is that the ratio for the metastable loss of H_2O and HDO from cis-cyclohexanol-4-d_1 following FI differs drastically from that following EI, although the metastable ions should have comparable lifetime windows and internal energies in both ionization modes. It has been argued that under FI conditions the cyclohexanol ring remains intact prior to decomposition while under EI conditions acyclic molecular ions take part in the metastable fragmentation.

FIK results on the McLafferty rearrangement in n-butyl acetate and several related compounds [31] have been interpretated in terms of the charge localization concept as mentioned in Section IV-1.4. Finally the decomposition of phenylhexanone has been studied using the FIK method [159].

2.7 Molecular Orbital Calculations

It has been demonstrated in Section IV-1 that energetic data such as activation energies or bond strengths may be used to predict or rationalize the fragmentation pattern of organic ions, at least qualitatively. If experimental data are not available they may be deduced from molecular orbital calculations which have been applied repeatedly in the past to the interpretation of mass spectra. Unfortunately the ions of interest are usually too large for ab initio molecular orbital (MO) calculations (see Section V-3.6). Therefore one generally has to rely on semiempirical MO calculations such as ETH, INDO, CNDO and MINDO. The accuracy of these methods is difficult to determine a priori and depends on the specific problem one is dealing with. For instance, it has been stated that the CNDO/2 method provides reliable estimates of the charge distribution of an ion, but gives unrealistic dissociation energies [22]. The success of these methods depends entirely on the parametrization employed. The problems and possible pitfalls which may arise when semiempirical calculations are used to rationalize the fragmentation of organic ions have been discussed by Krier et al. [22] using the ethylamine ion as example.

Semiempirical molecular orbital calculations have been employed to deduce charge distributions in ions [22,24-28,160], bond densities (or bond orders) [25,160,161], ionization potentials [162], rotational barriers [163,164] and activated complex configurations [81,165-169]. It will be shown below how this information can be used to rationalize mass spectrometric reaction mechanisms.

2.7.1 Qualitative Prediction of Fragmentation Pathways

Two approaches have been used to determine which bonds in a molecular ion are most easily cleaved. In the first, bond densities, bond orders or dissociation energies can be calculated for the equilibrium molecular ion*) in order to determine the weakest bonds, which are then assumed to be cleaved preferentially. Such bond densities allow predic-

---

*
 After removal of an electron upon ionization the remaining electrons are reorganized until they reach an equilibrium distribution.

tion of the gross features of fragmentation. Loew et al. [25], for instance, demonstrated that in ionized estrone (79) the aromatic ring and the carbonyl group have the strongest bonds and hence are not expected to fragment. Kao et al. [161] have shown that in o-fluoro-cis-β-methoxy styrene (80) the C-F and CH_3-O-bond are the weakest bonds in the molecular ion. Whilst loss of a methyl group is indeed observed experimen-

tally to be the lowest energy cleavage the predicted rupture of an aromatic carbon-fluorine bond is not a common process in mass spectrometry. Krier et al. [22] calculated the dissociation energy for the C-N and the C-C bond in the ethylamine molecular ion and found similar dissociation energies for both bonds although the experimental value for the dissociation energy of the C-C bond is considerably lower than that of the C-N bond. It is obvious from these examples that at present the criterion of bond densities is of limited value in predicting the mass spectrometric fragmentation behavior. Moreover, it is difficult to estimate the reliability of bond density data determined by semi-empirical MO methods. Finally, it has been pointed out by McMaster [170] that as the ion dissociates the electron distribution and thus the bond densities change, often in an unpredictable manner, so that the fragmentation may not be controlled by the properties of the ion in its equilibrium state.

In the other approach the total energy of the intermediate formed by a rupture of a certain bond is calculated [171,172]. It is argued that bond cleavage leading to the most stable intermediate will occur preferentially. Using the extended Hückel (EH) as well as the iterative extended Hückel (IEH) method Ichikawa and Ogata [171] and Kato et al. [172] concluded that in the molecular ions of pyridine (81) and pyrimidine (82) as well as pyrimidine derivatives the first bond scission of the heteroaromatic ring generally occurs at the bond β to the nitrogen. Again it is difficult to estimate the reliability of such semi-empirical MO calculations.

(81)   (82)

2.7.2 Hydrogen Scrambling and Skeletal Isomerization

The dependence of hydrogen scrambling in the benzoic acid molecular ion on energy and ion lifetime has been discussed in detail in Section III-2.3. The exchange process involves only the hydroxyl and the two ortho hydrogens. The salicylic acid molecular ion undergoes scrambling to a lesser extent, which may be explained by a reduction in the rate of rotation of the carboxyl group. Parker et al. [163] calculated the rotational barriers in both molecular ions. They demonstrated that the lowest pathways for scrambling in benzoic acid (rotation of the ring C-/ carboxylic C-bond) requires an activation energy of 21 kJ mol^{-1}, while the lowest pathway in salicylic acid (rotation about one side chain carbon-oxygen bond, followed by rotation of the whole side chain around the ring C-/ carboxylic C-bond) requires a much higher activation energy of 54 kJ mol^{-1}. Thus in salicylic acid steric interaction blocks the pathway that had the lowest energy in benzoic acid and the process that does occur in salicylic acid occurs at a slower rate, in agreement with experiment.

The same authors also studied the HD exchange mechanism in the ethane molecular ion prior to methyl loss [165] and demonstrated that the most stable transition state for this process has a diborane like structure.

SCF-calculations have been reported by Gallup et al. [166] on the hydrogen scrambling in the benzene molecular ion. The authors demonstrated that for this process interconversions between cyclic and acyclic isomers do not have to be invoked to account for the observed scrambling process. The hydrogen exchange can occur via a simple movement of one proton to an adjacent carbon. The cyclic intermediate has an energy 1.9 eV above the ionization potential of benzene, which is considerably lower than the energy for ring opening.

MINDO/3 calculations have been employed by Dewar et al. [167-169] to study the isomerization of $C_7H_7^+$ and $C_7H_8^{+\cdot}$ ions. The results suggest that the rearrangement of a benzyl ion [167] (as well as a substituted benzyl ion [168]) to a tropylium ion or its substituted analogue occurs via a norcaradienyl intermediate as discussed in detail in Section V-3.6.3. On the other hand, the toluene molecular ion can rearrange to the cycloheptatriene ion via several paths of similar overall activation energy [169].

Of special interest to organic mass spectrometrists are recent MINDO/3 calculations of the transition state for 1,1-and 1,2-H_2-eliminations from small organic ions [81] as discussed in detail in Section IV-2.3.4.

2.7.3 Substituent Effects

MO calculations were also applied to the study of substituent effects. As shown in Section III-6 the influence of the substituent on ionization potentials and, to a lesser extent, on appearance potentials predominantly determines the relative fragment abundance. It is not surprising that this conclusion is borne out by MO calculations [162]. More interesting is the application of MO calculations to anomalous substituent effects. An anomalous effect is observed for instance in acetophenones if a phenyl ring is substituted in para position. Whilst for most other substituents the ratio $\log[CH_3CO^+]/[CH_3COC_6H_4Y^{+\cdot}]$ shows a good linear correlation with Hammett's σ-values, the fragment abundance is too low in the case of the p-phenyl-substituent. Bursey and McLafferty [173] suggested that the phenyl rings become more nearly coplanar on going from the neutral molecule to the ion so that in the ion the p-phenyl substituent may enter into resonance more effectively with the reaction site. As a result bond cleavage becomes more difficult (Scheme IV-33). MO calculations corroborate this reasoning, show-

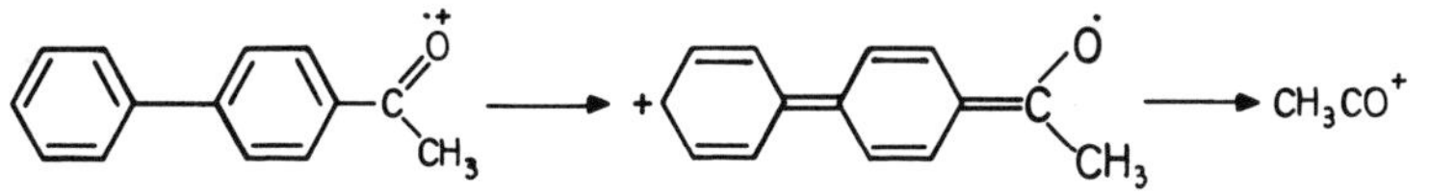

(83)

Scheme IV-33

ing an enhanced planarity of the phenyl rings in the ion compared with the neutral molecule and an increased bond density with increasing planarity [162,164].

2.8 Reaction Mechanisms and Ion Structures

Reaction mechanisms and ion structures are directly correlated, as any

IV   Reaction Mechanisms

postulated reaction mechanism includes assumptions on the structure of the fragment ion involved. Thus the determination of the ion structure(s) is a necessary step in confirming a postulated mechanism. Although a variety of reliable methods for ion structure elucidation are available today (see Chapter V), this approach is still little used by organic mass spectrometrists. Two examples have been selected to demonstrate the value of ion structure determination for the elucidation of reaction mechanisms.

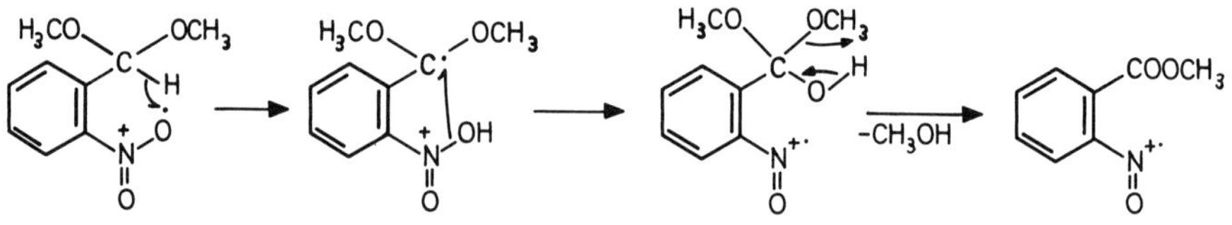

Scheme IV-34

Table IV-7. Collisional Activation Spectra[a] of $C_8H_7NO_3^{+\cdot}$ Ions from o-Nitrobenzaldehyde acetale and o-Nitrosobenzoic acid methylester [175]

m/e	Ion (84) $\xrightarrow{-CH_3OH}$ m/e 165	Ion (85) (m/e 165)
135	238	181
120	7.3	7.0
104	16	16
92	16	15
77	32	33
63	6.9	6.7
59	3.5	3.2
50	12	13
39	3.4	3.3
30	1.9	2.0
27	0.7	0.5
15	0.3	0.3

[a] Abundances relative to the sum of all collision induced fragments = 100 (except m/e = 135 which is also formed unimolecularly).

IV-2 *Methods for the Elucidation of Mechanisms*   201

(1) It has been postulated that methanol elimination from o-nitro benzaldehyde acetale leads in an intramolecular redox reaction to o-nitrosobenzoic acid methylester [174]. Comparison of the collisional activation spectra of the $(M - CH_3OH)^{+\cdot}$ fragment ion from (84) and the molecular ion of (85) (Table IV-7) confirms this result [175].

(2) Deuterium labelling revealed that loss of water from 6-phenylhexanol-1 involves to 80% the benzylic and to 15% the homobenzylic position which led to the conclusion [176] that 1-phenylhexene-1 is formed in this reaction according to Scheme IV-35a. Collisional activation spec-

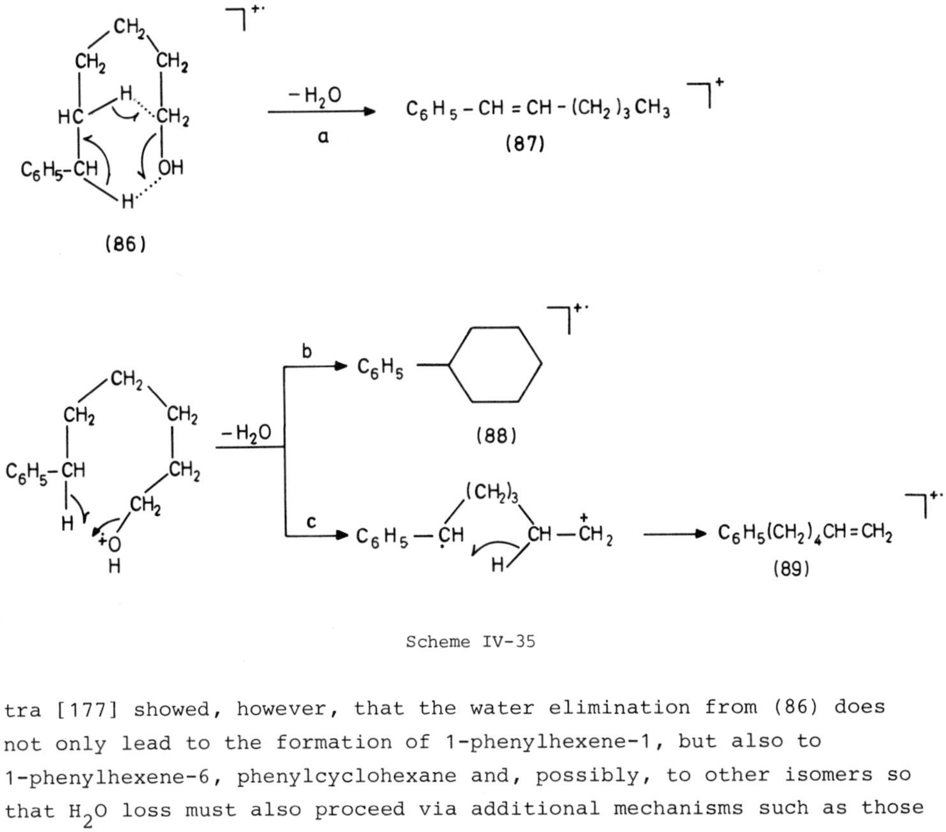

Scheme IV-35

tra [177] showed, however, that the water elimination from (86) does not only lead to the formation of 1-phenylhexene-1, but also to 1-phenylhexene-6, phenylcyclohexane and, possibly, to other isomers so that H_2O loss must also proceed via additional mechanisms such as those depicted in Scheme IV-35b and c.

Further examples demonstrating how reaction mechanisms may be deduced from ion structures have been presented in the previous sections (IV-2.5.1 and 2.5.2). For additional examples the reader is referred to the original publications [178-182].

2.9 Structure of Neutral Fragments

Knowledge of the structure of the neutrals is not only useful in order to confirm proposed reaction mechanisms, it is also of crucial importance when determining heats of formation of gaseous ions by appearance potential measurements*) (see Section V-3.3).

Two methods for the detection of neutrals have been proposed recently. Reeher et al. [183,184] developed a pulsed dual ion source in which the neutrals formed by fragmentation of the molecular ion are ionized by a second electron beam and detected mass spectrometrically. This method also allows the determination of the ionization potentials of neutral fragments, a valuable item of information in deducing their structure.

Burns and Mortan [185] constructed an "electron bombardment flow reactor" in which the products are collected in a cooled trap and analyzed by gas chromatography. Using this technique they were able to demonstrate that the electron impact induced hydrogen rearrangement in n-butylphenyl ether (see Section IV-2.6.3) leads to the formation of both 1-butene and 2-butene neutrals while by pyrolysis and photolysis exclusively 1-butene is formed.

3 References

1. F.W. McLafferty, "Interpretation of Mass Spectra", 2. edition, Benjamin, Reading, 1973.
2. R.W.A. Johnstone, "Mass Spectrometry for Organic Chemists", University Press, Cambridge, 1972, p. 39 and 78.
3. G.S. Hammond, J. Am. Chem. Soc., $\underline{77}$, 334 (1955).
4. D.H. Williams and I. Howe, "Principles of Organic Mass Spectrometry", McGraw-Hill, London, 1972.
5. K. Levsen, H. Heimbach, G.J. Shaw and G.W.A. Milne, Org. Mass Spectrom., $\underline{12}$, 663 (1977).
6. R.D. Bowen and D.H. Williams, J. Chem. Soc. Perkin Trans. II, 1479 (1976).
7. R.D. Bowen and D.H. Williams, Org. Mass Spectrom., $\underline{12}$, 453 (1977).

---

*
On the other hand appearance potentials can be used in turn to obtain information about the structure of the neutrals. Thus elimination of to H˙ radicals instead of a H_2 molecule leads to products whose heat of formation is 436 kJ mol^{-1} above that for H_2 elimination.

8. H. Budzikiewicz, C. Djerassi and D.H. Williams, "Mass Spectrometry of Organic Compounds", Holden-Day, San Francisco, 1967, p. 86.
9. H. Budzikiewicz, Adv. Mass Spectrom., $\underline{4}$, 313 (1967).
10. M. Bobrich, H. Schwarz, K. Levsen and P. Schmitz, Org. Mass Spectrom., $\underline{12}$, 549 (1977).
11. H. Schwarz, Org. Mass Spectrom., $\underline{9}$, 826 (1974).
12. D.H. Williams and R.D. Bowen, Org. Mass Spectrom., $\underline{11}$, 223 (1976).
13. D.P. Stevenson, Disc. Faraday Soc., $\underline{10}$, 35 (1951).
14. H.E. Audier, Org. Mass Spectrom., $\underline{2}$, 283 (1969).
15. A.G. Harrison, C.D. Finney and J.A. Sherk, Org. Mass Spectrom., $\underline{5}$, 1313 (1971).
16. R.G. Cooks, J.H. Beynon, R.M. Caprioli and G.R. Lester, "Metastable Ions", Elsevier, Amsterdam, 1973.
17. I. Howe, in "Mass Spectrometry", Vol. 2. (D.H. Williams, Ed.), The Chemical Society, London, 1973.
18. D.H. Williams, presented at the Euchem Conference on "Ion Beams", Nordwijk (Holland), 1977.
19. F.W. McLafferty, Chem. Commun., 78 (1966).
20. T.W. Bentley, R.A.W. Johnstone and F.A. Mellon, J. Chem. Soc. (B), 1800 (1971).
21. T.W. Bentley and R.A.W. Johnstone, Adv. Phys. Org. Chem., $\underline{8}$, 151 (1970).
22. C. Krier, J.C. Lorquet and A. Berlingin, Org. Mass Spectrom., $\underline{8}$, 387 (1974).
23. F.W. McLafferty in "Recent Developments in Mass Spectroscopy" (K. Ogata and T. Hayakawa, Eds.), University Park Press, Baltimore, 1970, p. 70.
24. J.A. Pople, Int. J. Mass Spectrom. Ion Phys., $\underline{19}$, 89 (1976).
25. G. Loew, M. Chadwick and D. Smith, Org. Mass Spectrom., $\underline{7}$, 1241 (1973).
26. M.M. Bursey, J.L. Kao and L. Pedersen, Org. Mass Spectrom., $\underline{10}$, 38 (1975).
27. D.R. Yarkony and H.F. Schaefer III, J. Chem. Phys., $\underline{63}$, 4317 (1975).
28. F. Jordon, J. Phys. Chem., $\underline{80}$, 76 (1976).
29. H. Budzikiewicz and R. Pesch, Org. Mass Spectrom., $\underline{9}$, 861 (1974).
30. C.G. van den Heuvel and N.M.M. Nibbering, Org. Mass Spectrom., $\underline{10}$, 250 (1975).
31. G. Wood, A.M. Falick and A.L. Burlingame, Org. Mass Spectrom., $\underline{8}$, 279 (1974).
32. D.H. Williams and J.H. Beynon, Org. Mass Spectrom., $\underline{11}$, 103 (1976).
33. I. Howe and F.W. McLafferty, J. Am. Chem. Soc., $\underline{93}$, 99 (1971).
34. A.A. Gamble, J.R. Gilbert and J.G. Tillett, Org. Mass Spectrom., $\underline{5}$, 1093 (1971).
35. M.M. Bursey, M.K. Hoffman and S.A. Benezra, Chem. Commun., 1417 (1971).
36. R.C. Dunbar and R. Klein, J. Am. Chem. Soc., $\underline{99}$, 3744 (1977).
37. K. Levsen, F.W. McLafferty and D.M. Jerina, J. Am. Chem. Soc., $\underline{95}$, 6332 (1973).
38. A.A. Gamble, J.R. Gilbert and J.G. Tillett, Org. Mass Spectrom., $\underline{3}$, 1223 (1970).
39. N.A. Uccella, I. Howe and D.H. Williams, Org. Mass Spectrom., $\underline{6}$, 229 (1972).
40. K.B. Tomer and C. Djerassi, Tetrahedron, $\underline{29}$, 3491 (1973).
41. P.C. Vijfhuizen, W. Heerma and G. Dijkstra, Org. Mass Spectrom., $\underline{10}$, 919 (1975).
42. H. Bosshardt and M. Hesse, Angew. Chem., $\underline{86}$, 256 (1974).
43. F.M. Benoit and A.G. Harrison, Org. Mass Spectrom., $\underline{11}$, 599 (1976).
44. K.H. Maurer, C. Brunnée, G. Kappus, K. Habfast, U. Schröder and P. Schulze, 19. Conference on Mass Spectrometry and Allied Topics, ASMS, Atlanta, 1971.
45. U.P. Schlunegger, Angew. Chem. Int. Ed., $\underline{14}$, 679 (1975).

46. J.H. Beynon, R.G. Cooks, J.W. Amy, W.E. Baitinger and T.Y. Ridley, Anal. Chem., 45, 1023 A (1973).
47. A.F. Weston, K.R. Jennings, S. Evans and R.M. Elliott, Int. J. Mass Spectrom. Ion Phys., 20, 317 (1976).
48. D.L. Kemp, R.G. Cooks and J.H. Beynon, Int. J. Mass Spectrom. Ion Phys., 21, 93 (1976).
49. R.K. Boyd and J.H. Beynon, Org. Mass Spectrom., 12, 163 (1977).
50. D.S. Millington and J.A. Smith, Org. Mass Spectrom., 12, 264 (1977).
51. J. Seibl, Helv. Chim. Acta, 50, 263 (1967).
52. E. Tajima and J. Seibl, Int. J. Mass Spectrom. Ion Phys., 3, 245 (1969).
53. K.R. Jennings, Chem. Commun., 283 (1968).
54. U. Löhle and Ch. Ottinger, Int. J. Mass Spectrom. Ion Phys., 5, 265 (1970).
55. J.H. Beynon, W.E. Baitinger, J.W. Amy and R.M. Caprioli, Int. J. Mass Spectrom. Ion Phys., 3, 309 (1969).
56. R.G. Cooks, M. Bertrand, J.H. Beynon, M.E. Rennekamp and D.W. Setser, J. Am. Chem. Soc., 95, 1732 (1973).
57. J.H. Beynon, M. Bertrand and R.G. Cooks, J. Am. Chem. Soc., 95, 1739 (1973).
58. J.H. Beynon, M. Bertrand and R.G. Cooks, Org. Mass Spectrom., 7, 785 (1973).
59. R.G. Cooks, K.C. Kim and J.H. Beynon, Int. J. Mass Spectrom. Ion Phys., 15, 245 (1974).
60. D.K. Sen-Sharma, K.R. Jennings and J.H. Beynon, Org. Mass Spectrom., 11, 319 (1976).
61. R.G. Cooks, J.H. Beynon, M. Bertrand and M.K. Hoffman, Org. Mass Spectrom., 7, 1303 (1973).
62. K.C. Kim, J.H. Beynon and R.G. Cooks, J. Chem. Phys., 61, 1305 (1974).
63. P.C. Vijfhuizen, W. Heerma, N.M.M. Nibbering, Org. Mass Spectrom., 11, 787 (1976).
64. P.C. Vijfhuizen and J.K. Terlouw, Org. Mass Spectrom., 11, 888 (1976).
65. P.C. Vijfhuizen, H. van der Schee and J.K. Terlouw, Org. Mass Spectrom., 11, 1198 (1976).
66. J.L. Holmes, D. McGillivray and N.S. Isaacs, Org. Mass Spectrom., 9, 510 (1974).
67. J.L. Holmes, A.D. Osborne and G.M. Weese, Org. Mass Spectrom., 10, 867 (1975).
68. K. Levsen and F.W. McLafferty, J. Am. Chem. Soc., 96, 139 (1974).
69. R.K. Boyd and J.H. Beynon, Int. J. Mass Spectrom. Ion Phys., 23, 163 (1977).
70. C.B. Theissling and N.M.M. Nibbering, in "Advances in Mass Spectrometry", Vol. 7 (N. Daly, Ed.), Heyden, London, 1977.
71. F. Borchers, K. Levsen, C.B. Theissling and N.M.M. Nibbering, Org. Mass Spectrom., 12, 746 (1977).
72. D. Russel, M.L. Gross, J. van der Greef and N.M.M. Nibbering, in preparation.
73. J.F. Elder, J.H. Beynon and R.G. Cooks, Org. Mass Spectrom., 10, 273 (1975).
74. J.R. Christie, P.J. Derrick and G.J. Rickard, J. Chem. Soc. Faraday Trans. II, 74, 304 (1978).
75. D.H. Williams and R.D. Bowen, J. Am. Chem. Soc., 99, 3192 (1977).
76. D.H. Williams and G. Hvistendahl, J. Am. Chem. Soc., 96, 6753 (1974).
77. D.H. Williams and G. Hvistendahl, J. Am. Chem. Soc., 96, 6755 (1974).
78. G. Hvistendahl and D.H. Williams, J. Am. Chem. Soc., 97, 3097 (1975).
79. R.D. Bowen and D.H. Williams, J. Chem. Soc. Chem. Commun., 378 (1977.
80. H. Schwarz, F. Borchers and K. Levsen, Z. Naturforsch. 31b, 935 (1976).

81. M.J.S. Dewar and H.S. Rzepa, J. Am. Chem. Soc., 99, 7432 (1977).
82. I. Howe and D.H. Williams, Chem. Commun., 1195 (1971).
83. I. Howe, N.A. Uccella and D.H. Williams, J. Chem. Soc., Perkin II, 76 (1973).
84. G. Hvistendahl and D.H. Williams, J.Chem. Soc., Chem. Commun., 4 (1975).
85. K. Levsen and H. Schwarz, J. Chem. Soc., Perkin Trans. II, 1231 (1976).
86. S. Hammerum and K.B. Tomer, Org. Mass Spectrom., 6, 1369 (1972).
87. K.B. Tomer, S. Hammerum and C. Djerassi, Tetrahedron Lett., 12, 915 (1973).
88. I. Howe, Org. Mass Spectrom., 10, 767 (1975).
89. R.H. Shapiro and T.F. Jenkins, Org. Mass Spectrom., 2, 771 (1969).
90. K.B. Tomer, J. Turk and R.H. Shapiro, Org. Mass Spectrom., 6, 235 (1972).
91. J.R. Dias and C. Djerassi, Org. Mass Spectrom., 6, 385 (1972).
92. N.M.M. Nibbering, T. Nishishita, C.C. Van de Sande and F.W. McLafferty, J. Am. Chem. Soc., 96, 5668 (1974).
93. C. Köppel and F.W. McLafferty, J. Am. Chem. Soc., 98, 8293 (1976).
94. F.W. McLafferty, D.J. McAdoo and J.S. Smith, J. Am. Chem. Soc., 91, 5400 (1969).
95. C.C. Van de Sande and F.W. McLafferty, J. Am. Chem. Soc., 97, 2298 (1975).
96. H. Schwarz, Org. Mass Spectrom., 10, 384 (1975).
97. H.J. Veith, A. Guggisberg and M. Hesse, Helv. Chim. Acta, 54, 653 (1971).
98. K. Sailer and M. Hesse Helv. Chim. Acta, 51, 1817 (1968).
99. C. Köppel, H. Schwarz and F. Bohlmann, Org. Mass Spectrom., 9, 567 (1974).
100. H. Schwarz, R.D. Petersen and C.C. Van de Sande, Org. Mass Spectrom., 12, 391 (1977).
101. J. Martens, K. Praefcke, U. Schulze, H. Schwarz and H. Simon, Z. Naturforsch., 32b, 657 (1977).
102. C.C. Van de Sande, Tetrahedron, 32, 1741 (1976).
103. C.C. Van de Sande, Org. Mass Spectrom., 11, 121 (1976).
104. C.C. Van de Sande, Org. Mass Spectrom., 11, 130 (1976).
105. C.C. Van de Sande, Bull. Soc. Chim. Belges, 84, 785 (1975).
106. C.C. Van de Sande, M. Vanhooren and F. Van Gaever, Org. Mass Spectrom., 11, 1206 (1976).
107. K. Levsen, H. Heimbach, C.C. Van de Sande and J. Monstrey, Tetrahedron, 33, 1785 (1977).
108. T.A. Molenaar-Langeveld and N.M.M. Nibbering, Org. Mass Spectrom., 9, 257 (1974).
109. A.P. Bruins and N.M.M. Nibbering, Tetrahedron, 30, 493 (1974).
110. H. Schwarz, K. Praefcke and J. Martens, Tetrahedron, 29, 2877 (1973).
111. F. Bohlmann, R. Herrmann, W. Mathar and H. Schwarz, Chem. Ber., 107, 1081 (1974).
112. M. Sheehan, R.J. Spangler and C. Djerassi, J. Org. Chem., 36, 3526 (1971).
113. B. Richter and H. Schwarz, Org. Mass Spectrom., 10, 522 (1975).
114. H. Schwarz, W. Mathar and F. Bohlmann, Org. Mass Spectrom., 9, 84 (1974).
115. A.P. Bruins and N.M.M. Nibbering, Org. Mass Spectrom., 11, 271 (1976).
116. H. Schwarz, F. Bohlmann, G. Hillenbrand and G. Altnau, Org. Mass Spectrom., 9, 707 (1974).
117. C.C. Van de Sande, C. De Meyer and A. Maquestiau, Bull. Soc. Chim. Belge, 85, 79 (1976).
118. H. Schwarz and M.T. Reetz, Angew. Chem. (Int. Ed.), 15, 704 (1976).

119. H. Schwarz, R. Herrmann and R. Wolfschütz, J. Heterocycl. Chem., 12, 633 (1975).
120. H. Schwarz, B. Richter and F. Bohlmann, Org. Mass Spectrom., 10, 1125 (1975).
121. H. Schwarz, "Topics in Current Chemistry", 73, 231 (1978).
122. R.H. Shapiro and K.B. Tomer, Org. Mass Spectrom., 3, 333 (1970).
123. H.D. Beckey, "Field Ionization Mass Spectrometry", Pergamon, Oxford, 1971.
124. H.D. Beckey, "Principles of Field Ionization and Field Desorption Mass Spectrometry", Pergamon, Oxford, 1977.
125. H.D. Beckey, K. Levsen, F.W. Röllgen and H.R. Schulten, Surface Science, 70, 325 (1978).
126. P.J. Derrick and A.L. Burlingame, Acc. Chem. Res., 7, 328 (1974).
127. P.J. Derrick, in "Mass Spectrometry", Int. Rev. Science, Phys. Chem. Series II, Vol. 5 (A. Maccoll,Ed.), Butterworths, London, 1975.
128. A.M. Falick, P.J. Derrick and A.L. Burlingame, Int. J. Mass Spectrom. Ion Phys., 12, 101 (1973).
129. H.D. Beckey, H. Hey, K. Levsen and G. Tenschert, Int. J. Mass Spectrom. Ion Phys., 2, 101 (1969).
130. K. Levsen and H.D. Beckey, Int. J. Mass Spectrom. Ion Phys., 15, 353 (1974).
131. J. van der Greef, F.A. Pinkse, C.W.F. Kort and N.M.M. Nibbering, Int. J. Mass Spectrom. Ion Phys., 25, 315 (1977).
132. J. van der Greef and N.M.M. Nibbering, Int. J. Mass Spectrom. Ion Phys., 25, 357 (1977).
133. P.J. Derrick, A.M. Falick and A.L. Burlingame, J. Am. Chem. Soc., 94, 6794 (1972).
134. P.J. Derrick and A.L. Burlingame, J. Am. Chem. Soc., 96, 4909 (1974).
135. P.J. Derrick, A.M. Falick and A.L. Burlingame,"Advances in Mass Spectrometry", Vol. 6 (A.R. West, Ed.), Institute of Petroleum, London, 1974, p. 877.
136. P.J. Derrick, A.M. Falick, S. Lewis and A.L. Burlingame, Org. Mass Spectrom., 7, 887 (1973).
137. P.J. Derrick, A.M. Falick, A.L. Burlingame and C. Djerassi, J. Am. Chem. Soc., 96, 1054 (1974).
138. F. Borchers, K. Levsen, H. Schwarz and C. Wesdemiotis, Int. J. Mass Spectrom. Ion Phys., in press.
139. P. Wolkoff, J. van der Greef and N.M.M. Nibbering, J. Am. Chem. Soc., 100, 541 (1978).
140. J. van der Greef, T.A. Molenaar-Langeveld and N.M.M. Nibbering, Int. J. Mass Spectrom. Ion Phys., in press.
141. A. Venema, N.M.M. Nibbering and Th. J. de Boer, Org. Mass Spectrom.,3, 583 (1970).
142. T.A. Molenaar-Langeveld, N.M.M. Nibbering and Th. J. de Boer, Org. Mass Spectrom., 5, 725 (1971).
143. S. Meyerson and G.J. Karabatsos, Org. Mass Spectrom., 8, 289 (1974).
144. R.P. Morgan, P.J. Derrick and A.G. Harrison, J. Am. Chem. Soc., 99, 4189 (1977).
145. A.M. Falick and T. Gäumann, 24th Annual Conference on Mass Spectrometry and Allied Topics, San Diego, Calif., May 1976.
146. T. Gäumann, private communication.
147. F. Borchers, K. Levsen, H. Schwarz, C. Wesdemiotis and H.U. Winkler, J. Am. Chem. Soc., 99, 6359 (1977).
148. A.N.H. Yeo and C. Djerassi, J. Am. Chem. Soc., 94, 482 (1972).
149. F.M. Benoit and A.G. Harrison, Org. Mass Spectrom., 11, 1056 (1976).
150. F. Borchers, K. Levsen and H.D. Beckey, Int. J. Mass Spectrom. Ion Phys., 21, 125 (1976).

151. F. Borchers, K. Levsen, G. Eckardt and G.W.A. Milne, in "Advances in Mass Spectrometry", Vol. 7 (N.R. Daly, Ed.) Heyden, London, 1977.
152. P.J. Derrick, A.M. Falick and A.L. Burlingame, J. Am. Chem. Soc., 95, 437 (1973).
153. P.J. Derrick, A.M. Falick and A.L. Burlingame, J. Am. Chem. Soc., 96, 615 (1974).
154. R.P. Morgan and P.J. Derrick, J. Chem. Soc., Chem. Commun., 836 (1974).
155. P.J. Derrick, in "Advances in Mass Spectrometry", Vol. 7, (N.R. Daly, Ed.), Heyden, London, 1977.
156. P.J. Derrick, J.L. Holmes and R.P. Morgan, J. Am. Chem. Soc., 97, 4936 (1975) and references herein.
157. M.M. Green, Topics in Stereochemistry, 9, 35 (1976).
158. D.G.I. Kingston, B.W. Hobrock, M.M. Bursey and J.T. Bursey, Chem. Rev., 75, 693 (1975).
159. D.G. Patterson, R.B. Scott and P. Brown, Org. Mass Spectrom., 12, 395 (1977).
160. C.E. Parker, J.R. Hass, M.M. Bursey and L.G. Pedersen, Org. Mass Spectrom., 7, 1189 (1973).
161. J.-L. Kao, C.E. Parker and M.M. Bursey, Org. Mass Spectrom., 9, 679 (1974).
162. G.H. Loew, R.F. Kirchner and J.G. Lawless, Org. Mass Spectrom., 11, 1158 (1976).
163. C.E. Parker, M.M. Bursey and L.G. Pedersen, Org. Mass Spectrom., 9, 204 (1974).
164. C.E. Twine, C.E. Parker and M.M. Bursey, Org. Mass Spectrom., 7, 1179 (1973).
165. C.E. Parker, M.M. Bursey and L.G. Pedersen, Org. Mass Spectrom., 7, 1077 (1973).
166. G.A. Gallup, D. Steinheider and M.L. Gross, Int. J. Mass Spectrom. Ion Phys., 22, 185 (1976).
167. C. Cone, M.J.S. Dewar and D. Landmann, J. Am. Chem. Soc., 99, 372 (1977).
168. M.J.S. Dewar and D. Landmann, J. Am. Chem. Soc., 99, 4633 (1977).
169. M.J.S. Dewar and D. Landmann, J. Am. Chem. Soc., 99, 2446 (1977).
170. B.N. McMaster, in "Mass Spectrometry", Vol. 3 (R.A.W. Johnstone, Ed.), The Chemical Society, London, 1975, Chapter 1.
171. H. Ichikawa and M. Ogata, J. Am. Chem. Soc., 95, (1973).
172. T. Kato, H. Yamanaka, H. Abe, S. Sasaki and H. Ichikawa, Org. Mass Spectrom., 9, 981 (1974).
173. M.M. Bursey and F.W. McLafferty, J. Am. Chem. Soc., 88, 529 (1966).
174. M.M. Bursey, Tetrahedron Lett., 981 (1968).
175. H. Schwarz, R. Sezi, K. Levsen, H. Heimbach and F. Borchers, Org. Mass Spectrom., 12, 569 (1977).
176. S. Meyerson and L.C. Leitch, J. Am. Chem. Soc., 93, 2244 (1971).
177. K. Levsen, H. Heimbach, M. Bobrich, J. Respondek and H. Schwarz, Z Naturforsch., 32b, 880 (1977).
178. H. Schwarz, C. Wesdemiotis, B. Hess and K. Levsen, Org. Mass Spectrom., 10, 595 (1975).
179. K. Levsen and H. Schwarz, Org. Mass Spectrom., 10, 752 (1975).
180. H. Bornowski, V. Feistkorn, H. Schwarz, K. Levsen, and P. Schmitz, Z. Naturforsch., 32b, 664 (1977).
181. K. Levsen, G.E. Berendsen, N.M.M. Nibbering and H. Schwarz, Org. Mass Spectrom., 12, 125 (1977).
182. F. Borchers, H. Heimbach, K. Levsen, H. Morhenn and H. Schwarz, Org. Mass Spectrom., 12, 573 (1977).

183. J.R. Reeher, G.D. Flesch and H.J. Svec, Int. J. Mass Spectrom. Ion Phys., 19, 351 (1976).
184. J.R. Reeher, G.D. Flesch and H.J. Svec, Org. Mass Spectrom., 11, 154 (1976).
185. F.B. Burns and Th. Hellman Morton, J. Am. Chem. Soc., 98, 7308 (1976).

Chapter V. The Structure of Gaseous Ions

1 Introduction

In the study of the properties of an organic ion the most important question is undoubtedly that of its structure. The structure of organic ions in solution has been extensively studied in the past in particular by Olah and his coworkers [1]. While the synthesis of such ions in solution is no easy undertaking, the corresponding gaseous ions can be readily generated in a mass spectrometer. At the pressures typical for the ion source of a conventional mass spectrometer ($< 10^{-5}$ torr) such ions are isolated species. Thus the influence of the solvent does not need to be taken into account, which makes the properties of gaseous organic ions directly amenable to comparison with the results of molecular orbital calculations.

In the early stage of organic mass spectrometry the main emphasis was laid on the elucidation of the decomposition mechanisms of excited organic ions, and metastable transitions and isotopic labelling were the most powerful tools for the determination of these reaction mechanisms. Supported by chemical intuition the structures involved in the decomposition of organic ions were most frequently deduced from the mechanism leading to their formation. To illustrate this approach and its possible pitfalls the decomposition of the 2-hexanone molecular ion will be discussed. Three important fragmentation routes of this ion are outlined in Scheme V-1. α-Cleavage leads to the $C_2H_3O^+$ ion which, as

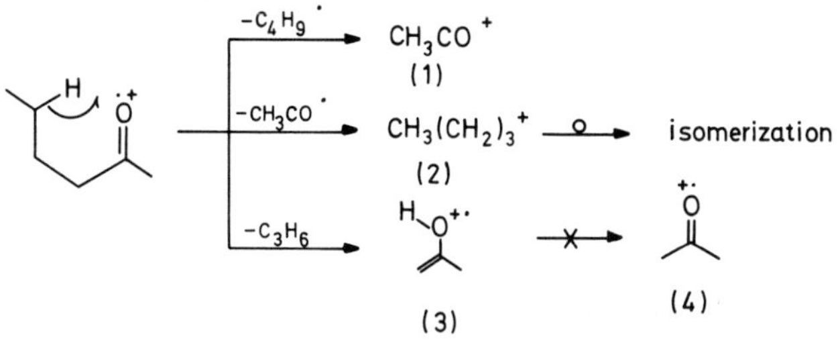

Scheme V-1

suggested by the mechanism of formation, has the structure of the acetyl ion. The corresponding cleavage with charge migration initially gives rise to the n-butyl ion. However it has been shown by a variety of additional experiments that excited n-butyl ions isomerize to a mixture of interconverting primary, secondary and tertiary ions [2-6] a piece of information which cannot be extracted directly from knowledge of the mechanism. Finally, $C_3H_6O^{+\cdot}$ ions are formed by a McLafferty rearrangement [7]. Although deuterium labelling suggests, but does not prove, the formation of the enolic ion (3) via a six-membered transition state, the elucidation of this mechanism alone does not allow the exclusion of a subsequent isomerization to the acetone molecular ion (4). Additional studies revealed that this rearrangement does not occur [8-11].

This example demonstrates that the elucidation of the mechanism leading to the formation of a fragment ion gives at most only information on the initial, but not on the final structure. Furthermore, if only the primary fragmentation is considered, one should keep in mind that isotopic labelling can only establish which atom in the original molecule is transferred in a rearrangement reaction, no direct information being available about the site to which this atom migrates. Additional problems which may arise if ion structures are assigned from the mechanism leading to their formation are discussed in Section V-3.2.

It is evident from this discussion that the numerous ion structure assignments based on the reaction mechanism and chemical intuition as reported in the early mass spectrometric literature are often only speculative. This view is reinforced when it is borne in mind that it is not always possible to extrapolate conventional views of bonding in neutral organic molecules to ions in the gas phase where the existence of nonclassical structures such as the CH_5^+ ion is well known [12]. Thus it has been stated by Bentley and Johnstone in 1970 "that the knowledge of ion structures at present is very poor" [13]. During the last decade several powerful techniques for the elucidation of ion structures have been developed (vide infra) and at present the structures of a variety of gaseous organic ions seem to be well established. However, our knowledge of ion structures today is not as good by far as in normal chemistry.

2 Fundamental Considerations

2.1 Definitions

The discussion of the fragmentation of the 2-hexanone molecular ion in the previous section demonstrated that the term ion structure needs to be defined more precisely.

First, all experimental methods for ion structure elucidation give, in general, only information on the constitution, but hardly any information on the configuration (e.g. on bond angles and bond lengths), if the ion is formed by a fragmentation process. Qualitative information on the configuration is, however, often available for molecular ions through stereospecific fragmentations and ion-molecule reactions. In the following only the ion constitution will be discussed.

Second, one has to differentiate between the *initial* structure, i.e. the structure of a molecular ion immediately after ionization or a fragment ion immediately after formation, and the *final* structure which the ion reaches after a given lifetime. Although most techniques for ion structure elucidation give information on the final structure, i.e. on ions with lifetimes $\geq 10^{-6}$ s, some methods are available which allow conclusions to be drawn about the initial structure. However, when discussing the apparent initial structure of an ion formed by a fragmentation process it should be remembered that one might only be dealing with the structure of the transition state.

Third, the ion structure depends on its internal energy (and lifetime). For experimental reasons it is usual to differentiate between ions with sufficient energy to decompose*) (in the following termed "de-

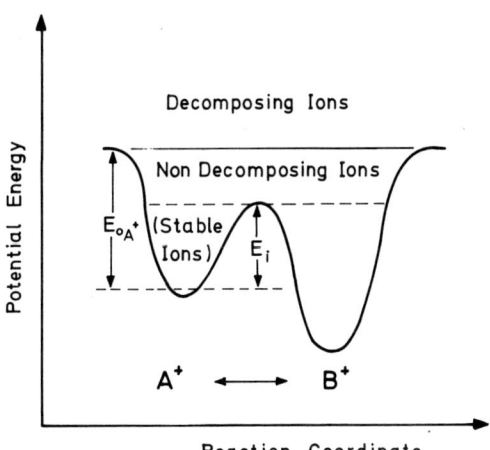

Fig. V-1. Schematic potential energy diagram illustrating the definitions "decomposing ions", "non-decomposing ions" and "stable ions".

---

* The term "reactive" ion has been frequently used in the literature. Note, however, that "non-decomposing" ions may also react in an ICR cell.

composing ions") and those which do not decompose on the mass spectrometric time scale ("*non-decomposing ions*"). Non-decomposing ions may still have sufficient energy to rearrange to an isomeric structure (see Fig. V-1) if the threshold for isomerization is below that for decomposition. Non-decomposing ions with insufficient energy to overcome the isomerization barrier will be termed "*stable ions*"*).

Unfortunately the differentiation between decomposing and non-decomposing ions is rather arbitrary with respect to the ions' internal energy. Thus non-decomposing benzene ions have a range of internal energies from zero to 4.5 eV (H-elimination), non-decomposing isobutane ions a range from zero to 0.1 eV (CH_4 elimination), which explains inter alia why the former but not the latter show extensive skeletal reorganization before decomposition**).

The apparently arbitrary distinction between decomposing and non-decomposing ions has, however, also a more basic significance: in decomposing ions the fragmentation and the isomerization channels are in competition with each other (see following section) which is of fundamental importance for the structure of these ions. As a wide range of internal energies (up to at least 10 eV) is transferred to a molecule upon ionization by 70 eV electrons it is possible in principle for the molecular ion to rearrange to various isomeric structures (including a large variety of non-classical ones) as shown by molecular orbital calculations. However, in most cases the decomposition from the original structure will be much faster than any isomerization, so that even in decomposing ions the situation is less complex than one might expect.

2.2 Isomerization of Gaseous Ions

It is evident from the above discussion that possible isomerization reactions complicate the structure elucidation of organic ions. For this reason the parameters determining the isomerization behavior of gaseous organic ions will be discussed in a qualitative fashion [14,15].

The relative ratio of the energy barriers for decomposition (E_{oA}^+) and isomerization (E_i) is the parameter which principally determines whether and to what extent an ion A^+ rearranges to an isomeric ion B^+ at a given internal energy E, as illustrated in Fig. V-2 for four sche-

---

*
 Note, that in previous publications the term "stable ions" often refers to all non-decomposing ions.

**
 The extent of skeletal reorganization is, however, also influenced by the height of the isomerization barrier (see Section V-2.2).

matic potential energy diagrams (the lowest threshold for decomposition is represented as solid curve, competing decomposition channels as dashed curves and the energy range for metastable decomposition in-

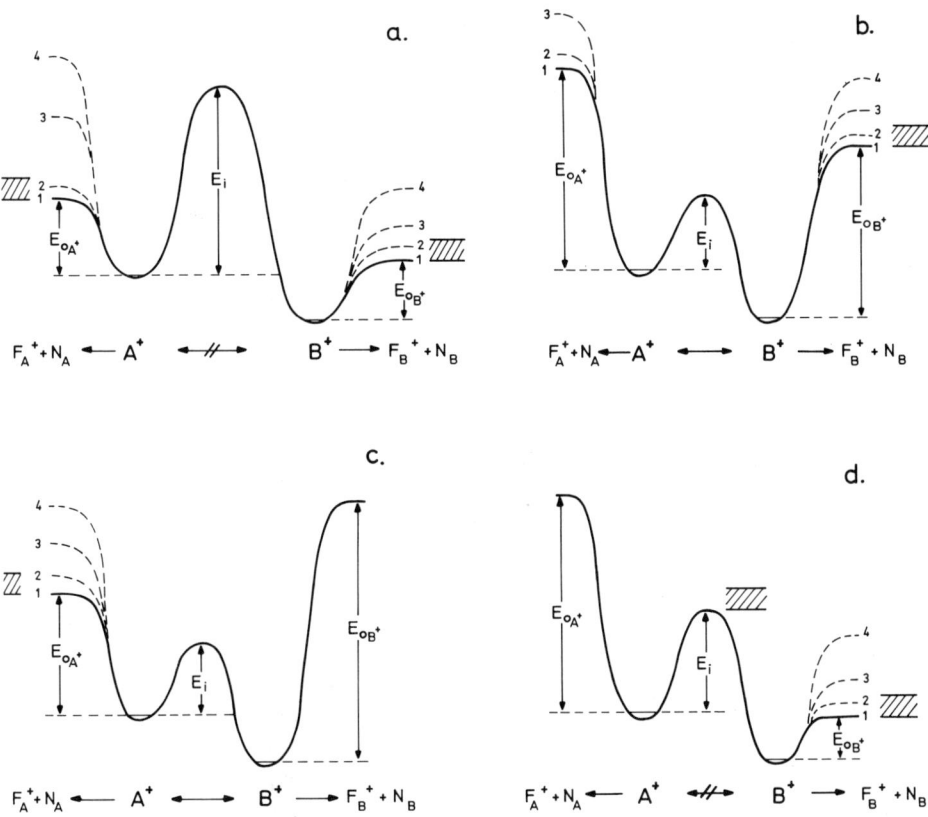

Fig. V-2. Schematic potential energy diagrams illustrating the parameters which influence the isomerization of organic ions (see text).

dicated as hatched area). The most important cases are shown in Fig. V-2a and b where the isomerization barrier is either much higher (a) or much lower (b) than the lowest threshold for decomposition.

In Fig. V-2a ($E_i \gg E_{oA^+}$) no isomerization is possible at internal energies below the threshold for isomerization (excluding tunnelling). At higher energies ($E > E_i$) isomerization is possible in principle, but in general decomposition will be much faster than isomerization, since the density of states in the activated complex for decomposition is considerably higher than that in the activated complex for isomerization as a result of the higher excess energy, $(E - E_{oA^+}) > (E - E_i)$ and (in general) the looser geometry of the transition state*). As shown later the potential energy diagram in Fig. V-2a is representative for many heteroatom-containing ions. The situation is more complex in Fig. V-2b ($E_i < E_{oA^+}$). Again no isomerization is possible at $E < E_i$, whilst at internal energies $E_i < E < E_{oB^+}$ A^+ can rearrange to B^+ without decomposition. However, in the absence of any mechanism for energy relaxation (e.g. by collision or radiation) A^+ will not rearrange completely to B^+ (as expected for thermal systems), but there will be a mixture of rapidly interconverting structures A^+ and B^+. At internal energies $E_{oA^+} > E > E_{oB^+}$ decomposition is possible after rearrangement to B^+, preceded by several interconversions between A^+ and B^+ which may lead to a partial or complete randomization of skeletal atoms.

For $E > E_{oA^+}$ the relative ratio of the rate constants for direct decomposition and rearrangement with subsequent decomposition through B^+ will determine whether and to what extent isomerization is observed. As a result of the much larger excess energy of the activated complex for the isomerization, $(E - E_i) > (E - E_{oA^+})$, decomposition through B^+ will prevail at low internal energies, whilst at high energies ($E \gg E_{oA^+}$) direct decompositions from A^+ with loose activated complexes will compete effectively with the isomerization. However, the degree of isomerization (detectable for instance as the extent of carbon randomization) may vary considerably for different decomposition processes depending on the activation energy and the geometry of the activated complex. The potential energy diagram in Fig. V-2b is representative for (inter alia) even-electron hydrocarbon ions (see Section V-4.1.1).

More special situations are illustrated in Fig. V-2c and d. In Fig. V-2c the lowest threshold for decomposition of the rearranged ion B^+ is considerably higher than the lowest threshold for decomposition from A^+ although B^+ has a lower heat of formation than A^+. In this particular case the initial ion A^+ with an internal energy $E_{oB^+} > E > E_{oA^+}$ will decompose through its original structure, but such decomposition may be preceded by several interconversions between A^+ and B^+ leading to partial or complete randomization of the skeletal atoms. It has been tentatively proposed that the system styrene-cyclooctatetraene may be represented by such a potential energy diagram [16]. In Fig. V-2d the lowest threshold for decomposition of the initial ion A^+ is also considerably higher than the isomerization barrier, which is in turn higher

---

*) Note however that the ion A^+ may have a decomposition channel whose threshold is higher than that for isomerization such as the dashed curve 4 in Fig. V-2a. If this decomposition occurs via a tight transition state, the isomerization to B^+ may compete effectively with the decomposition although both processes will be largely suppressed by decomposition channels of lower activation energy.

than several decomposition thresholds of the rearranged ion B^+. All initial ions A^+ with energy $E_{oA^+} > E > E_i$ will decompose through structure B^+. However, in contrast to Fig. V-2b hardly any interconverting structures $A^+ \rightleftharpoons B^+$ will be observed in this case: once the ions A^+ have overcome the isomerization barrier E_i decomposition through B^+ will be much faster than rearrangement back to A^+. It has been demonstrated that a variety of heteroatom-containing ions such as $C_2H_4O_2^{+\cdot}$ [17], $C_2H_6N^+$ [18,19], $C_3H_7O^+$ [20-22] and $C_4H_9O^+$ [23] are best represented by such a potential energy diagram (see Section V-4.2.2).

The discussion centered on Fig. V-2 clearly demonstrates that in the absence of any mechanism for energy relaxation the isomerization of organic ions is mainly determined by the internal energy and the relative ratio of the thresholds for isomerization and decomposition, and not (or only indirectly) by the thermodynamic stability of the ion.

Furthermore it has been argued above that in the absence of any mechanism for energy relaxation isomerization of an ion A^+ often does not lead to a given structure B^+, but rather to a mixture of interconverting structures $A^+ \rightleftharpoons B^+$. Is this description of the system correct? It is obvious that at the typical pressure of a conventional electron impact mass spectrometer one can exclude collisional deactivation. Moreover, until 1975 it was generally assumed that in large organic ions radiative deactivation can also be excluded as in excited ions radiationless transitions to the electronic ground state were expected to be much faster than radiative transitions. Since then, however, Maier et al. [24-28] have observed several larger organic ions which show fluorescence spectra (see Section II-2.1.1) so that radiative stabilization cannot be excluded a priori. Moreover, the quasi-equilibrium of interconverting structures $A^+ \rightleftharpoons B^+$ can be shifted predominantly towards B^+. For instance, it has been concluded that non-decomposing cycloalkane ions with three- to five-membered rings undergo ring opening to form the isomeric 1-alkene ions rather than a mixture of interconverting structures [29]. It is obvious that for entropic reasons the rate for the opening of a cycloalkane ring is much larger than that for the back reaction (ring formation out of an alkene ion) so that in this case it may be justified to speak of an isomerization of A^+ to B^+.

3 Methods for Ion Structure Elucidation

3.1 Internal Energy and Ion Lifetime

The considerable success achieved in the past in the structure elucidation of neutral organic compounds extracted, for instance, from natural products is to a large extent based on the use of spectroscopic techniques such as IR, UV and NMR spectroscopy. So far the application of spectroscopic methods to the structure elucidation of gaseous organic ions has been precluded by the low concentrations of ions in an ion beam, with the exception of the photodissociation technique, which has been used to obtain structural information on molecular ions (see Section V-3.5.2).

Thus it has been pointed out by Williams and Howe [30] that the mass spectrometrist has to rely on techniques which correspond to those used in early organic chemistry for structure determination, i.e. the degradation of ions and the comparison of their reactivity.

The most important methods employed for ion structure determination are the following:
(1) Decomposition pathways
(2) Heats of formation (ΔH)
(3) Metastable ion characteristics (MI)
(4) Kinetic energy release (T)
(5) Collisional activation spectra (CA)
(6) Ion-molecule reactions studied in an ion cyclotron resonance spectrometer (ICR)
(7) Molecular orbital calculations (MO)

In most of the techniques the unknown ion structure is determined by comparison with a reference ion: If the heats of formation, the metastable ion characteristics, the kinetic energy release values, the collisional activation spectra or the ion-molecule reactions of two ions of identical composition are identical one concludes that their structures are the same and vice versa. However, reference ions[*]) of known structure are not always on hand so that additional information such as characteristic decomposition pathways is used to infer the correct structure.

Before discussing the various techniques in detail, it is important to remember that with these methods ions of different internal energies and ion lifetimes are sampled. If ion structures are inferred from metastable ion characteristics or kinetic energy release data, one is sampling decomposing ions with a narrow range of internal energies just above the threshold for decomposition, whilst collisional activa-

---

[*] Molecular ions are often used as reference ions for ion structure determination of odd-electron fragment ions, fragment ions, formed by simple cleavage reactions or protonated molecular ions as reference for even-electron fragments with the tacit assumption that they retain their original structural features upon ionization, protonation or fragmentation.

tion spectra and ion-molecule reactions studied in an ICR cell refer to non-decomposing ions with internal energies ranging from zero up to the lowest threshold for decomposition*). Finally, heat of formation data and molecular orbital calculations characterize ions in their electronic and vibrational ground state. Table V-1 summarizes for these six methods approximate values for the ion's internal energy, E, and ion lifetime, t.

Table V-1. Internal Energies and Lifetimes of Ions Studied with Various Techniques

Method	Energy^a	Lifetime (s)
ΔH	0	∞
MI	$\sim E_o{}^b$	$10^{-6} - 10^{-5}$
T	$\sim E_o{}^b$	$10^{-6} - 10^{-5}$
CA	$0 - E_o$	$> 10^{-5}$
ICR	$0 - E_o$	$\geq 10^{-3}$
MO	0	∞

^{a} E_o = lowest threshold for decomposition

^{b} The internal energy of metastable ions is not identical with the lowest threshold for decomposition. Rather, metastable ion may possess a range of internal energies up to $\sim$ 1 eV above the decomposition threshold.

It is apparent from Table V-1 that as the various techniques are sampling ions with different internal energies and ion lifetimes, one does not necessarily get identical results with each technique. Whether all methods give the same information or not depends on the relative ratio of the thresholds for isomerization and decomposition as illustrated in Fig. V-3 [32] where again schematic potential energy diagrams are shown in which the energy ranges corresponding to each technique are indicated. (For the sake of simplicity only one decomposition channel is considered). If the isomerization barrier is much higher than the threshold for decomposition (a) all methods will yield the same information. If on the other hand the isomerization barrier is lower than the lowest threshold for decomposition (b), the result obtained with the various techniques may differ. Heat of formation data, colli-

---

* However, an opposite statement regarding the ICR technique has recently been published by Hass et al. [31].

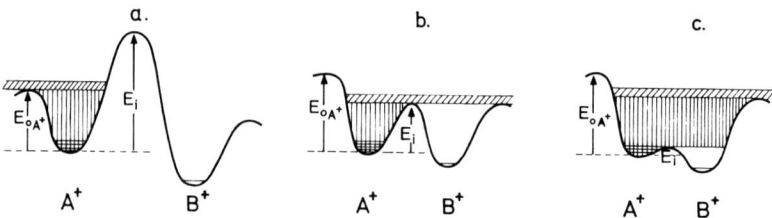

Fig. V-3. Influence of internal energy on structural information obtained by different methods about an ion A^+ that may rearrange to an ion B^+ after overcoming an isomerization threshold E_i (the excitation energy region for decomposing metastable ions is hatched diagonally, in CA and ICR investigations vertically, and in measurements of the heats of formation horizontally).
(a) $E_i \gg E_{oA^+}$; (b) $E_i < E_{oA^+}$; (c) $E_i \ll E_{oA^+}$

sional activation spectra and ion-molecule studies will demonstrate the existence of two distinct structures while metastable ion characteristics, and kinetic energy release data will show that A^+ and B^+ rearrange to a mixture of interconverting structures (isomerization). Finally it might be difficult to detect very small isomerization barriers (as shown in c) using collisional activation spectra and ion-molecule reactions as the number of ions with close to zero energy represents only a small fraction of the total number of non-decomposing ions.

It follows that if those methods which sample ions with sufficient energy to decompose (MI, T) show that two isomeric ions have distinct structures all other methods sampling non-decomposing ions (ΔH, CA, ICR) must yield the same result, whereas the reverse conclusion is not permissible.

It is evident from Table V-1 that most techniques used in ion structure work today sample ions with low to medium internal energies (ranging from zero to roughly the lowest threshold for decomposition) and long ion lifetimes ($t \geq 10^{-6}$ s). Only few attempts have been made to develop techniques for ion source reactions, as both the lifetime and the internal energy of ions generated in the source are poorly defined.

In the following sections, the most important methods used for ion structure elucidation will be discussed and illustrated by a variety of examples. If available, examples have been chosen which have been studied by several methods so that a direct comparison of the results is possible.

3.2 The Mechanism leading to the Formation of an Ion

It has been pointed out earlier (Section V-1) that the elucidation of the mechanism leading to the formation of a fragment ion yields, at most, only information on the initial ion structure. Other problems arise from the fact that the reaction mechanism is not always well established, even if isotopic labelling is used. Isotope effects and isotopic scrambling may make it difficult to assign an unambiguous mechanism. Competing decompositions (i.e. a non-specific fragmentation) may lead to various structures of identical composition. Although this approach is still used today to assign ion structures no reliable results can be expected.

3.3 Thermochemical Properties (Heats of Formation)

3.3.1 The Principle

Ion structures may be inferred from heat of formation data by comparison with reference ions [33] or molecular orbital calculations. Identical heats of formation are taken to imply identical structures and vice versa. Heat of formation values for ions can be calculated from appearance and ionization potentials if the thermochemical data of the neutrals are known (Fig. V-4). The heats of formation of various neutrals, including radicals, have been compiled [34]. For the reaction $AB^{+\cdot} \rightarrow A^+ + B^\cdot$ the heats of formation for the molecular and fragment ion, respectively, are given as

$$\Delta H_f(AB^{+\cdot}) = IP + \Delta H_f(AB) \tag{V-1}$$

$$\Delta H_f(A^+) = AP + \Delta H_f(AB) - \Delta H_f(B^\cdot) - \varepsilon_{excess} \tag{V-2}$$

Moreover, if A can be produced as radical, $\Delta H_f(A^+)$ can be determined with considerably higher accuracy via the ionization potential of this radical. In addition ionic heats of formation are available in some instances from proton affinities (PA), as determined from gas phase ionic equilibria studies in an ICR cell or a high pressure mass spectrometer. For instance, ions containing an OH-group can be formed by proton transfer to aldehydes and ketones for which proton affinities have been reported [35]. For the reaction $AB + H^+ \rightarrow ABH^+$ we have:

$$\Delta H_f(ABH^+) = \Delta H_f(AB) + \Delta H_f(H^+) - PA \tag{V-3}$$

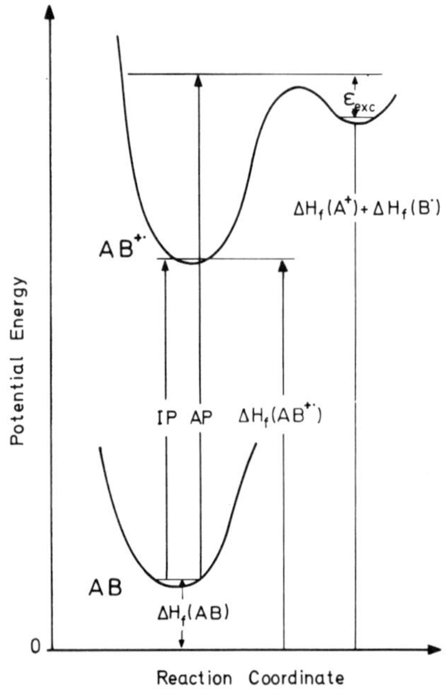

Fig. V-4. Schematic potential energy diagram illustrating the determination of heats of formation of gaseous organic ions.

3.3.2 Possible Sources of Error

Since the early days of organic mass spectrometry heat of formation data have repeatedly been used to assign ion structures. However, in recent years several possible sources of error associated with this technique have been recognized:

(1) The structure assignment is based on a single item of information, the heat of formation. It is quite possible that two isomeric ions of distinct structure coincidentally have identical heats of formation. (Examples will be given below).

(2) If heats of formation of ions formed by a fragmentation process are determined, the structure of the ejected neutral must be known. Although the identity of these neutrals may be determined experimentally (see Section IV-2.9) plausibility arguments are generally used to infer their structures.

(3) If heat of formation data are derived from appearance potential or ionization potential measurements, one has to be aware of the limited accuracy of such measurements especially if non-monoenergetic electrons are used in conjunction with one of the extrapolation methods. This

problem was discussed in Section I-4.2.

(4) The most serious problem, however, arises from the fact that the excess energy in eq. V-2 is usually not known. This excess energy has two sources: the reverse activation energy, ε_o^r, (see Section III-7.3) and the kinetic shift, E_s, (see Section III-4). To obtain reliable heat of formation data one either has to prove that the excess energy is negligible or one has to correct for both contributions which is, at present, no easy undertaking. Although the kinetic energy released upon fragmentation reflects part of the excess energy, the partitioning quotient of the excess energy, i.e. the fraction of the total excess energy funnelled into the translational channel, is not known a priori. The reverse activation energy can only be determined if the heat of formation of the products is known, i.e. the answer is required before the question can be cleared. Similarly it is at present difficult to measure the kinetic shift, although experimental methods which allow the determination of at least a fraction of the kinetic shift are being developed (see Section III-4.2). Furthermore in measuring the appearance potential of a fragment ion it is important to be aware of the competitive shift (Section III-4.3). Even the thermal shift (Section III-4.4) cannot be neglected, especially if a fragment of given elemental composition is formed from a precursor of either small or large molecular weight.

If no information about the excess energy is available, the appearance potential should at least be corrected for that fraction of the excess energy*) which is released as kinetic energy and can be determined from the metastable peak width (Section III-7.2).

3.3.3 Examples

A convincing example demonstrating how unreliable ion structure assignments based on uncorrected heat of formation data may be has been reported by Elder et al. [36]. The authors determined the uncorrected heats of formation of $C_6H_5CO^+$ ions generated from compounds of the general type C_6H_5COX from appearance potentials and observed variations from 712 kJ mol^{-1} (if X = NH_2) to 929 kJ mol^{-1} (if X = CF_3). As it is hardly disputed that $C_6H_5CO^+$ ions have a unique structure, i.e. that of the benzoyl ion, the authors conclude that the differences result mainly from variations of the excess energy. (An earlier, alternative

---

* It should, however, be kept in mind that in addition to the reverse activation energy the excess energy available upon metastable decomposition contains the total non-fixed energy, $\varepsilon^{\neq}$, only part of which results from the kinetic shift.

explanation [37], involving isolated electronic states is less convincing). This example demonstrates that caution should be excercised if heats of formation are used for ion structure determination. There is a variety of examples in the literature where ion structure assignments based on uncorrected heats of formation data proved to be wrong. For instance, a heat of formation reported for the $C_{12}H_{10}O^{+\cdot}$ ion formed by CO_2 loss from ionized diphenyl carbonate was 155 kJ mol^{-1} in excess of that for the isomeric molecular ion of diphenyl ether [38], suggesting that both ions have distinct structures. On the other hand, the same heat of formation was determined for $C_3H_3^+$ ions from cyclopropene, propyne, and allene molecular ions [39] pointing at a common structure (or interconverting structures). It will be shown in the following sections that these early results are clearly wrong: the two $C_{12}H_{10}O^{+\cdot}$ ions have an identical structure, whereas two distinct $C_3H_3^+$ ions exist.

As an example of a more recent attempt to assign ion structures from heat of formation data the study of $C_2H_5O^+$ ions by Solka and Russell [40] and the reinvestigation of this system by Lossing [41] will be discussed. This example has been chosen because on the one hand this study illustrates both the advantages and limitations of the method, and on the other hand the structures of the $C_2H_5O^+$ ion have been studied repeatedly using various techniques, so that a comparison of the results is of interest.

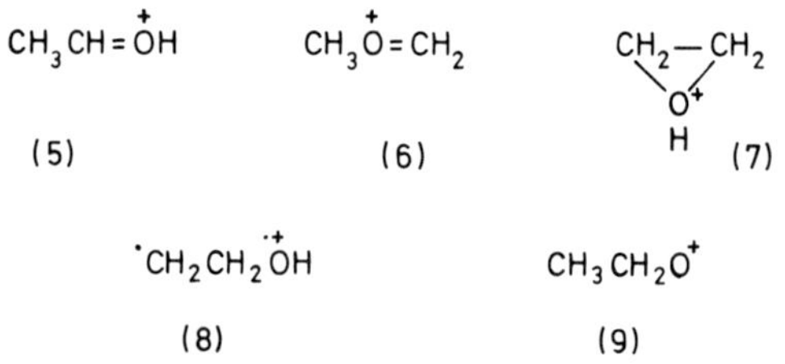

$$CH_3CH=\overset{+}{O}H \qquad CH_3\overset{+}{O}=CH_2 \qquad \underset{H}{\overset{CH_2-CH_2}{\underset{O^+}{\diagdown\diagup}}}$$

(5) (6) (7)

$$^\cdot CH_2CH_2\overset{+}{O}H \qquad CH_3CH_2O^+$$

(8) (9)

Scheme V-2

The appearance potentials for formation of $C_2H_5O^+$ obtained by Solka and Russell are listed in Table V-2 and contrasted, where available, with data obtained by photoionization, monoenergetic electron impact or via proton affinities*).

---

* Lossing reported evidence based on energetic considerations that (6) is better represented by the formula $CH_3O-CH_2^+$ [41].

Table V-2. Energetics of $C_2H_5O^+$ Ion Formation[a]

	Precursor	Neutral Fragment	Appearance Potential (eV)	ΔH (kJ/mol)	Assumed Structure
(I)	CH_3CH_2OH	H	10.75 ± 0.03[b]	586 ± 4	(5)
			10.67[c]	582	
			10.78[d]	582	
(II)	CH_3OCH_3	H	11.23 ± 0.04[b]	682 ± 4	(6)
			10.99[c]	657	
			11.04[f]	664 ± 4	
(III)	$CH_3CH_2CH_2OH$	CH_3	11.35 ± 0.04[b]	690 ± 4	(7)
				712 ± 17[e]	
(IV)	$BrCH_2CH_2OH$	Br	10.47 ± 0.05[b]	695 ± 4	(7)
(V)	CH_3CH_2ONO	NO	10.62 ± 0.07[b]	829 ± 8	(9)

[a] Further reference data are given in Ref. 40, [b] Solka and Russel (RPD-technique) [40], [c] Lossing (monochromatic electrons) [41], [d] Refaey and Chupka (photoionization) [42], [e] Beauchamp and Dunbar (proton affinity of ethylene oxide) [43], [f] Botter et al. (photoionization) [44].

As appearance potentials are obtained with the RPD technique (see Section I-4.2.2) reliable heat of formation values are expected, especially as the structures of the ejected neutrals are well known. The excess energy was assumed to be negligible which may be a reasonable assumption for such direct cleavage processes with low activation energy.

The authors assume that the precursors I, II and V yield the ion structures (5), (6), and (9) which are expected according to the precursor ion structures, if fragmentation occurs by direct cleavage. Further, they conclude that structure (8) which is expected to be formed by direct cleavage from precursors III and IV does not exist as the observed heat of formation (~ 695 kJ mol^{-1}) is too low for the biradicalic ion (8). Rather, they assume that the precursors III and IV undergo a rearrangement reaction to form the pronated ethylene oxide, whose heat of formation is almost identical with the observed values[*]. This reasoning demonstrates that not only the heat of formation data, but also additional information as well as plausibility arguments are

---

[*] This conclusion is consistent with the potential energy surface of $C_2H_5O^+$ ions reported by Bowen and Williams [18].

used to deduce the ion structure. Structures (5) and (6) have been confirmed to be distinct species by various other techniques (see below). Moreover it was recently shown both by an ICR [45] and a CA study [46] that protonated ethylene oxide ions are stable species and, at least partially, are formed in a displacement reaction from the precursor ion IV [46] (see Scheme V-3). However, the existence of (9) as distinct

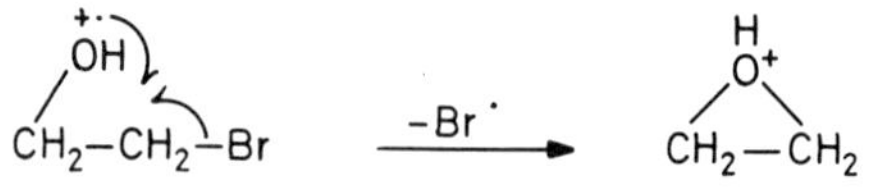

$$\underset{(10)}{\overset{+\cdot}{O}H} \underset{CH_2-CH_2-Br}{} \xrightarrow{-Br\cdot} \underset{(7)}{\overset{H}{\underset{CH_2-CH_2}{O^+}}}$$

Scheme V-3

structure has not been confirmed by any other technique. It was argued that this ion may not have an energy minimum on the $C_2H_5O^+$ energy surface which is separated from that of $CH_3CH=OH^+$, i.e. (9) isomerizes to (5) [41].

Before passing on two observations are of interest:
(1) The heats of formation of ions (6) and (7) are identical within the reproducibility so that one would infer identical structures in the absence of any additional information.
(2) The heat of formation determined for ion (6) using the RPD [40] and electron monochromator technique [41] respectively still differ by a substantial amount (0.24 eV = 25 kJ mol^{-1}) which exceeds the cited reproducibility by far.

Another approach to determine ion structures from heat of formation values may be illustrated using an example reported recently by Benoit et al. [47]. The authors studied $RCOOH_2^+$ ions generated from aliphatic esters, $RCOOR'^{+\cdot} \rightarrow RCOOH_2^+ + (R' - 2H)^{\cdot}$. This ion may have the following structures

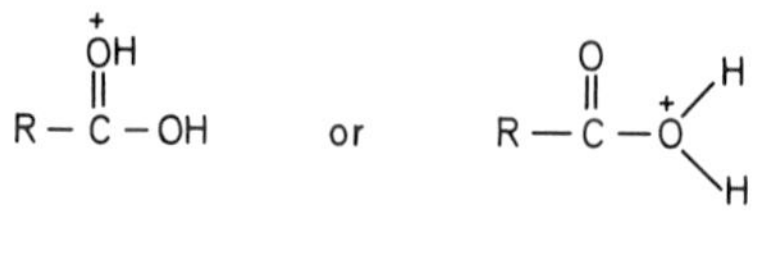

$$R-\overset{\overset{+}{O}H}{\underset{\parallel}{C}}-OH \quad \text{or} \quad R-\overset{O}{\underset{\parallel}{C}}-\overset{+}{O}\overset{H}{\underset{H}{\diagdown}}$$

By comparing the experimental appearance potentials (determined with monoenergetic electrons) with appearance potentials calculated from published thermodynamic data it was possible to demonstrate unequivocally that the $RCOOH_2^+$ ion has the protonated acid structure with the hydrogen bound to the carbonyl and not the hydroxyl group.

For other, more recent ion structure studies based on heat of formation values the reader is referred to the original publications [48-62].

Whilst the above discussion demonstrated that reliable ion structure assignments based on heat of formation values are difficult it should be emphasized that this technique yields important information about the energetics of organic ions in the gas phase. Such data can be used in conjunction with other methods to corroborate proposed ion structures. Moreover, this information can, for instance, be used to predict the occurrence of metastable transitions (see Section IV-1.1), functional group interaction (Section IV-2.5.4) or to construct potential energy surfaces (Section V-4.2.2). High quality heat of formation data for a large variety of hydrocarbon ions obtained from ionization potential measurements of radicals have been reported by Lossing [63-70][*]. Moreover, an extensive compilation of heat of formation data was recently published [34].

3.4 Degradation Reactions

3.4.1 Decomposition Mechanisms

Whilst the elucidation of the mechanism leading to the formation of a fragment ion is of little value for the determination of its structure (see Section V-1 and V-3.2), the determination of the mechanisms for further secondary decompositions may yield more direct information about the ion structures involved. The interdependence between ion structure and reaction mechanism has already been mentioned in Section IV-2.8.

The mechanisms for secondary decomposition of fragment ions may be elucidated using metastable transitions or collision-induced dissociations (see Section III-8.1) in conjunction with isotopic labelling. Collision-induced dissociations give more valuable information on ion structures, as they include the structural significant direct bond cleavages, whilst with metastable transitions rearrangement processes prevail. If molecular ions are studied instead of fragment ions, it is found that ion source reactions (i.e. the normal mass spectrum) characterize their structure.

Ion structure assignments based exclusively on decomposition mech-

---

[*] See Ref. 34, p. 691, for additional references.

anisms are rarely unequivocal. However, when used in conjunction with
other techniques, valuable complementary information on the ion structure may be extracted from the mechanism. Several examples of ion
structure assignments based on this method have been reported in Chapter IV and will also be mentioned in the following sections. An especially valuable approach to the detection of cyclic structures is the
use of symmetry arguments supported by isotopic labelling, provided
that there is no scrambling of the label. This concept has been discussed in detail in Section IV-2.5.5. It should, however, be emphasized
that, strictly speaking, symmetry arguments based on metastable ions
or collision-induced dissociations in conjunction with isotopic labelling allow only the inference of a cyclic transition state, but not of
a final cyclic structure. Possible pitfalls which may arise from structure assignments based on symmetry arguments have been discussed by
Bentley and Johnstone [13].

3.4.2 Comparison of Mass Spectra

The similarity of normal mass spectra has been repeatedly used to infer
structural similarities. Thus from the fact that the mass spectrum of
o-nitrobenzylalcohol is remarkably similar to that of o-nitrosobenzaldehyde (except for the molecular ions) it has been concluded that both
molecular ions decompose via a common intermediate [71] (see Fig. V-5).
Later collisional activation studies demonstrated that the $(M-H_2O)^+$
fragment from o-nitrobenzylalcohol and the molecular ion of o-nitrosobenzaldehyde indeed rearrange to a common structure, most likely the
2,1-benzisoxazoline-3-one [72] (see Scheme V-4).

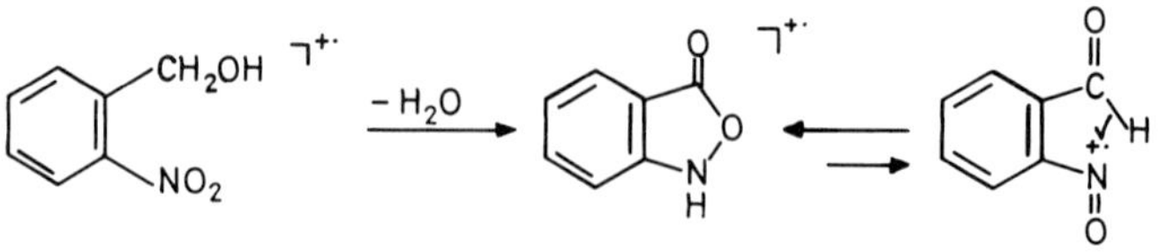

Scheme V-4

As the mass spectrum of o-nitrobenzylalcohol is a superposition of decomposing $(M-H_2O)^+$ and M^+ ions*) no complete correspondence of this
spectrum with that of o-nitrosobenzaldehyde is expected. Today more
accurate methods are available for ion structure assignments so that

---

* Not all molecular ions will decompose via an initial $(M-H_2O)^+$ fragment.

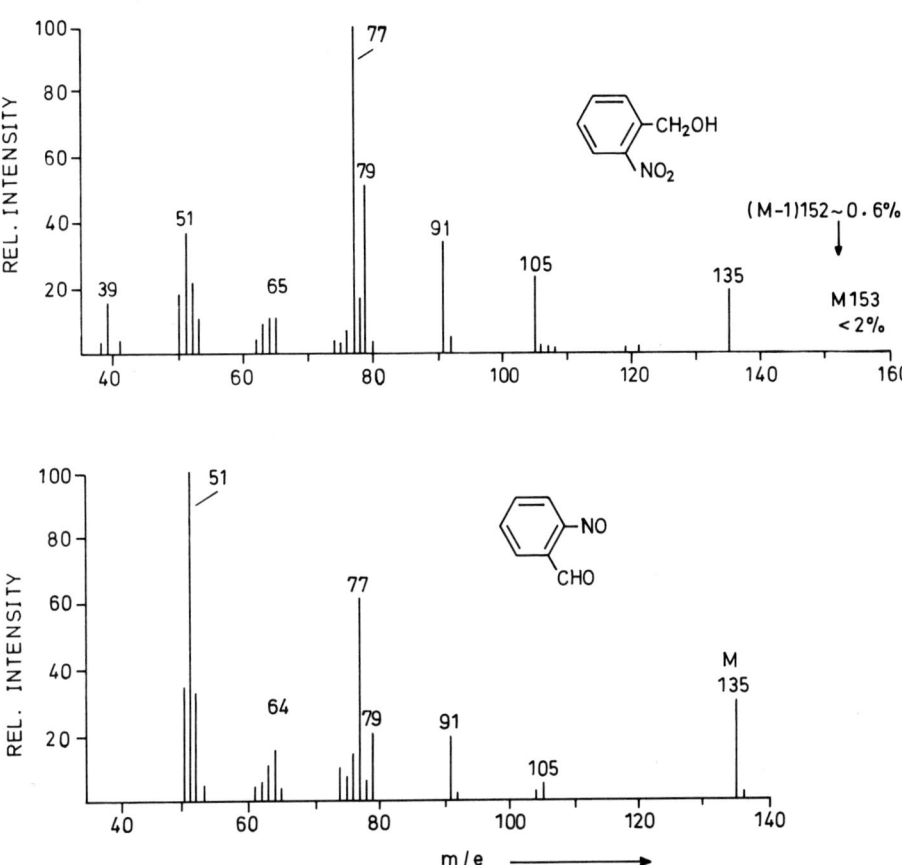

Fig. V-5. Comparison of the 70 eV mass spectra of o-nitrobenzyl alcohol and o-nitroso benzaldehyde [71]. (By courtesy of Heyden & Son Ltd.)

there is no need to rely on the more qualitative comparison of spectra except if information on the molecular ion structure is sought.

In the latter case the normal mass spectrum reflects the molecular ion structure directly. Thus identical electron impact spectra point to the same molecular ion structure. This criterion was used by Momigny et al. [73] to conclude that the benzene and the hexadiene-(1,3)-yne-5 molecular ions decompose via common intermediates. In using this approach it is important to remember that the relative fragment abundances in a 70 eV electron impact spectrum reflect differences in the internal energy distribution by far less than, for instance, metastable ion spectra do (as discussed in the following section). This is due to the fact that in a 70 eV spectrum the abundance of each fragment is the result of an integration over a broad range of energies (in contrast to metastable ions where only a narrow energy window contributes to the decay). This insensitivity of the 70 eV spectra with respect to a variation of the internal energy distribution has already been emphasized whilst discussing the propane spectrum (Section II-2.2.1).

3.4.3 Metastable Ion Characteristics

3.4.3.1 The Principle

If two ions of the same composition have not only the same types of metastable fragments but also identical metastable abundance ratios one concludes that they must have an identical structure and vice versa.

This is a direct consequence of the QET: As shown in Section II-1.7 the intensity of a given fragment is a function of the internal energy distribution and the rate constant. The rate constant is in turn a direct function of the activation energy for further decomposition as well as the vibrational and rotational frequencies and moments of inertia in the reactant and the activated complex. Thus as two isomeric ions differ in general in either their activation energies for further decompositions and (or) the vibrational and rotational frequencies, they should also differ in their rate constants for further decomposition. Hence given that the internal energy distributions for two ions are identical they should yield identical abundance ratios for the metastable transitions if the structures are the same, but different metastable abundance ratios for different structures. Metastable ion characteristics reflect the structure of *decomposing ions* with lifetimes of 10^{-6} to 10^{-5} s and an average internal energy slightly above the lowest threshold for decomposition.

Metastable ion characteristics were first used by Rosenstock et al. [74] in 1964 to demonstrate that hexyl ions, $C_6H_{13}^+$, from various n-hexyl derivatives give identical abundance ratios and thus should

Table V-3. Metastable Abundance Ratio of Decomposing $C_2H_5O^+$ Ions [75]

Precursor Structure	Ion Structure	Relative Abundance (Loss of C_2H_2/CH_4)
CH_3OCH_2Y	$CH_3\overset{+}{O}=CH_2$	< 0.01
$HOCH_2CH_2Y$	$CH_3CH=\overset{+}{O}H$	1.8 $\pm$ 0.2
$CH_3CH(OH)Y$	$CH_3CH=\overset{+}{O}H$	1.9 $\pm$ 0.1
CH_3CH_2OY	$CH_3CH=\overset{+}{O}H$	2.0 $\pm$ 0.1

have the same structure throughout. Two years later Shannon and McLafferty [75] used the same approach to demonstrate that there are two distinct $C_2H_5O^+$ isomers (see Table V-3)*).

3.4.3.2 Critical Evaluation of the Method

As already mentioned, the use of metastable abundance ratios for ion structure assignments is based on the assumption that the internal energy distribution does not affect the relative abundance. This assumption is of course doubtful, as metastable ions originate from a rather narrow energy window (0 - 1 eV above threshold) so that differences in the internal energy caused for instance by any structure in the energy distribution are directly reflected in the relative abundances.

The influence of the internal energy distribution on the metastable abundance ratio has been studied both theoretically (using the classical approximation of the QET) and experimentally by Occolowitz [76], by Yeo and Williams [77] and by Tsang and Harrison [78]. The authors demonstrated that pronounced differences in the internal energy of ions of identical structure may lead to variations in the relative abundance by a factor of two to five. Whilst a variation of a factor of five should be a safe criterion in general, there may be exceptions where the effect is even more pronounced. Table V-4 compares the metas-

Table V-4. Metastable Ion Spectra of $C_2H_6N^+$ Ions Generated from 1-Methylalkylamines [79][a]

m/e	Loss of	1-Methyl-ethylamine	1-Methyl-butylamine	1-Methyl-octylamine
43	H	7	21	60
42	H_2	26	24	19
18	C_2H_2	67	55	21

[a] Abundances relative to the sum of all fragments.

---

*) It was originally assumed that these $C_2H_5O^+$ ions have the structure of protonated acetaldehyde (5) and protonated ethylene oxide (7). Although there is unequivocal evidence that the cyclic ion (7) exists as a stable species [45,46] it is more likely that the decomposing $C_2H_5O^+$ ions originally studied by Shannon and McLafferty possess the structures (5) and (6), i.e. that (7) rearranges to (5) prior to further metastable decomposition. This conclusion is supported by a recent study of the potential energy surface of $C_2H_5O^+$ ions reported by Bowen and Williams [18].

table ion spectra of three $C_2H_6N^+$ ions generated from 1-methylalkyl amines [79]. Although the CA spectra unequivocally demonstrate that these $C_2H_6N^+$ ions have identical structures [79] the metastable abundance ratio (e.g. m/e 43/18) varies by a factor of up to 30.

Summarizing one may confidently conclude that two ions with sufficient energy to overcome the decomposition threshold have the same structure if the metastable ion spectra are identical within the reproducibility. The opposite conclusion is less safe: Pronounced differences in the abundance ratio (e.g. by a factor of five) point to different structures. Small changes in the relative abundances should, however, be interpreted with caution. The differentiation between two distinct ions can be substantiated if both ions are generated from a large variety of precursors and each of the two sets of metastable ion spectra shows identical ratios. Finally, an unequivocal distinction between two different structures is possible if the metastable spectra of two isomeric ions differ not only in the relative abundances but also in the types of metastable fragments.

Apart from the poorly-known influence of the internal energy on the abundance ratio there are two further disadvantages to the use of metastable ion characteristics.
(1) The structure assignment is based on very few peaks. In some instances there is only a single metastable ion of measureable intensity so that no abundance ratio can be formed.
(2) The observed metastable fragments often originate from rearrangement reactions which give less specific information on the ion structure than direct bond cleavages.

On the other hand several advantages of the method should be mentioned.
(1) The method is simple. No special instrumentation is necessary, although the use of instruments with reversed geometry (magnetic sector preceding the electric sector) or a linked scan of the magnetic and the electrostatic field is of advantage.
(2) Both the internal energy ($\sim 0 - 1$ eV above the lowest threshold for decomposition) and ion lifetime ($10^{-6} - 10^{-5}$ s) are well defined.
(3) Isotopic labelling (i.e. the elucidation of the decomposition mechanisms) can be used to substantiate any structure assignment based on metastable ion characteristics.

3.4.3.3 Examples

Various investigations have been reported in which metastable ion characteristics have been used to deduce ion structures. The more important studies are included in Table V-9 (Section V-4.3). In the light of the ambiguities discussed above, which result from the dependence of the abundance ratio on the internal energy, not all structure assignments are conclusive. Thus the criterion of metastable abundances has been used to obtain information about the structure of the $C_{12}H_{10}O^{+\cdot}$ ion formed on the one hand by loss of CO_2 from the diphenyl carbonate molecular ion and on the other hand by direct ionization of diphenyl ether [80]. The ions differ in their metastable abundance ratios. As pointed out by the authors these variations could result from differences in the internal energy or from different structures so that no unambiguous conclusion is possible.

Thus for a reliable ion structure assignment it is important to use isotopic labelling in addition to metastable abundance ratios. To illustrate this approach which has been used repeatedly by Williams and co-workers two examples have been selected.

$C_3H_8N^+$ *Ions.* The metastable ion spectra of $C_3H_8N^+$ ions have been studied by Uccella et al. [81] and were reinvestigated by Levsen and McLafferty [79]. $C_3H_8N^+$ ions, which according to the mechanism of formation have the five initial structures (11) to (15) yield four distinct metastable

$$CH_2 = \overset{+}{N}HCH_2CH_3 \qquad (CH_3)_2C = \overset{+}{N}H_2 \qquad CH_3CH_2CH = \overset{+}{N}H_2$$

(11) (12) (13)

$$CH_3CH = \overset{+}{N}HCH_3 \qquad (CH_3)_2\overset{+}{N} = CH_2$$

(14) (15)

Scheme V-5

ion spectra [79,81] demonstrating that the $C_3H_8N^+$ ions in (11) to (13) with sufficient energy to decompose and a lifetime of 10^{-6} s have distinct structures while the ions (14) and (15) rearrange to a common structure or interconverting structures prior to further decomposition. In support of the postulated interconverting structures the initial ions $(CH_3)_2N=CH_2^+$ (15) when specifically labelled with deuterium or ^{13}C, show complete randomization of all deuterium and carbon atoms prior to ethylene loss, whilst, for instance, ethylene loss from $CH_2=NHC_2H_5^+$ (11) occurs via a specific mechanism (see Scheme V-6) without any atom randomization, which is consistent with the proposed structure [81]. It will be shown later that in contrast to the decomposing $C_3H_8N^+$ ions the non-decomposing ions (14) and (15) are stable species [79].

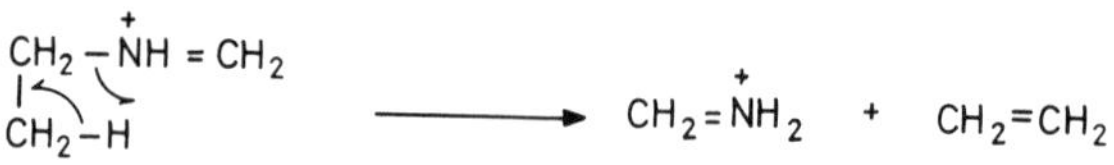

$$CH_2-\overset{+}{N}H=CH_2 \atop CH_2-H \quad \longrightarrow \quad CH_2=\overset{+}{N}H_2 \;+\; CH_2=CH_2$$

(11)

Scheme V-6

$C_4H_9^+$ *Ions*. Davis et al. [2] demonstrated that the metastable abundance ratios of fragmenting $C_4H_9^+$ ions initially formed as n-, sec-, tert- and iso-butyl ions are similar (although not identical) suggesting that butyl ions with sufficient internal energy to overcome the decomposition threshold and an ion lifetime of 10^{-6} s have largely isomerized to a mixture of interconverting structures. The remaining differences observed for the metastable abundance ratio (loss of methane to loss of ethylene) could be interpreted as resulting from incomplete interconversion (different stable structures) or from differences in the internal energy distribution, the latter being more likely. In support of the postulated interconverting butyl ions complete randomization of all carbons and hydrogens in n- and iso-butyl ions prior to fragmentation was observed by several authors [2,3,5,82].

3.4.3.4 Rate-determining Isomerization prior to Decomposition

Recently Hvistendahl and Williams [20] concluded that metastable abundance ratios cannot be used as criterion for determining ion structures in those cases where a rate-determining isomerization is involved. In the meantime several examples for such a slow isomerization have been reported [18-23]. Consider the two isomeric oxonium ions (16) and (17)

$$CH_3CH_2\overset{+}{O}=CH_2 \qquad CH_3CH=\overset{+}{O}-CH_3$$

(16)   (17)

Scheme V-7

The metastable ion spectra of these two precursors are summarized in Table V-5. Based on the pronounced differences in metastable abundance ratios Tsang and Harrison [78] concluded earlier that both ions "do not fragment through an intermediate which is common to both". Moreover, the authors demonstrated by QET calculations that even large differences in the internal energy of both ions will not affect the abundance

Table V-5. Metastable Ion Spectra of Isomeric $C_3H_7O^+$ Ions [20][a]

m/e	Loss of	$CH_3CH_2-\overset{+}{O}=CH_2$ (16)	$CH_3CH=\overset{+}{O}-CH_3$ (17)
41	H_2O	70	1.4
31	C_2H_4	30	77
29	CH_2O	0	22

[a] Normalized to the sum of all fragments = 100

ratio by more than a factor of two.

Hvistendahl and Williams, however, deduced from collision-induced decompositions, deuterium labelling and appearance potential measurements, that the ion (17) undergoes a slow (rate-determining) isomerization to (16) prior to further (fast) decomposition as illustrated by

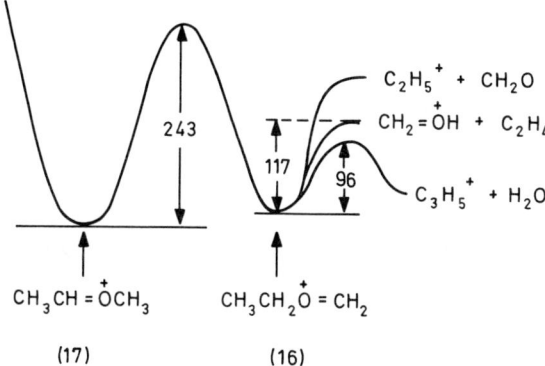

Fig. V-6. Potential surface for the isomerization of (17) to (16) and unimolecular decomposition of (16). Activation energies are given in kJ mol^{-1} [20]. (By courtesy of the American Chemical Society.)

the potential energy diagram in Fig. V-6*). As the most convincing piece of evidence for the proposed rearrangement of (17) to (16) prior to decomposition the authors observed identical activation energies for loss of H_2O, C_2H_4 and CH_2O, which obviously do not correspond to the individual thresholds for decomposition, but to an isomerization bar-

---

* This diagram corresponds to the potential energy curve in Fig. V-2d, discussed in Section V-2.2.

234   V   *The Structure of Gaseous Ions*

rier of 243 kJ mol^{-1} between (17) and (16). In $C_3H_7O^+$ ions (16) with sufficient energy to decompose only the loss of H_2O and C_2H_4 gives rise to a metastable peak as both processes have comparable activation energies and can thus compete effectively with each other. On the other hand after overcoming the isomerization barrier $C_3H_7O^+$ ions (17) have a high excess energy and thus do not only lose H_2O and C_2H_4, but also CH_2O in a metastable decomposition. Hence although both ions decompose via the common structure (16) they show pronounced differences in their metastable abundance ratio.

However, this result does not invalidate the use of metastable abundance ratios for ion structure assignments. Apart from the fact that collisional activation spectra demonstrate that the non-decomposing $C_3H_7O^+$ ions (16) and (17) are stable species, as is also apparent from the potential well in Fig. V-6, $C_3H_7O^+$ ions (16) and (17) with sufficient energy to decompose and an ion lifetime of $\geq 10^{-6}$ s also have distinct structures prior to decomposition: once the ion (17) has overcome the isomerization barrier it will decompose, as for both energetic and entropic reasons the rate for direct decomposition is much faster than for the back reaction (16) → (17), i.e. there will be no quasiequilibrium (17) ⇌ (16) prior to decomposition.

To further substantiate this point let us consider $C_2H_4O_2^{+\cdot}$ ions formed by a McLafferty rearrangement or by direct ionization of acetic acid [17]. According to the mechanism of formation the former should have the enolic structure $CH_2=C(OH)_2^{+\cdot}$ (18), the latter that of the acetic acid molecular ion (20). The pronounced differences observed between the metastable abundance ratios for (18) and (20) support the conclusion that both $C_2H_4O_2^{+\cdot}$ ions with sufficient energy to decompose have distinct structures. However, deuterium labelling revealed that H_2O loss from (18) occurs from the original enolic structure, but $OH^\cdot$

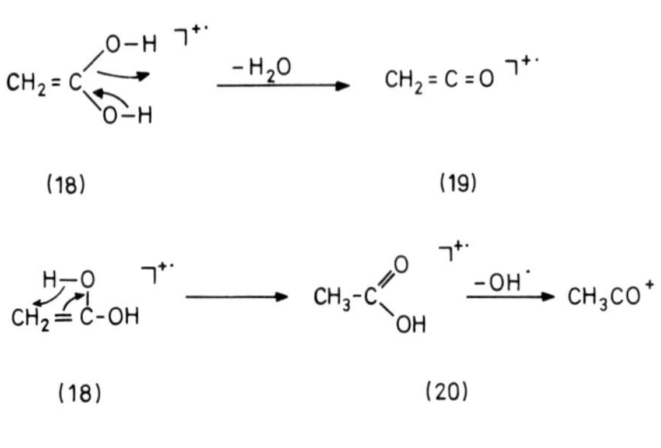

$$CH_2=C\begin{matrix}O-H\\O-H\end{matrix}\quad\rceil^{+\cdot} \xrightarrow{-H_2O} CH_2=C=O\quad\rceil^{+\cdot}$$

(18)   (19)

$$CH_2=\overset{H-O}{\underset{}{C}}-OH\quad\rceil^{+\cdot} \longrightarrow CH_3-C\begin{matrix}O\\OH\end{matrix}\quad\rceil^{+\cdot} \xrightarrow{-OH^\cdot} CH_3CO^+$$

(18)   (20)

Scheme V-8

loss only after rearrangement to acetic acid (see Scheme V-8). In contrast to the $CH_3CH=OCH_3^+$ ion (17) where all metastable decompositions occur after rearrangement to (16), $CH_2=C(OH)_2^{+\cdot}$ (18) decomposes either from the original structure or after rearrangement to the acetic acid molecular ion. This example shows even more clearly that the decomposing $C_2H_4O_2^{+\cdot}$ ions (18) and (20) have distinct structures prior to fragmentation, although one decomposition channel is common to both ions.

Both examples demonstrate that metastable abundance ratios should be interpreted with caution. In general, they can be used to determine whether isomeric ions with sufficient energy to decompose interconvert or not[*]. However, in contrast to previous conclusions they do not give any direct information whether decomposition occurs via common or distinct intermediates. It has become evident from the above discussion that apart from their value in ion structure work, metastable abundance ratios are an important piece of information in deducing potential energy surfaces of gaseous organic ions (see Section V-4.2.2).

3.4.3.5 Time-dependent Metastable Ion Characteristics

The field ionization technique (see Section II-2.1.3.3 and IV-2.6) allows the metastable abundance ratio[**] of isomeric ions to be studied as a function of the ion lifetime from 10^{-11} to 10^{-5} s after ionization and thus offers a valuable extension to the method of competing metastable transitions. Such time-dependent metastable ion characteristics reflect structural changes with time, i.e. isomerization reactions, provided the internal energy difference between the isomeric ions under study is small or negligible.

The experimental procedure makes use of the fact that ions which suffer delayed decomposition within the acceleration field between the emitter and counterelectrode of a field ionization source have a deficit of translational energy which is the larger the later the decomposition occurs. This energy deficit can be compensated by increasing the high voltage at the emitter stepwise at constant electric sector voltage, which allows ions with successively longer ion lifetime to be sampled. Thus scanning the magnetic field at each high voltage step over the mass range of interest yields a spectrum of metastable ions

---

[*] Note, however, that Bowen et al. reported similar abundance ratios for metastable decompositions from isomeric $C_2H_6N^+$ ions although there is only partial interconversion between the isomers [18].

[**] Here metastable ions are defined as those ions which as a result of their delayed decomposition do not contribute to the normal field ionization spectrum.

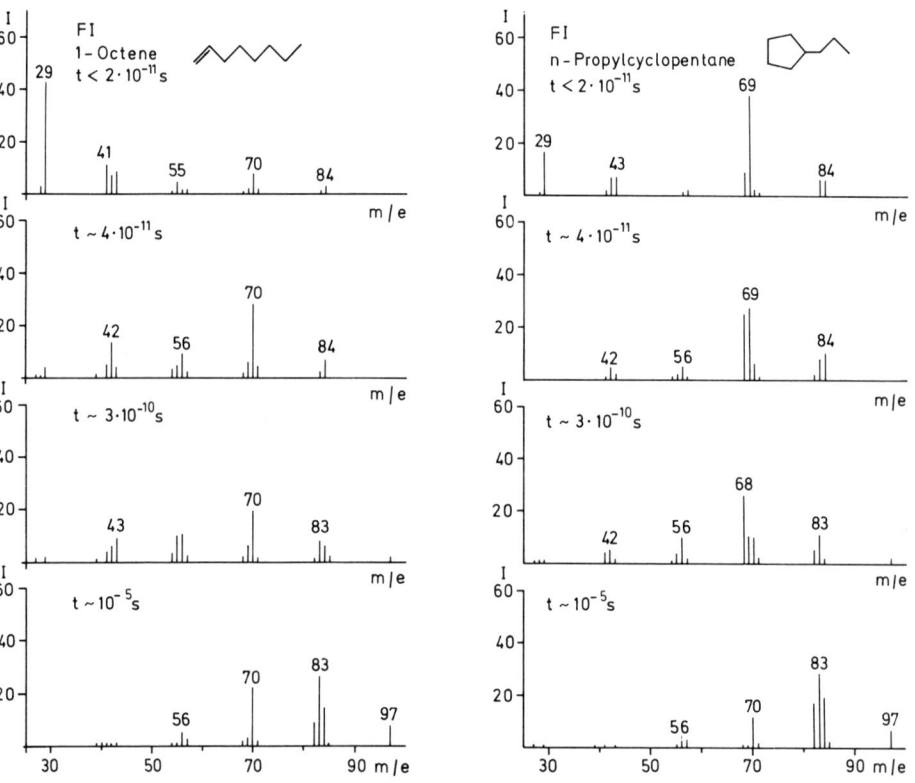

Fig. V-7. Time-dependent metastable ion spectra of the molecular ions of 1-octene and n-propylcyclopentane (t < 2 x 10^{-11} s, ∼ 4 x 10^{-11} s, ∼ 3 x 10^{-10} s and ∼ 10^{-5} s) [29].

at a given average decomposition time[*]. For reasons of principle the method is restricted to molecular ions. Fig. V-7 compares the metastable ion spectra of the isomeric 1-octene and n-propylcyclopentane molecular ions after a precursor ion lifetime of < 2 x 10^{-11} s, ∼ 4 x 10^{-11} s, ∼ 3 x 10^{-10} s and ∼ 10^{-5} s [29]. There is additional experimental evidence that the internal energy distribution is rather similar for both ions so that differences in ion abundances reflect structural differences. It is apparent that the abundance ratios show pronounced differences at the shortest times, indicating that both molecular ions re-

---

[*] The decomposition time depends also on the fragment mass. Thus the time calculated for a given metastable ion spectrum is not exactly constant but increases slightly with increasing mass. This time variation within one spectrum is, however, the same for all isomeric ions.

tain their original structure prior to décomposition. In support of this conclusion n-propylcyclopentane shows an abundant fragment at m/e 69 which corresponds to the loss of the side chain, consistent with the conclusion that the five-membered ring is intact prior to decomposition. With increasing ion lifetime the metastable ion spectra become more similar and are identical at $\sim 10^{-5}$ s after ionization, demonstrating that both ions have isomerized to a common structure. In addition, collisional activation spectra suggest that prior to decomposition n-propylcyclopentane molecular ions undergo a ring opening to form the 1-octene molecular ion. Furthermore, field ionization kinetic measurements demonstrate that the isomerization process is complete at $\sim 10^{-9}$ s after ionization. Additional experimental data show that alkyl cycloalkane molecular ions with three-, four- and five-membered rings undergo ring opening to form the 1-alkene ion initially, while those with six-, seven- and eight-membered rings retain the intact ring prior to decomposition. This result is in agreement with earlier electron impact and field ionization studies of methylcyclopentane and methylcyclohexane reported by Stevenson [83], Meyerson et al. [84], and Falick and Burlingame [85]. Time dependent metastable ion characteristics have also been reported for linear octene ions [86] (see Section V-4.1.3) and have been shown to give valuable insight into isomerization reactions.

3.4.4 Kinetic Energy Release

3.4.4.1 The Principle

Not only the relative abundance of metastable peaks, but also their peak width, i.e. the kinetic energy released upon decomposition can be used for ion structure determination. This technique is again a comparative method: Identical kinetic energy release data point to identical structures and vice versa. The justification of this approach will be discussed using Fig. III-21, which shows a hypothetical potential energy diagram for the reaction $AB^{+\cdot} \to A^+ + B^{\cdot}$. As illustrated in this figure the kinetic energy release has two sources: the activation energy for the reverse reaction, ε_0^r, and the non-fixed energy in the transition state, $\varepsilon^{\neq}$, as discussed in detail in Section III-7 [87]. If two ions have identical structure the potential energy surfaces and thus the reverse activation energies will be the same. Moreover the rate constant for the reaction, $k(E)$, and thus the kinetic shifts will be identical. However, the kinetic shift represents only a fraction of the non-fixed energy in the activated complex. The remainder need not necessarily be the same for two identical ions of different internal energy distributions. Thus a dependence of the kinetic energy release might be expected in those cases where the reverse activation energy is small or negligible. Indeed a strong dependence of the average kinetic energy release on the internal energy has been observed by Baer et al. [88] using the PEPICO technique (see Section III-7). As far as metastable ions are concerned the kinetic energy release has been studied by Jones et al. [89] as function of the degree of freedom (which leads to a variation of the energy distribution in the precursor ion) and as function of the electron energy. No effect on the metastable peak width was observed although the authors state that the kinetic energy release cannot be measured very precisely at low electron energies. In view of the importance of this question with respect to ion structure determinations a reinvestigation of the influence of the internal energy on the kinetic energy release is desirable. If there is any effect at all, it will certainly be small.

The kinetic energy release method again samples *decomposing ions* with a range of internal energies of 0-1 eV and a lifetime of $10^{-6} - 10^{-5}$ s.

It should be kept in mind that the method gives only information about the ion structure prior to decomposition, but does not allow a safe conclusion whether the decomposition occurs via a common intermediate or not. If one of the isomeric ions under study decomposes via a slow, rate-determining isomerization to a second isomer substantial differences in the kinetic energy release values for metastable decompositions from both isomers are expected and observed, as shown by Hvistendahl and Williams for the case of $C_3H_7O^+$ ions [20].

3.4.4.2 Critical Evaluation of the Method

There is little doubt that if substantial differences in the kinetic energy released upon metastable decomposition of isomeric ions (e.g. a factor of two) are observed this is a safe criterion for distinct structures. Similarly the same kinetic energy release values point to identical structures. Unfortunately this assignment is often based on the analysis of a single metastable peak. Thus, although unlikely, it cannot be ruled out completely that two distinct ions decomposing via a common channel give by accident identical kinetic energy release values: On the other hand it is conceivable that two distinct, non-interconverting structures show completely different metastable decomposition channels so that the kinetic energy release method cannot be applied at all.

Although the existing experimental evidence suggests that the influence of the internal energy on the kinetic energy release is far less pronounced than on metastable abundance ratios [89], it is not at present clear how small differences in the kinetic energy should be interpreted. For instance Jones et al. [89] report that the kinetic energy release in acetylene loss from $C_6H_5^+$ ions generated from various sources differs by a factor of 1.6, and that for acetylene loss from a variety of $C_6H_6^{+\cdot}$ ions by a factor of almost two, while in both cases identical collisional activation spectra were observed [16]*).

Of special advantage is the fact that composite metastable peaks may reveal the presence of ion mixtures. One should, however, keep in mind that composite metastable peaks observed upon decomposition of a fragment ion may either result from the presence of two non-interconverting structures formed by two competing decomposition channels from the precursor, or from subsequent decomposition of an ion with a unique structure via two distinct transition states.

Finally, it is obvious that double focussing mass spectrometers with good energy resolution and electronic stability are necessary to obtain reproducible high quality kinetic energy release data.

In spite of some limitations of the method outlined above there is little doubt that the use of kinetic energy release values constitutes one of the most reliable and powerful techniques for ion structure elucidation.

As information on ion structures obtained through the elucidation of decomposition mechanisms by isotopic labelling, from metastable abundance ratios and kinetic energy release values are all based on metastable ions, i.e. decomposing ions with a lifetime of $10^{-6} - 10^{-5}$ s, it is strongly recommended to use all three methods together to obtain supplemental information on a given ion structure.

3.4.4.3 Examples

In the early studies of $C_2H_5O^+$ ions by Shannon and McLafferty [75] and $C_3H_8N^+$ ions by Uccella et al. [81] (mentioned in Section V-3.4.3) kinetic energy release data were already used in addition to metastable abundance ratios to infer ion structures. The first extensive investigation of kinetic energy release data as tool for ion structure elucidation has been reported by Jones et al. [89]. The authors demonstrated that the influence of various experimental parameters such as source

---

* As CA spectra sample non-decomposing ions, even more pronounced differences should be observed if distinct structures were present.

temperature*⁾, ion source residence time, acceleration voltage and electron energy on the kinetic energy release is usually small or even negligible. Furthermore they report a variety of examples illustrating the usefulness of the method for ion structure elucidation. Thus $C_{12}H_{10}O^{+\cdot}$ ions formed by CO_2 loss from diphenyl carbonate as well as by direct ionization of diphenyl ether show within the reproducibility the same kinetic energy release of 0.435 - 0.438 eV during CO loss consistent with an identical structure for both ions, whilst heat of formation data [38] and metastable abundance ratios [80] erroneously suggested the presence of two distinct structures. On the other hand the kinetic energy release values for methyl loss from $C_8H_8O^+$ ions formed by a McLafferty rearrangement from the butyrophenone molecular ion (21) or by direct ionization of acetophenone differ considerably (see Scheme V-9) demonstrating that the former ion has the enolic structure (22), the latter the keto structure (23).

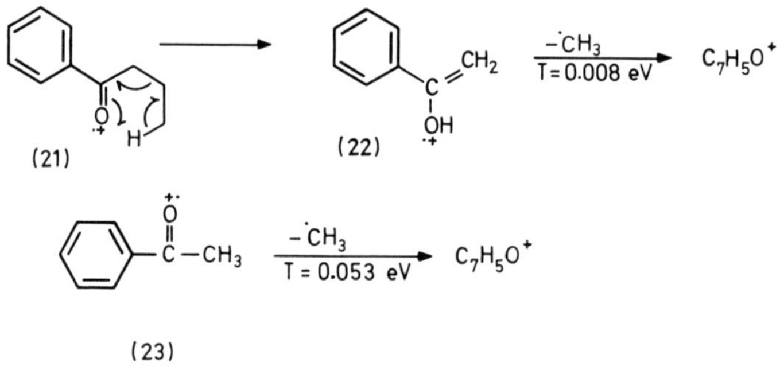

Scheme V-9

In addition to Beynon, Cooks and coworkers [11,89,92-94] Holmes et al. [95-101] have repeatedly used kinetic energy release data. $C_2H_4O^{+\cdot}$ ions studied by Holmes and Terlouw will be discussed in more detail [97] as these results can be compared with a collisional activation study by Van de Sande and McLafferty [102]. Fig. V-8 compares the metastable peak shapes for H· loss from $C_2H_4O^{+\cdot}$ ions generated by direct ionization of ethylene oxide (VI) and acetaldehyde (VII) or by fragmentation from ethyl vinyl ether (VIII), dimethyl ether (IX) and isopropanol (X). It is apparent that the metastable peaks from VI, VII and VIII differ

---

* A pronounced temperature dependence has so far only been observed for H loss from n-propanol [90] and methane [91].

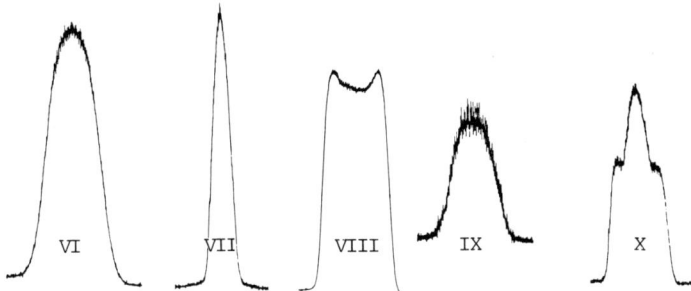

Fig. V-8. Metastable peak shapes for the reaction $C_2H_4O^{+\cdot} \rightarrow C_2H_3O^+ + H^{\cdot}$. Precursor: ethylene oxide (VI), acetaldehyde (VII), ethyl vinyl ether (VIII), dimethylether (IX) and isopropanol (X) [97]. (By courtesy of National Research Council of Canada.)

considerably from each other both in their shape and half width (VI = 0.430 eV, VII = 0.043 eV and VIII = 0.580 eV) suggesting that these decomposing ions have distinct non-interconverting structures (24), (25) and (26) (see Scheme V-10). $C_2H_4O^{+\cdot}$ ions from IX show a peak of similar shape but smaller half width (0.320 eV) than that from VI that may result from ion (27), but may also be due to a composite peak with different proportions of several isomeric ions.

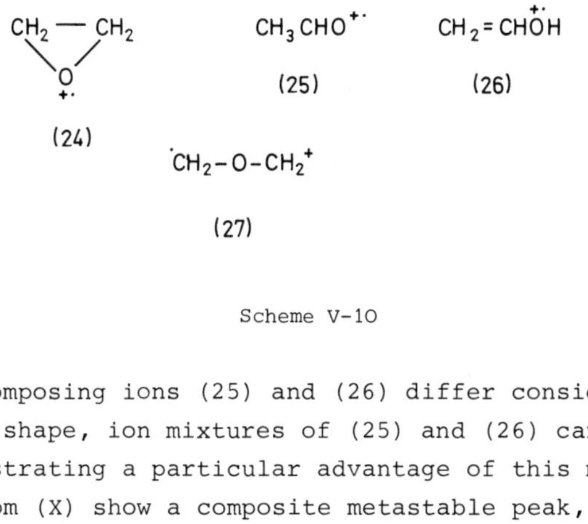

$$
\begin{array}{cccc}
\underset{(24)}{\overset{CH_2 - CH_2}{\underset{O}{\diagdown \quad \diagup}}^{+\cdot}} & \underset{(25)}{CH_3CHO^{+\cdot}} & \underset{(26)}{CH_2 = C\overset{+\cdot}{H}OH} \\
& \underset{(27)}{\overset{\cdot}{C}H_2-O-CH_2^+} &
\end{array}
$$

Scheme V-10

As the decomposing ions (25) and (26) differ considerably in their metastable peak shape, ion mixtures of (25) and (26) can be easily recognized, demonstrating a particular advantage of this method. Thus $C_2H_4O^{+\cdot}$ ions from (X) show a composite metastable peak, revealing the simultaneous presence of both ions (25) and (26) in a ratio of about 1:6.

It should be mentioned that consideration of metastable ion characteristics would probably have led to the same conclusion, as the authors reported that $C_2H_4O^{+\cdot}$ ions (25) and (26) show a metastable loss of HCO$^{\cdot}$ which is absent in (24), whilst loss of CH$_3^{\cdot}$ was strong in (24), weak in (25) and barely detectable in (26). The above results agree well with a collisional activation study in which the non-decomposing ions (24), (25) and (26) were identified as distinct, non-interconverting species [102]. In this context it should be mentioned again that if the kinetic energy release method demonstrates that two decomposing ions are distinct, the collisional activation method (characterizing non-decomposing ions) must give the same result. The reverse conclusion does not necessarily hold.

Further examples where kinetic energy release data were used for ion structure assignment are included in Table V-9 (Section V-4.3).

3.4.4.4 Kinetic Energy Release and Molecular Orbital Symmetry Considerations

The concept of molecular orbital symmetry conservation has been applied to mass spectrometric elimination reactions by Williams and Hvistendahl [103-104] (see Section IV-2.3.4). The authors conclude that concerted symmetry-forbidden 1,2-eliminations lead to repulsive molecular orbitals and thus to a large release of kinetic energy. The mechanistic consequences of this conclusion have been discussed in Section IV-2.3.4. However, the approach can also be applied to ion structures. As H_2 loss from $C_6H_7^+$ ions is associated with a small kinetic energy release, Williams and Hvistendahl [104] concluded that this reaction occurs via a symmetry-allowed 1,1-elimination from a protonated benzene structure, which has been confirmed by studying the kinetic energy release associated with H_2 loss from $C_6H_7^+$ ions generated by protonation of benzene in a chemical ionization source. On the other hand, H_2 loss from $C_7H_9^+$ is associated with a large release of kinetic energy. Thus it was concluded that this reaction does not correspond to a 1,1-elimination of protonated toluene, but rather to a symmetry-forbidden 1,2- or 1,3-elimination from a dihydrotropylium ion. The same conclusion was reached earlier by Cooks et al. [94]. This rationalization was later used by Brady et al. [105] to deduce the ion structures of $C_6H_7^+$, $C_7H_9^+$, $C_8H_{11}^+$ and $C_9H_{13}^+$ generated from alloocimene and α-pyronene by ring closure and ring contraction*).

---

*) Strictly speaking the kinetic energy values give only information on the constitution of the transition state prior to H_2 loss [106].

3.4.5 Collisional Activation

3.4.5.1 The Principle

Fundamental studies related to collision-induced fragmentations in a mass spectrometer have been discussed in Section III-8.1. The potential of collision-induced fragments for ion structure work was first recognized by McLafferty [107,108] and various ion structure studies based on this approach have been reported by his group (see Table V-9, Section V-4.3).

The method, which has been recently reviewed [32], closely resembles that of metastable ion characteristics (Section V-3.4.3): The relative abundance ratio of all collision-induced fragments of a given ion (i.e. the collisional activation (CA) spectrum) reflects its structure. Identical CA spectra point to identical structures and vice versa. The justification of this approach is the same as that used for metastable ion abundances. In addition, the kinetic energy released upon collision-induced decomposition can also be used as structural criterion.

The processes of lowest activation energy, recognizable by the presence of intense unimolecular dissociations in the metastable ion spectra, are usually excluded from the normalization [108] as they may depend on the internal energy distribution. This energy dependence of processes with lowest activation energy is not fully understood, but may simply result from the fact that these peaks include contributions from unimolecular dissociations which are energy dependent. Quantitative corrections for these contributions are difficult. Moreover processes with low activation energies may originate from a rather narrow range of internal energies (see breakdown graphs) and thus may be affected more strongly by internal energy variations whilst the abundance of processes with higher activation energy usually results from an integration over a wide range of internal energies which reduces the influence of smaller differences in the internal energy distribution. Thus any structure assignments based on collision-induced peaks with unimolecular contributions, where such are at all possible, must be made with caution.

If after the exclusion of the processes with lowest activation energy the intensity ratio is found to depend on the electron energy, then it can be concluded that a **mixture** of non-interconverting structures is present, these being formed simultaneously from the precursor ion by competing mechanisms with different activation energies. Because of the different activation energies the composition of the mixture changes as a function of the excitation energy. On the other hand, if the intensity ratio changes little or not at all with the electron energy it cannot be definitely concluded that only a single structure (or interconverting structures) is involved. The absence of any energy dependence may be due to the presence of a mixture of non-interconverting structures which happen to have similar energetic and kinetic parameters.

CA spectra reflect the structure of *non-decomposing ions* with a range of internal energies from zero up to the lowest threshold for decomposition. At first sight this statement appears to be surprising: although it is true that the large majority of ions entering the collision cell have insufficient energy to decompose these ions are highly excited after collision. They do not only decompose through various channels, but may also undergo isomerization reactions which are not accessible to the non-decomposing ions. However, as with such highly excited ions the direct decomposition from the original structure is at least as fast or even faster than isomerization reactions which are theoretically possible on energetic grounds, the CA spectra indeed largely reflect the stable ion structure, i.e. even if the threshold for isomerization of an ion A^+ to B^+ is below that for decomposition (see Fig. V-3b) the CA spectra of A^+ and B^+ will be distinct.

3.4.5.2 The Technique

Since the region in which collision-induced fragments are formed in the mass spectrometer is the same as for metastable ions (one of the field free regions of the instrument) they can be detected by the same methods. Collision-induced fragments can be observed with almost any type of mass spectrometer. However, double focussing instruments either with "reversed Nier-Johnson" geometry or operating with a linked scan of the electric and magnetic sector field are particularly suited for this purpose. If an instrument of reversed geometry is used the ion under study is first selected by means of the magnetic field and passes with high translational energy (3 - 10 keV) into the collision chamber, into which a collision gas (e.g. helium) is admitted at a pressure of $10^{-4} - 10^{-5}$ torr. The collision induced fragments are subjected to energy and thus mass analysis by scanning the electric sector potential. The CA spectrum obtained in such a manner is a superposition of collision-induced and unimolecular dissociations.

3.4.5.3 Critical Evaluation of the Method

The advantages and disadvantages of the method can be summarized as follows:

Advantages

(1) Collisional activation leads to a larger variety of fragments, i.e. the structure assignment is not based on a single or few peaks.

(2) The collision-induced fragments are usually more intense than metastable ions, which leads to an enhanced reproducibility and allows even weak precursors to be studied.

(3) Both the relative abundance and the kinetic energy release can be used as structural criterion.

(4) If processes with unimolecular contributions are excluded (vide supra) the collisional activation spectrum is independent of the internal energy distribution. Thus even small variations in the relative abundance (if clearly beyond the limit of reproducibility*[)] are structurally significant.

(5) Ion mixtures can be detected by varying the electron energy (vide supra). If the CA spectra of the pure reference ions are known the relative composition of the mixture can be determined at least semiquantitatively.

(6) CA spectra include structurally significant direct bond cleavages which, if combined with isotopic labelling, give valuable additional information on the structure.

Disadvantages

(1) The internal energy of the ions under study is not well defined, ranging from zero up to the lowest threshold for decomposition.

(2) If the threshold for isomerization is considerably lower than that for decomposition (see Fig. V-3c) not only partial isomerization prior to collision, but also a possible isomerization after collision obscures structurally significant differences in the CA spectra. Thus two stable ions which have a low threshold for interconversion may give rather similar CA spectra. This may explain why especially with hydrocarbon ions other techniques which sample non-decomposing ions, such as ion cyclotron resonance or charge stripping, often allow the identification of stable isomers more readily than the collisional activation method does.

(3) The instrumental equipment is not yet generally available to all mass spectrometrists.

3.4.5.4 Examples

Several examples which illustrate the use of collisional activation

---

* The reproducibility is usually better than $\pm$ 10 % and can be improved considerably if computer averaging is employed.

spectra for ion structure work will be discussed. These examples have also been studied with other techniques so that a comparison of the results is possible.

$C_2H_5O^+$ *Ions*. Although the existence of four distinct $C_2H_5O^+$ ions has been postulated on the inconclusive basis of their heat of formation data [40], both early reports on metastable abundance ratios [75] and kinetic energy release values [75] point to two distinct decomposing $C_2H_5O^+$ ions, $CH_3CH=OH^+$ (5) and $CH_3O=CH_2^+$ (6). Considerable efforts have also been made to elucidate the structure of the non-decomposing ions using the collisional activation technique. Whilst in an earlier CA study [108] only two distinct $C_2H_5O^+$ ions (5) and (6) were identified, a reinvestigation of this system [46] revealed the existence of three stable non-decomposing $C_2H_5O^+$ ions. Partial CA spectra of $C_2H_5O^+$ ions from a variety of precursors are summarized in Table V-6. The table

Table V-6. Partial Collisional Activation Spectra of $C_2H_5O^+$ Ions [46]

Precursor	Ion Structure		Relative Abundance					
			m/e 24	25	26	28	30	31
$CH_3OCH_2CH_2OCH_3$	$CH_3O=CH_2^+$	(6)	< 1	< 1	4	41	43	12
$CH_3OCH_2CH_2CN$	$CH_3O=CH_2^+$	(6)	< 1	2	4	39	43	12
CH_3CHO^a	$CH_3CH=OH^+$	(5)	4	16	45	18	14	3
$(CH_3)_2CHOH$	$CH_3CH=OH^+$	(5)	4	14	49	15	14	4
$CH_3CH_2(CH_3)CHOH$	$CH_3CH=OH^+$	(5)	5	16	47	17	12	3
$\overline{CH_2CH_2O}^a$	$\overline{CH_2CH_2OH}^+$	(7)	4	15	42	13	11	16

[a] By protonation with H_2O at high pressure

demonstrates that not only the ions (5) and (6), but also the protonated ethylene oxide (7) is a stable species in the gas phase. The fragmentation pathways and isotopic labelling support this conclusion. Protonated ethylene oxide shows an abundant loss of CH_2 (m/e 31), which is consistent with the proposed structure, whilst deuterium labelling demonstrates that the hydrogen originally on the oxygen is not lost as methylene. It was further shown that part of the $C_2H_5O^+$ ions formed by a displacement reaction from 2-bromo- and 2-nitroethanol have the cyclic structure [46].

$C_3H_8N^+$ *Ions*. The metastable ion spectra of this ion have been discussed in detail [79,81] (Section V-3.4.3.3). They clearly demonstrated that the decomposing ions $CH_2=NHCH_2CH_3^+$ (11), $(CH_3)_2C=NH_2^+$ (12) and

$CH_3CH_2CH=NH_2^+$ (13) have distinct structures, whilst the ions $CH_3CH=NHCH_3^+$ (14) and $(CH_3)_2N=CH_2^+$ (15) interconvert prior to metastable decomposition, and this is supported by isotopic labelling. The non-decomposing $C_3H_8N^+$ ions have been studied using the collisional activation technique [79]. The CA spectra (see Table V-7) unequivocally show that (14) and (15) are stable species. This example illustrates that

Table V-7. CA Spectra of $C_3H_8N^+$ Ions [79][a]

Compound	Ion Structure	Relative Abundances							
		m/e → 15	27	28	29	41	42	43	54	55
N-Methylethyl-amine	$CH_3CH=NHCH_3^+$ (14)	5	10	24	15	12	16	25	3	2
2,5-Dimethyl-piperazine	$CH_3CH=NHCH_3^+$ (14)	6	11	28	13	14	19	17	4	3
N-Methyliso-propyl amine	$CH_3CH=NHCH_3^+$ (14)	6	10	26	13	13	18	23	3	2
Trimethyl-amine	$(CH_3)_2N=CH_2^+$ (15)	7	3	6	8	<6	52	21	2	1
Tetramethyl-diaminomethane	$(CH_3)_2N=CH_2^+$ (15)	9	4	6	7	<6	62	19	2	1
Dimethylamino-acetone	$(CH_3)_2N=CH_2^+$ (15)	8	4	6	7	<6	49	22	2	1

[a] Fragments with contributions from unimolecular dissociations are omitted. Abundance relative to sum of all fragments except m/e 41.

methods for ion structure elucidation which sample either decomposing or non-decomposing ions give valuable complementary information.

$C_{12}H_{10}O^{+\cdot}$ *Ions.* Not only heat of formation data [38], metastable abundance ratios [80] and kinetic energy release values [89] (vide supra), but also collisional activation spectra have been used to characterize this ion [109,110] generated by loss of CO_2 from the diphenyl carbonate ion and by direct ionization of diphenyl ether. The CA spectra of the two ions are identical within the limits of the reproducibility (but differ from those of the isomeric phenylphenols), demonstrating in accordance with kinetic energy release values that loss of CO_2 from the diphenyl carbonate ion leads to the diphenyl ether molecular ion.

$C_2H_4O^{+\cdot}$ *Ions.* The kinetic energy release values for H˙ loss from isomeric $C_2H_4O^{+\cdot}$ ions have been discussed in Section V-3.4.4.3 [97]. They established that the decomposing ions $\overline{CH_2CH_2O}^{+\cdot}$ (24), $CH_3CHO^{+\cdot}$ (25) and $CH_2=CHOH^{+\cdot}$ (26) have distinct non-interconverting structures. Thus the non-decomposing ions (24) - (26) should also be distinct species, as indeed confirmed by an earlier CA study by Van de Sande and McLafferty [102].

$C_7H_7^+$ *Ions.* Hardly any gaseous organic ion has attracted more interest than the $C_7H_7^+$ ion since Rylander et al. [111] concluded from isotopic labelling that this ion should have the tropylium ion (28) rather than the benzyl ion (29) structure. Despite the numerous experimental [94, 111-137] and theoretical studies [139-141] which followed this initial proposal, until recently there was no conclusive evidence as to whether

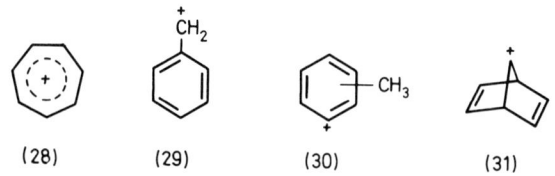

Scheme V-11

this species exists as tropylium (28) or benzyl (29) ion. An extensive collisional activation study by McLafferty and Winkler [123,124] demonstrated that although the CA spectra of $C_7H_7^+$ ions from 60 precursors are almost identical (indicating a substantial isomerization of the non-decomposing ions) they differ characteristically in the structurally significant mass range from m/e 74 - 77. Thus at low electron energies $C_7H_7^+$ ions from 1,2-diphenyl-ethane show an abundant collision-induced loss of CH_2 which, according to isotopic labelling data, includes predominantly the original methylene group - consistent with the benzyl structure. The authors report evidence that not only in the non-decomposing $C_7H_7^+$ ions from toluene but also in those from cycloheptatriene which were seen as prototype for the tropylium ion (28) an appreciable fraction of benzyl ions (29) is present. Furthermore, the authors showed that, besides these structures, stable o-, m- and p-tolyl cations (30) and probably even norbornadienyl ions (31) can be detected when suitable precursors are used. This example demonstrates again that the decomposition pathway established by isotopic labelling gives valuable additional information on ion structures when used in conjunction with collisional activation. Further evidence for the existence of both stable benzyl and tropylium ions will be presented in the following sections.

There is no doubt that the collisional activation method has contributed considerably to our knowledge on gaseous ion structures. In

Table V-9 (Section V-4.3) the important CA studies on ion structures published so far are summarized.

3.5 Ion-Molecule Reactions

3.5.1 Ion-Molecule Reactions Studied by Ion Cyclotron Resonance Spectrometry

3.5.1.1 The Principle

While most methods discussed so far used degradation reactions to obtain structural information a quite different approach is used if ion-molecule reactions are employed for ion structure assignments. Here the reactivity of an ion with neutral molecules is used as criterion for its structure. Three different approaches have been used to deduce ion structures:

(1) The usual approach is to compare ion reactivities. If the ion molecule reactions of two isomeric ions are identical one infers that they have identical structures and vice versa.

(2) Proton transfer reactions to a base in conjunction with isotopic labelling may yield additional information on the ion structures.

(3) Finally the equilibrium constants for ion-molecule reactions may differ substantially for isomeric ions. Examples illustrating these approaches are given below.

In an ion cyclotron resonance (ICR) cell ions of a wide range of internal energies and thus ion lifetimes are generated. However, those with longest ion lifetimes ($t \geq 10^{-3}$ s) have the highest chance of undergoing a collision with a neutral molecule which may lead to a reaction. Thus the ICR method samples predominantly *non-decomposing ions* with lifetimes $\geq 10^{-3}$ s and, except for the different lifetime window, results should be comparable to those obtained with the collisional activation method.

3.5.1.2 The Technique

Although ion-molecule reactions can be studied with various types of instruments such as tandem mass spectrometers [142], chemical ionization [143] or other types of high pressure mass spectrometers [144] and ion trapping in the space charge of a pulsed electron impact source [145] the ion cyclotron resonance mass spectrometer [146-154] is especially suited for this purpose as it allows product and reactant ions to be distinguished from one another.

A typical ICR cell is shown in Fig. V-9. It is placed between the

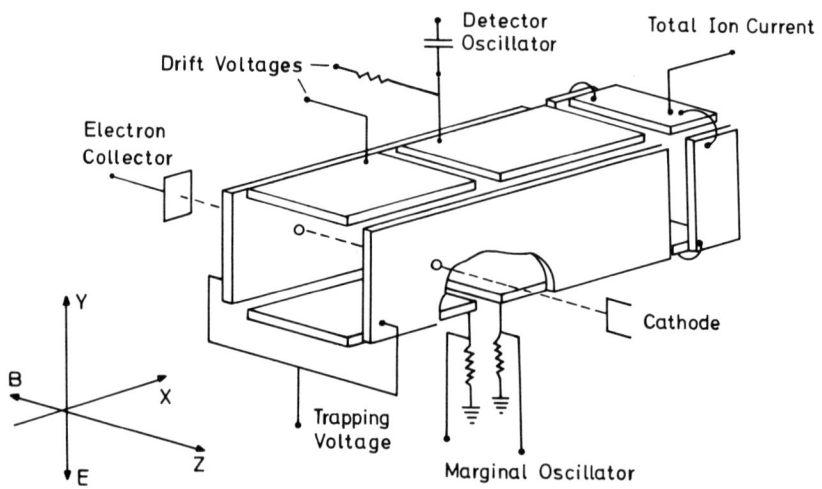

Fig. V-9. Cutaway view of an ion cyclotron resonance cell.

poles of an electromagnet, the field of which is directed as indicated. Ionization occurs by electron impact, where the electrons traverse the cell in the direction of the magnetic field. The ions are forced by this field into a circular path. The angular frequency, ω_c, of this motion is given by

$$\omega_c = eB/m \qquad (V-4)$$

where B is the electromagnetic flux density. Thus the mass to charge ratio m/e of the resonating ions is determined by the values of both the angular frequency, ω_c, and the magnetic flux, B. A small electric field, E, (∼ 0.5 V/cm) is applied to the upper and lower plate perpendicular to B and causes the ions to move into a direction at right angle to both fields towards a total ion collector. The drift velocity is given by

$$v = E/B \qquad (V-5)$$

and is typically 50 m/s. Thus the ion spends a time of the order of 10^{-3} s in the cell before being collected. This long ion residence time allows ion-molecule reactions to be studied even at low pressures (e.g. 10^{-5} torr). For mass-selected ion detection a high frequency field from a marginal oscillator is applied to the upper and lower plates (i.e. normal to B). Resonance is achieved if the angular frequency of the oscillator, ω_1, matches that of the ion, ω_c. This leads to an absorption of energy by the ion from the HF field which creates an electrical signal in the circuit. A complete mass spectrum is obtained by scanning the magnetic field at constant oscillator frequency.

For the study of ion-molecule reactions the double resonance technique is especially useful. Consider the reaction $m_1^+ + N_1 \rightarrow m_2^+ + N_2$. First, the magnetic field is adjusted to detect the product ion m_2^+ at its resonance frequency $\omega_2 = \omega_c$. Then, a second HF field is applied to the cell and its frequency, ω_1, varied until resonance for the reactant ion m_1^+ is achieved. The reactant ion gains translational energy which affects its further reaction with N_1, and thus the yield of product ions m_2^+. It is apparent from eq. V-4 that at constant magnetic field the double resonance condition is fulfilled if

$$\frac{\omega_1}{\omega_2} = \frac{m_2}{m_1} \qquad (V-6)$$

Thus the double resonance experiment allows the product ion to be linked to its precursor.

3.5.1.3 Critical Evaluation of the Method

When the ICR technique was first applied to ion structure work in 1969 by Diekmann et al.[8] it was heralded as a particularly powerful method for the structure elucidation of gaseous organic ions. While this is still true, the number of publications using this technique for structure determination is declining. This does not only result from the fact that the necessary experimental equipment is only available to few organic mass spectrometrists, but also reflects the problems inherent in this method. Thus it is often difficult to find suitable ion-molecule reactions to characterize a given ion.

If the comparative method is used, different ion-molecule reactions reflect different ion structures. The opposite conclusion is less convincing. Thus it is conceivable that two ions of distinct structure still undergo identical ion-molecule reactions. For instance, both the phenol molecular ion (34) and the $C_6H_6O^{+\cdot}$ ion formed by ketene loss from bicyclo[2,2,2]-oct-2-en-5,7-dione (32) (which according to the mechanism of formation should be initially formed as cyclohexadienone ion (33)- see Scheme V-12) show the same ion-molecule reactions with 1-methylcyclobutanol [155] whilst collisional activation spectra demonstrate that the ions are distinct species [156,157].

Finally, it has been repeatedly pointed out that the occurrence and extent of ion-molecule reactions may in some instances depend on the ion's internal energy, and this may handicap ion structure assignments. Thus a dependence of the ion reactivity on the electron energy has been

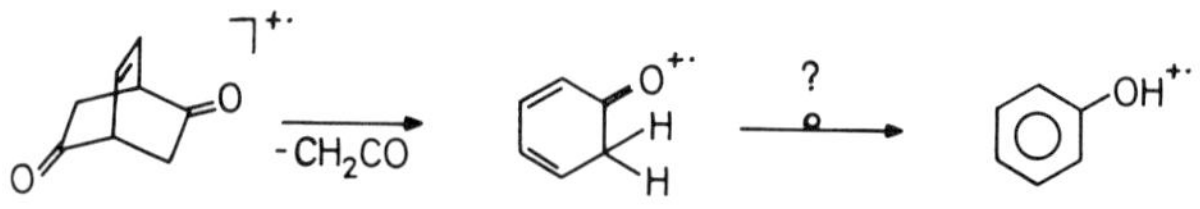

Scheme V-12

frequently observed [43,158,159]. Moreover it has recently been shown by Gross et al. [158] that $C_6H_6^{+\cdot}$ ions formed e.g. from tropone differ in their ion-molecule reactions from $C_6H_6^{+\cdot}$ ions assumed to have either a cyclic or an acyclic structure. These differences were tentatively explained in terms of differences in internal energy, although no conclusive evidence could be obtained. Similarly Blair and Harrison [160] stated that the observed differences in the ion-molecule reactions of $C_2H_4O^{+\cdot}$ ions initially having the acetaldehyde or ethylene oxide structure may arise either from structural or from energetic effects. It is highly desirable to study the influence of the internal energy on ion structure assignments based on the ICR technique further. In any case this possible source of error should be kept in mind if ion structures are determined using this method.

3.5.1.4 Examples

Again examples have been selected which were also studied using other techniques and which illustrate the three different approaches used for ion structure assignments.

The Comparative Method.
$C_2H_5O^+$ *Ions.* Not only heats of formation data [40], metastable ion characteristics [75], kinetic energy release values [75] and collisional activation spectra [46,108] (vide supra), but also ion-molecule reactions have been used to identify isomeric $C_2H_5O^+$ ions. In a first attempt Beauchamp and Dunbar [43] were able to differentiate the $CH_3O=CH_2^+$ ion (6) from protonated acetaldehyde (5) and protonated ethylene oxide (7) whilst the ions (5) and (7) could not be distinguished. The ion-molecule reactions are summarized in Scheme V-13 which demonstrates that the ion (6) shows a hydride abstraction and methyl cation transfer reaction which are not observed for ions (5) and (7). The authors conclude that the protonated ethylene oxide (7) rearranges to the protonated acetaldehyde (5) upon formation as the latter is more stable than the former by 113 kJ mol^{-1}.

Isomer	Process	Abundance
(6) $CH_3\overset{+}{O}=CH_2 + CH_3OC_2H_5$	$\rightarrow CH_3\overset{+}{O}=CHCH_3 + CH_3OCH_3$	0.94
	$\rightarrow (CH_3)_2\overset{+}{O}C_2H_5 + CH_2O$	0.06
(7) $CH_3\overset{+}{CH}=OH + (CH_3)_2CHOH$	$\rightarrow (CH_3)_2\overset{+}{C}HOH_2 + CH_3CHO$	0.44
	$\rightarrow CH_3CH=\overset{+}{O}-CH(CH_3)_2 + H_2O$	0.20
	$\rightarrow C_5H_9^+ + 2H_2O$	0.19
	$\rightarrow CH_3CH=O--\overset{+}{H}--OH_2 + C_3H_6$	0.16
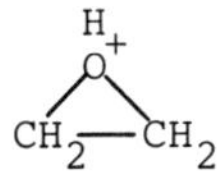 (8) $\underset{CH_2-CH_2}{\overset{H_+\atop O}{\triangle}} + (CH_3)_2CHOH$	$\rightarrow (CH_3)_2\overset{+}{C}HOH_2 + C_2H_4O$	0.38
	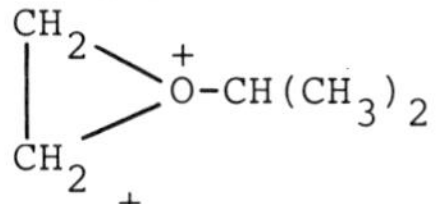 $\rightarrow \underset{CH_2}{\overset{CH_2}{\triangle}}\!\!O\text{-}CH(CH_3)_2^+ + H_2O$	0.20
	$\rightarrow C_5H_9^+ + 2H_2O$	0.24
	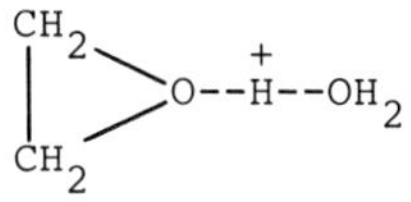 $\rightarrow \underset{CH_2}{\overset{CH_2}{\triangle}}\!\!O\text{--}\overset{+}{H}\text{--}OH_2 + C_3H_6$	0.18

Scheme V-13

A distinction between the protonated ethylene oxide (7) and the two other isomers was, however, achieved in a second study from the same laboratory. Staley et al. [45] demonstrated that protonated ethylene oxide (7) reacts with phosphine and hydrogen sulfide with water elimination (Scheme V-14) whilst this reaction is not observed with the isomers (5) and (6), demonstrating that protonated ethylene oxide is a stable ion in the gas phase.

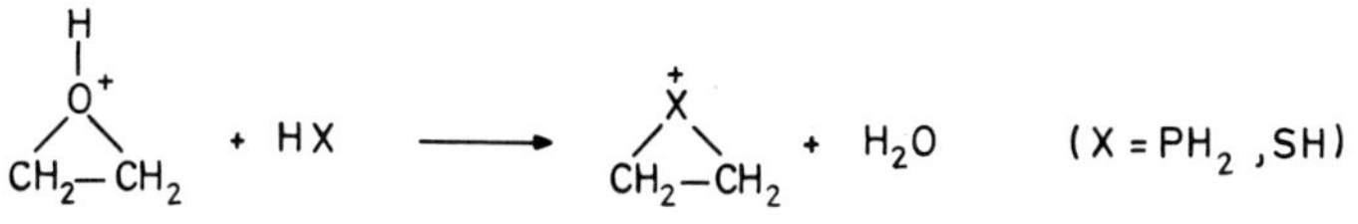

$$\underset{CH_2-CH_2}{\overset{H\atop |\atop O^+}{\triangle}} + HX \longrightarrow \underset{CH_2-CH_2}{\overset{\overset{+}{X}}{\triangle}} + H_2O \qquad (X = PH_2, SH)$$

(7)

Scheme V-14

Using ion trapping in the space charge of a pulsed electron impact source some indication for the existence of a cyclic $C_2H_5O^+$ ion (7) was also observed by Blair and Harrison [160] when studying proton transfer reactions from (5) and (7) to $(CD_3)_2O$, although the identification of (7) was less conclusive.

The unambiguous existence of protonated ethylene oxide as a stable species has also been demonstrated by the collisional activation technique [46] (vide supra).

$C_7H_7^+$ *Ions.* Venema and Nibbering [125] demonstrated that $C_7H_7^+$ ions generated from benzylmethyl ether $C_6H_5CH_2OCH_3$ (35) react with neutral dimethyl amine whilst $C_7H_7^+$ ions from 7-methoxy-cycloheptatriene (36) do not react (Scheme V-15). The authors concluded that the non-reactive

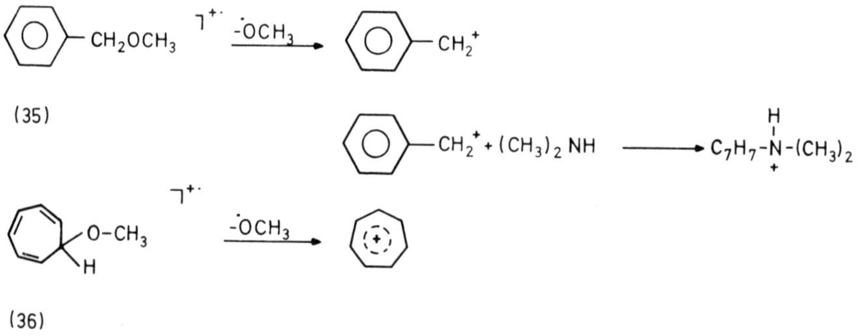

Scheme V-15

$C_7H_7^+$ ion has the tropylium ion structure and the reactive one the benzyl ion structure. The existence of both stable benzyl and tropylium ions has also been inferred from collisional activation spectra [123, 124] although the CA spectra seem to indicate that both the ions (35) and (36) give a mixture of tropylium and benzyl ions in a similar ratio.

The existence of stable benzyl cations in the gas phase was also inferred from an ICR study by Abboud et al. [141] by comparing calculated and experimental heats of formation of $C_7H_7^+$ ions formed from benzyl chloride.

Proton Transfer Reactions

$C_3H_8N^+$ *Ions.* It has been shown earlier (Section V-3.4.3.3 and V-3.4.5.4) that the decomposing ions $CH_3CH=NHCH_3^+$ (14) and $(CH_3)_2N=CH_2^+$ (15) interconvert prior to decomposition [81], while the non-decomposing ions (14) and (15) are stable species [79]. An ICR study by Tomer [161] supports this conclusion. The author demonstrated that the N-deuterated ion (14a) transfers only deuterium to 3,5-lutidine (Scheme V-16) whilst transfer of both hydrogen and deuterium is expected if the ions (14) and (15) interconvert freely. Proton transfer reactions were also used by Nibbering [162] to demonstrate that the large majority of the $C_6H_6O^{+\cdot}$ ions generated from phenetole have the phenol ion structure.

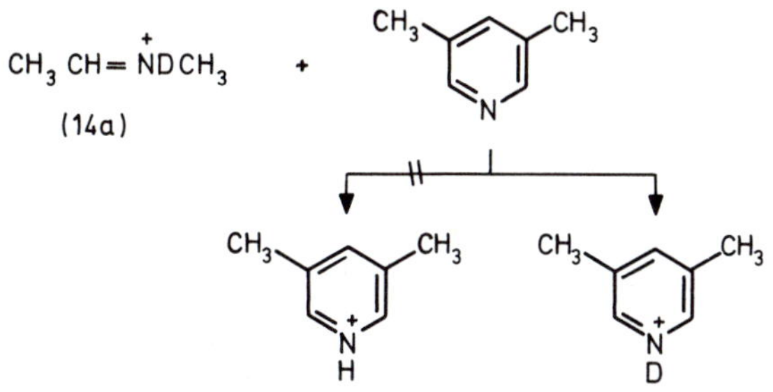

$$CH_3 CH = \overset{+}{N}DCH_3$$
(14a)

Scheme V-16

Table V-8. Comparison of Electron Impact Mass Spectra and Pulsed Ion Cyclotron Resonance Data for C_8H_{10} Isomers [163]

	Electron-impact mass spectra;[a] rel. int. of major fragments at 70 eV				
C_8H_{10} Isomers	m/e 91	m/e 106	m/e 105	m/e 39	m/e 51
1,2-Dimethylbenzene	100	57.2	23.8	16.3	15.7
1,3-Dimethylbenzene	100	62.7	28.2	18.6	15.4
1,4-Diemethylbenzene	100	61.5	29.6	15.7	15.9

	Pulsed ICR equilibrium data; $-\Delta G^{o}_{298}$ (kJ mol^{-1}) for proton-transfer reactions
C_8H_{10} Isomers	Reaction $(C_9H_{13})^+$ + M = $(M+1)^+$ + n-propyl benzene
1,2-Dimethylbenzene	4.6 ± 0.8
1,3-Dimethylbenzene	8.8 ± 0.8
1,4-Diemethylbenzene	2.1 ± 0.8

[a] American Petroleum Institute Project 44, 1972.

Equilibrium Constants*)

It was shown by McIver [163] that the equilibrium constant of ion-molecule reactions is highly sensitive to structural features. Thus although the electron impact spectra of the isomeric dimethylbenzenes are almost identical, pointing to an isomerization of these $C_8H_{10}^{+\cdot}$ ions,

---

* Strictly speaking, the reported example demonstrates that the neutrals have distinct structures. However, the same approach can be used to characterize ions.

the equilibrium constant for the reaction $C_9H_{13}^+ + C_8H_{10} \rightarrow C_8H_{11}^+ + C_9H_{12}$ varies by a factor of four between the three isomers (see Table V-8). Further ion structure studies employing the ICR technique are summarized in Table V-9 (Section V-4.3).

3.5.2 The Ion Cyclotron Resonance Photodissociation Technique

The ICR photodissociation technique was developed by Dunbar [164-182]. This is the only spectroscopic method used extensively in ion structure work today. With this technique, cations (usually molecular ions) are produced by electron impact and trapped in an ICR cell. These cations are subsequently irradiated by a photon source of variable wavelength whose energy is sufficient to cause photodissociation of the ion. A plot of the cross section for disappearance of the trapped ions (due to photodissociation) as function of the irradiating wavelengths constitutes the photodissociation spectrum of the ion involved. The position and intensity of the peaks in the spectrum is characteristic of the ion structure in much the same way as the optical absorption spectrum would be.

This approach has been repeatedly employed by Dunbar to investigate the structure of ions such as $C_7H_8^{+\cdot}$ [127,165-167], $C_7H_7^+$ [126], $C_8H_8^{+\cdot}$ [168], $C_9H_{10}^{+\cdot}$ [168], $C_{10}H_{12}^{+\cdot}$ [168], the molecular ions of mono- and dialkyl-substituted benzenes [169], benzylhalides [170], halotoluenes [170], hexatrienes [171], hexamethylbenzene and hexamethyl (Dewar benzene) [172]. In these studies, the identity of the photodissociation spectra of two isomeric cations is used as a criterion for identical structures and vice versa.

The examples reported so far demonstrated that in several instances stable isomers could be distinguished more readily with the photodissociation technique than with the collisional activation method. Thus the CA spectra of $C_7H_8^{+\cdot}$ ions from both n-butylbenzene and 2-phenylethanol are almost indistinguishable from that of $C_7H_8^{+\cdot}$ ions from cycloheptatriene [108] while the photodissociation spectra allowed the unequivocal differentiation of the two sets of isomers [167]. Similarly, the photodissociation spectra of the styrene and cyclooctatetraene cation do not reveal any interconversion between the two isomers [168] whilst the collisional activation spectra of the two $C_8H_8^{+\cdot}$ ions have been interpreted in terms of partial equilibration below the threshold for decomposition [16].

Another approach employed by Dunbar uses a time resolved photodissociation technique to obtain information on the structure of gaseous ions. Here the most interesting investigation is that of $C_7H_7^+$ ions generated from toluene molecular ions [126]. In this study the toluene molecular ions were produced by electron impact with electron energies slightly above the ionization threshold. The parent ions were trapped in the ICR cell for several minutes and irradiated with UV light at 4200 Å, which led to the exclusive formation of $C_7H_7^+$ ions by gas phase photodissociation. These $C_7H_7^+$ ions reacted with neutral toluene to give $C_8H_9^+$ ions:

$$C_7H_7^+ + C_7H_8 \rightarrow C_8H_9^+ + C_6H_6$$

The production of $C_8H_9^+$ ions was studied as function of time by commencing the irradiation when the electron beam was turned off. The intensities of the $C_7H_8^{+\cdot}$, $C_7H_7^+$ and $C_8H_9^+$ ions as function of time are shown in Fig. V-10. It is evident from this figure that $C_7H_8^{+\cdot}$ ions rapidly

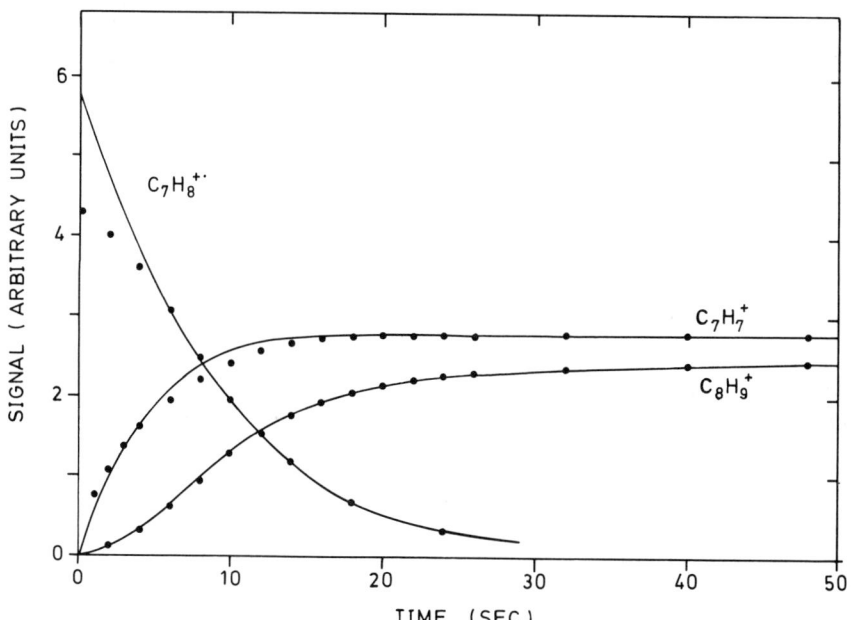

Fig. V-10. Time-dependent photodissociation of the toluene molecular ion [126]. At time zero the electron beam was turned off and the monochromatic irridation (4200 Å) was initiated. (By courtesy of the American Chemical Society.)

dissociate to $C_7H_7^+$ which in turn react to give $C_8H_9^+$. The point of interest is that a major fraction of the $C_7H_7^+$ ions does not react, even after very long time. Rather, a constant ratio of $C_8H_9^+/C_7H_7^+$ is reached after 25 s. The most straightforward explanation of this result is that photodissociation of toluene leads to two structural distinct $C_7H_7^+$ ions. Based on earlier labelling studies [127], the author concluded that the reactive ion has the benzyl structure, the non-reactive ion the tropylium structure. The simultaneous formation of two $C_7H_7^+$ ions of distinct structures was explained by assuming that the excited $C_7H_8^{+\cdot}$ precursor exists in a dynamic equilibrium between the toluene and cycloheptatriene structure. With decreasing internal energies, the production of benzyl ions decreases and reaches zero near the dissociation threshold, i.e. at threshold only tropylium ions are formed, in agreement with recent photoionization appearance potential measurements [136].

The same conclusion was reached in a later time-resolved ICR study by Jackson et al. [137] who also reported quantitative data on the relative abundance of benzyl and tropylium ions in $C_7H_7^+$ ions generated from various other compounds.

As mentioned earlier (see Section V-3.4.5.4), the simultaneous formation of both tropylium and benzyl ions from dissociating toluene ions has also been postulated by McLafferty and Winkler [123,124] based on collisional activation spectra. The quantitative agreement between the CA and the ICR studies is, however, poor. Moreover, Jackson et al. [137] could not find any evidence for the existence of tolyl ions in the gas phase as postulated by Winkler and McLafferty [124].

3.6 Molecular Orbital Calculations

3.6.1 Molecular Orbital Calculations and Ion Structure

Strictly speaking molecular orbital (MO) calculations do not represent a method for the elucidation of gaseous ion structures. Rather, they allow the energetics of gaseous ions to be estimated. This in turn gives valuable information about the structures of these ions. This information may be used to support or even supplement experimental evidence. MO calculations provide three main pieces of information:
(1) Relative heats of formation of ions.
(2) Equilibrium geometries.
(3) Energy barriers for isomerization and fragmentation.
Independent of its usefulness for ion structure elucidation the heat of formation is one of the most important properties of an ion and its theoretical determination, if reliable (vide infra), is highly desirable. Thus energetic data can be directly compared with experimental values, and, if the latter are of high quality, may in turn be used to test the accuracy of the computation. Moreover, such MO calculations usually include various non-classical structures which the mass spectrometrist trained in thinking along conventional chemical lines may have overlooked when discussing ion structures.

However, such calculated relative heats of formation are of limited value for ion structure assignment as they do not allow a prediction of which ions are actually observed in the gas phase to be made. Relatively unstable isomers may exist in the gas phase if the isomerization barrier is sufficiently high, whilst on the other hand two isomers of almost identical heats of formation may have a vanishingly small energy barrier for interconversion and thus will not be observed as distinct species.

In principle ion structure assignments based on calculated relative heats of formation can be made by comparison with experimental values provided the experimental and the calculated data are of sufficient reliability. However, as pointed out in Section V-3.3, one cannot exclude the case of two isomeric ions which coincidentally have identical heats of formation. It will be shown below that this is the case for instance with isomeric $C_2H_3^+$ ions.

MO calculations do not only provide relative heats of formation, but also equilibrium geometries of ions. This information is of particular value as the experimental methods for ion structure elucidation, at least as far as fragment ions are concerned, give no information on the configuration of an ion, i.e. the bond lengths and angles. Examples for equilibrium geometries of ions will be given below.

Of special interest with regard to ion structures are MO calculations which are not restricted to the estimation of relative heats of formation, but include calculations on the potential energy surfaces, i.e. the energy barrier for isomerization and decomposition. So far unfortunately only relatively few calculations include this information. Whilst the decomposition threshold may be readily determined experimentally, information about the isomerization threshold is difficult to obtain although this information is of crucial importance when discussing ion structures. It has been pointed out in Section V-2.2 that the ratio of the decomposition and the isomerization thresholds determines whether and to what extent an ion isomerizes. Thus knowledge of the isomerization barrier will allow a prediction to be made of whether non-decomposing isomeric ions or even two ions with sufficient energy to decompose may be detected as distinct species in the gas phase.

Finally not only the overall energy barrier for decomposition or isomerization, but also its "fine structure", i.e. the detailed potential energy surface of an ion is of interest. Most notably, potential energy surfaces for a variety of organic ions have recently been deduced by Williams et al. from thermochemical data, kinetic energy release values and isotopic labelling experiments (see Section V-4.2.2). Several MO calculations of at least partial potential surfaces have been published and will make an important contribution to enhancing our knowledge of gaseous organic ions.

3.6.2 The Method

Computational methods used to calculate energetic data on gaseous organic ions have been critically reviewed by McMaster [183] who also summarized the requirements necessary for calculations of predictive value. Both ab initio and semiempirical methods are used to calculate ion structures.

Most ab initio calculations which have been applied to the structure of positive ions have been on the Hartree-Fock level. In this approximation each electron moves in a field which is the sum of the field due to the nucleus and the averaged field of all other electrons, i.e. direct electron-electron interaction (electron correlation) is not taken into account. Thus it is assumed that the full wave function can be written as a single determinant of one-electron functions. The orbitals in the Hartree-Fock Self Consistent Field (HF-SCF) procedure are determined by minimizing the energy integral

$$E = \int \Psi^* \hat{H} \Psi \, d\tau \qquad (V-7)$$

where $\hat{H}$ is the many-electron Hamiltonian. For this purpose each orbital is approximately expressed as a linear combination of basis functions of Gaussian type, i.e. $\phi = \sum c_i \phi_i$. The larger the basis set the more accurate the data which can be expected. Thus it has been pointed out that the minimum basis set (STO-3G) although widely used gives results of hardly any predictive value. The accuracy of such calculations can be improved by including polarization functions (e.g. contributions of the d-function to the p-function of a C atom, etc.)*[)].

As the geometry of the ion is usually unknown the calculations start with a geometrical search for the equilibrium structure for which the simplest basis set (STO-3G) is in general sufficient.

Ab initio calculations on the HF level are still incorrect by the amount of the correlation energy (due to individual interactions of the electrons-vide supra). Thus only if the correlation energy of several isomeric ions is identical will comparison at the HF-level be quite accurate. A number of techniques have been developed for estimating the correlation energy including the method of configuration interaction (CI). Unfortunately, most ab initio calculations on organic ions do not take any correlation energy into account. However, such neglect will lead in general to unreliable conclusions regarding the relative stabilities of different isomers if the energy differences between the

---

* This is often indicated by a superscript asterisk, e.g. 6-31G*.

isomers are small. An example illustrating such a case will be discussed below.

From the various semiempirical MO calculations which usually take electron correlation effects into account only Dewar's MINDO/3 [184] method seems to yield results of sufficient reliability to predict relative stabilities. Although there has been a lively debate about the apparent deficiencies of this method by the protagonists of the ab initio [185,186] and the MINDO method [187], respectively, MINDO/3 calculations seem to reflect the experimental results on the relative stabilities of various isomeric propyl ions better than ab initio calculations with a large basis set do, if electron correlation is neglected [188]. For a comparison of both methods it is important to keep in mind that *ab initio calculations underestimate the stability of non-classical structures owing to the neglect* of electron correlation, while *MINDO/3 calculations still seem to overestimate the stability of non-classical structures* (especially cyclic structures) to some extent.

3.6.3 Examples

Examples are chosen which highlight the problems involved in obtaining sufficiently reliable data for a correct prediction of the relative stabilities whilst simultaneously demonstrating the wealth of information which can be extracted from such calculations for ion structure work.

$C_2H_3^+$ *Ions.* The vinyl ion is the simplest carbonium ion for which both a classical structure (37) or a non-classical structure, a π-protonated acetylene (38) can be written. The first semiempirical MO calculations

(37)   (38)

[189,190] on the structure of $C_2H_3^+$ ions, reported in 1969, predicted the bridged structure (38) to be more stable whereas the earliest ab initio SCF-calculations indicated that the classical form (37) was more stable than the bridged form by 80 kJ mol^{-1} [191]. Subsequent extension of the basis set reduced this difference to 24 kJ mol^{-1} although the classical structure remained most stable [192]. However, when estimates on the correlation energy were included it was now found that the bridged structure was more stable by 29 kJ mol^{-1} [193]. Finally a thorough reinvestigation by Weber et al. in 1976 [194] using the approach of

configuration interaction (which takes electron correlation into account) and a further extension of the basis set demonstrated that the two structures have the same energy to within 4-8 kJ mol^{-1} with the bridged structure probably more stable. The conflicting earlier results reemphasize that small differences in energy should be interpreted with caution, when electron correlation is not taken into account.

It was found that both structures are planar. Weber et al. [194] also calculated the optimum rearrangement path for an isomerization of the linear to the bridged structure. This transition state is again planar. The barrier for this rearrangement is low, less than 4-12 kJ mol^{-1}. Hence there is no chance of identifying two distinct $C_2H_3^+$ ions by any of the above discussed mass spectrometric techniques as at room temperature almost all $C_2H_3^+$ ions have isomerized to a mixture of interconverting structures.

$C_3H_3^+$ *Ions.* There have been several ab initio calculations on $C_3H_3^+$ ions. Radom et al.[195] predict that the propargyl ion (39) and cyclopropenyl ion (40) are the most stable isomers, cyclopropenyl being more stable by 142 kJ mol^{-1} which agrees closely with a difference of 130 kJ mol^{-1} calculated by Kebabcioglu and Dyczmons [196]. Experimentally a difference of 105 kJ mol^{-1} was determined by Lossing [68]. The discrepancy between the experimental and theoretical value may result from the neglect of correlation energy.

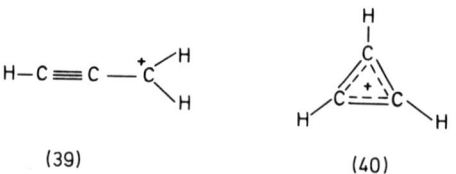

(39)   (40)

More interesting than the relative heats of formation is the energy barrier for interconversion between (39) and (40) which has been calculated to be surprisingly high (348 kJ mol^{-1}). Thus it should be possible to observe both ions as distinct species by mass spectrometric methods. Indeed the existence of two stable $C_3H_3^+$ ions has been confirmed by kinetic energy release [197-199], collisional activation [201] and photoelectron-photoion-coincidence studies [200,202].

$C_7H_7^+$ *Ions.* In view of the numerous mass spectrometric studies devoted to the structure of $C_7H_7^+$ ions and the recent conclusive evidence of the existence of both stable tropylium (28) and benzyl (29) ions in the gas phase [123-127,136,137] it is not surprising that this was also a challenging problem for MO calculations. As the $C_7H_7^+$ ion still seems to be too large for accurate ab initio calculations, one has at present to rely on MINDO/3 calculations reported by Cone et al. [140]. Again the calculated energy barriers for interconversion between both structures are more interesting than the relative heats of formation[*]. From their calculations the authors concluded that the easiest path for a rearrangement of the benzyl ion to the tropylium ion involves a norcaradienyl cation (41) as stable intermediate, as shown in Scheme V-17, with an overall isomerization barrier of 137 kJ mol^{-1}. Fig. V-11 shows

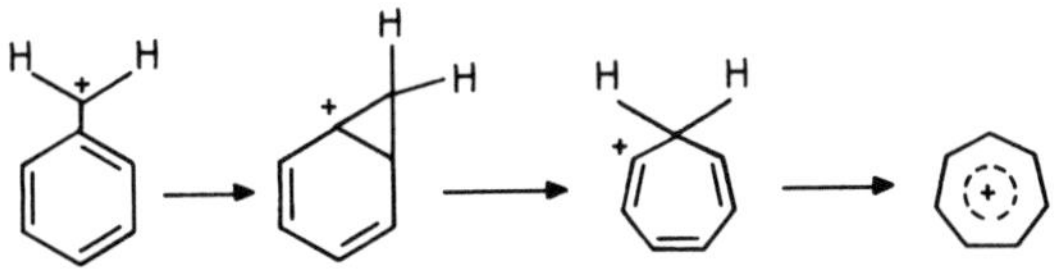

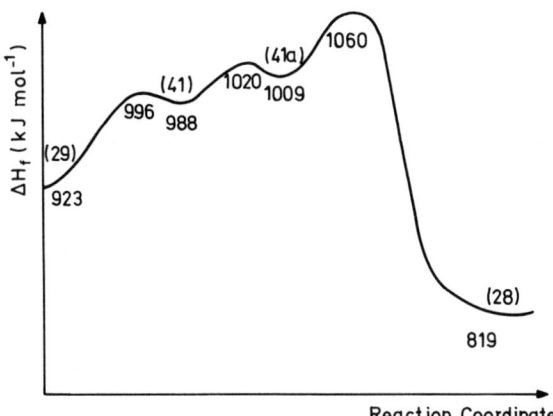

Scheme V-17

the partial potential energy surface for a rearrangement of the benzyl to the tropylium ion.

Fig. V-11. Potential energy surface for the rearrangement of benzyl to tropylium ions (energies in kJ mol^{-1}) [140]. (By courtesy of the American Chemical Society.)

---

[*] At present the heats of formation values for (28) and (29) deduced from experiment [34,136], ab initio [141] and MINDO/3 calculations [140] differ substantially (tropylium ion: ΔH_f = 867 ± 33 kJ mol^{-1}, benzyl ion: ΔH_f = 908 ± 17 kJ mol^{-1}).

It has been stated [124] that the lowest threshold for decomposition of $C_7H_7^+$ ions (C_2H_2 loss) is 398 kJ mol^{-1}. Even if the limited reliability of these data is kept in mind it is apparent that the large majority of the non-decomposing $C_7H_7^+$ ions have equilibrated to a mixture of benzyl and tropylium ions which explains why complete scrambling of all carbons is observed prior to *metastable* loss of acetylene [112-115], consistent with *interconverting decomposing $C_7H_7^+$ ions*. Nevertheless, the isomerization barrier is still sufficiently high to enable stable benzyl and tropylium ions to be detected using the CA [123,124] or ICR [125,126,137] method.

Additional MO calculations on organic ions are summarized in Table V-9 (Section V-4.3).

3.7 Other Techniques

A variety of other methods have been proposed and occasionally used to determine the structure of gaseous ions. The more important ones will be discussed briefly.

3.7.1 Isotope Effects

It has been shown in Section IV-2.4 how isotope effects can be used to elucidate the mechanisms of a variety of rearrangement reactions. As a result of the direct correlation between the mechanism leading to the formation of a fragment ion and the ion's initial structure, isotope effects may provide information about this initial structure of an ion. For instance, Howe and Williams [203] demonstrated that ethylene loss from p-bromophenetole proceeds via a four-membered transition state to form a p-bromophenol, but not a p-bromocyclohexadienone ion (see Scheme IV-13).

There is still another approach to using hydrogen-deuterium isotope effects as structural criterion: complete randomization of all hydrogens and an identical isotope effect for H/D loss in two partially labelled isomeric ions indicate, but do not unambiguously prove, that a single structure or a mixture of interconverting structures is present. Thus not only complete hydrogen randomization but also an almost identical isotope effect for H/D loss was observed for long living (t = 10^{-5} s) $C_7H_8^{+\cdot}$ ions from a variety of precursors [204,205].

3.7.2 Charge Stripping

Collisions of fast moving ions with a neutral target gas may lead not only to collision-induced dissociation, but also to a variety of charge transfer reactions as discussed in Section III-8.2. For instance, in a charge stripping reaction a doubly-charged ion is formed. It has been shown by Cooks et al. [206] that the cross sections for charge stripping reactions differ considerably for isomeric $C_6H_6^{+\cdot}$ ions, whilst the differences in the collisional activation spectra are less pronounced [207] (although the spectra are not identical, as initially suggested [206]).

A comparison of collisional activation and charge stripping spectra has also been reported by Maquestiau et al. [208] for $C_3H_5^+$, $C_3H_6^{+\cdot}$, $C_5H_5^+$ and $C_5H_6^{+\cdot}$ ions. At least in one case ($C_5H_6^{+\cdot}$) the authors demonstrated that charge stripping spectra are better suited than collisional activation spectra to identify stable isomers.

There are two explanations for these observations: first, the cross section of a collision process is an inverse function of the energy transferred [209]. Thus the collisional activation method samples predominantly those non-decomposing ions with energies near the threshold for decomposition which may have isomerized to a large extent. On the other hand during a charge stripping reaction an excitation energy of at least 15 eV must be transferred to reach the second ionization potential of $C_6H_6^{+\cdot}$ isomers. Hence the charge stripping process samples all non-decomposing ions with energies ranging from 0 - 4.5 eV more uniformly, i.e. a larger fraction of non-isomerized ions. Second, an isomerization after collision may further obscure the structurally significant differences in the collisional activation spectrum. As a collisional activation spectrum often includes charge stripping processes it is recommended that this additional information be used in distinguishing between isomeric ions.

3.7.3 Experimental Determination of Rate Constants

As shown in Section II-2.2.3 ion lifetime measurements of energy-selected ions, e.g. using the photoelectron-photoion-coincidence technique, allow a direct determination of rate constants as function of the internal energy to be made. Using this method Werner and Baer [210] demonstrated that the five isomeric $C_4H_6^{+\cdot}$ molecular ions show the same $k(E)$ curves for methyl loss within the limits of reproducibility, supporting earlier conclusions that these ions isomerize prior to decomposition.

4 Experimental Results

In describing the techniques employed in present day ion structure work a variety of examples illustrating the use of these techniques have been discussed. The following sections contain a more general survey on the structure of gaseous organic ions, their potential energy surfaces and in particular their isomerization behavior, whilst studies on specific ion structures are summarized in Table V-9 (Section V-4.3).

4.1 Hydrocarbon Ions

4.1.1 Aliphatic and Alicyclic Hydrocarbon Ions

Although the ion chemistry of alkanes has been described as perhaps the most complex [211] more recent results employing metastable ion and collisional activation spectra as well as isotopic labelling [212,213] demonstrated that our knowledge at least of the ion structures of linear and branched alkanes is much better than often assumed in the past, whilst the decomposition mechanisms of these ions still remain largely obscure.

The in part contradictory literature on the fragmentation of labelled n-alkanes has recently been critically reviewed [213]. According to the labelling data larger n-alkane ions (> C_7) eliminate alkyl radicals and alkane molecules mainly in specific mechanisms from the terminal positions and only with a low probability from central parts of the ion, demonstrating that alkane molecular ions predominantly retain their original structure prior to decomposition. Moreover, the small fraction of molecular ions undergoing non-specific fragmentation need not have isomerized to interconverting structures prior to decomposition. Rather, the non-specific fragmentation most probably occurs from the original structure via skeletal rearrangements during decomposition. The same conclusion also holds for smaller n-alkane ions where a non-specific methyl loss is observed with high abundance [214,215].

Metastable ion [216] and collisional activation spectra [212] support this conclusion: Isomeric alkane molecular ions with a lifetime of 10^{-6} s which differ in the branching of the carbon skeleton have the same structure (or more precisely: constitution) as the corresponding neutrals. This is illustrated in Fig.V-12 where the CA spectra of four isomeric octane molecular ions are contrasted [212]. The spectra do not only show pronounced differences in the relative abundances, but also a structure specific fragmentation.

The opposite behavior is observed for the even-electron alkyl ions, $C_nH_{2n+1}^+$. Isotopic labelling revealed that all carbons and hydrogens are completely randomized in small alkyl ions (such as butyl ions [2,3,5,82]). Atom randomization is still substantial although not complete in larger alkyl ions [213,217] indicating an equilibration of the various branched isomers to a mixture of interconverting structures.

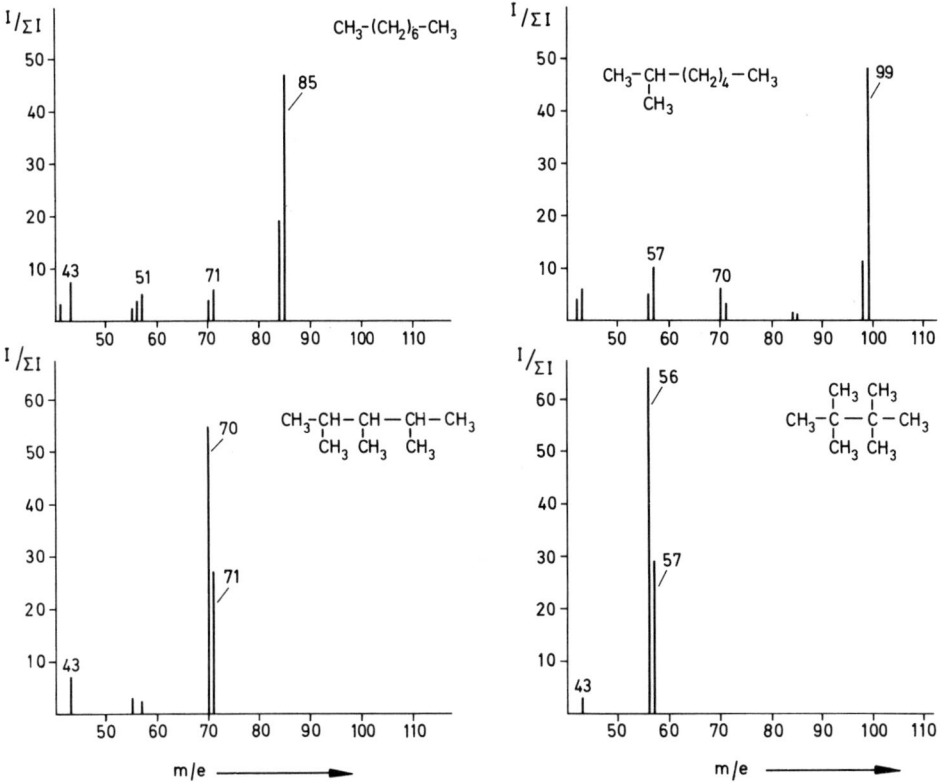

Fig. V-12. Collisional activation spectra of four octane isomers [212].

In support of this conclusion almost identical metastable ion [2,218] and collisional activation spectra [212] have been observed for a large variety of isomeric alkyl ions. Fig. V-13 compares the CA spectra of four $C_5H_{11}^+$ ions generated from differently-branched octane molecular ions [212]. It is obvious that within the limits of reproducibility of the measurement the CA spectra of these pentyl ions are identical, demonstrating that not only the decomposing but also the large majority of the non-decomposing pentyl ions have isomerized to a mixture of interconverting structures.

In contrast, open-shell (odd-electron) fragment ions such as isomeric $C_nH_{2n}^{+\cdot}$ ions generated by a specific mechanism, e.g. by H_2O loss from the corresponding alcohols do not give identical collisional activation spectra [14], the differences becoming the more pronounced the larger the fragment ion, indicating that such odd-electron fragments retain their original structure at least to a larger extent.

The most straightforward explanation to rationalize the different

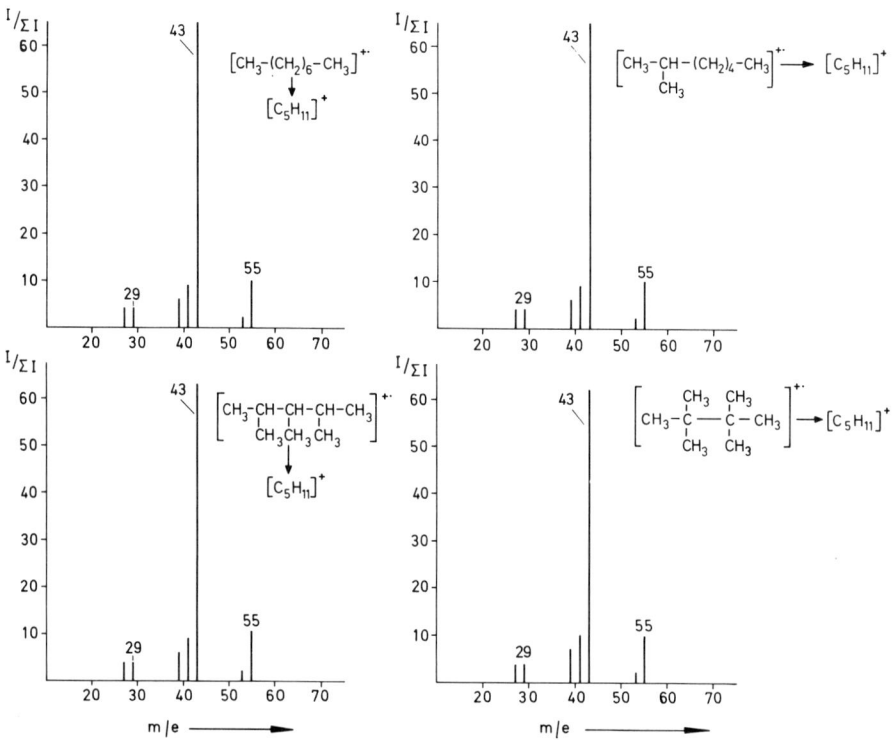

Fig. V-13. Collisional activation spectra of pentyl ions generated from four isomeric octane ions [212].

isomerization behavior of open-shell and closed-shell aliphatic hydrocarbon ions is the following: the decomposition thresholds of open-shell ions are considerably lower than those of closed-shell ions since odd-electron ions are less stable (with respect to fragmentation) than even-electron ions. Thus it is conceivable that in open-shell hydrocarbon ions (such as molecular ions) the decomposition threshold is much lower than the isomerization threshold (see Fig. V-2a) so that decomposition is much faster than isomerization, whilst in closed-shell hydrocarbon ions the isomerization barrier is below that for decomposition (see Fig. V-2b) leading to complete or extensive equilibration (and thus atom randomization) between the different isomeric ions [14].

It should, however, be kept in mind that not only the decomposition threshold but also the height of the energy barrier for interconversion determines the extent of isomerization. This isomerization barrier may also differ for corresponding open- and closed-shell ions.

The isomerization of n-, sec-, iso- and tert-butyl ions has been discussed in more detail by Bowen and Williams [15,219]. The authors make the plausible assumption that the interconversion of isomeric butyl ions occurs by rapid, symmetry-allowed 1,2-methyl and 1,2-hydrogen shifts (which, as known from solution chemistry, do not involve activation energies where these processes are exothermic). This leads to a mechanism for interconversion outlined in Scheme V-18. The heats of

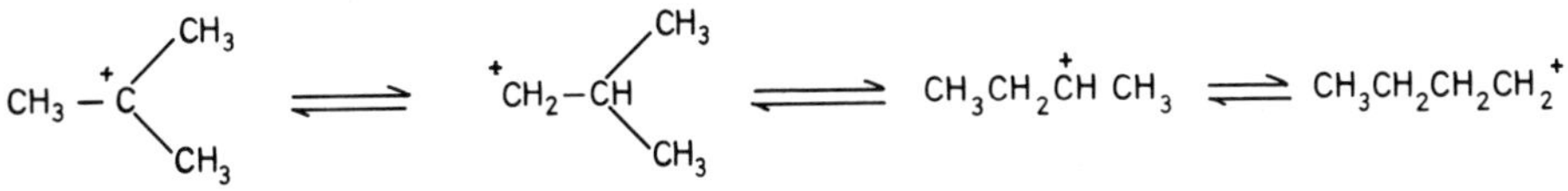

$$CH_3-\overset{+}{C}(CH_3)_2\text{-}CH_3 \rightleftharpoons \overset{+}{C}H_2-CH(CH_3)_2 \rightleftharpoons CH_3CH_2\overset{+}{C}HCH_3 \rightleftharpoons CH_3CH_2CH_2\overset{+}{C}H_2$$

(42)　　　　　(43)　　　　　(44)　　　　　(45)

Scheme V-18

formation of these ions as well as those of the products have been determined earlier with high accuracy and allow the potential energy surface shown in Fig. V-14 to be constructed. In agreement with the above

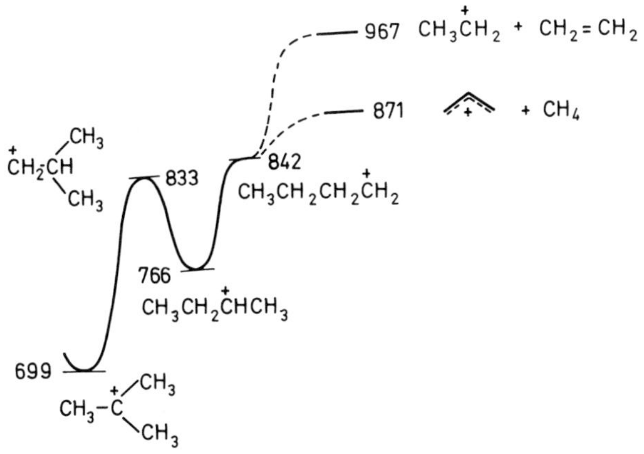

- 967　$CH_3\overset{+}{C}H_2 + CH_2=CH_2$
- 871　$+ CH_4$
- 842　$CH_3CH_2CH_2\overset{+}{C}H_2$
- 833
- 766　$CH_3CH_2\overset{+}{C}HCH_3$
- 699　$CH_3-\overset{+}{C}(CH_3)_2$
- $\overset{+}{C}H_2CH(CH_3)_2$

Fig. V-14. Potential surface for the unimolecular reactions of butyl ions (energies in kJ mol^{-1}) [15]. (By courtesy of the American Chemical Society.)

conclusion the isomerization barrier between sec- and tert-butyl ions, for instance, is below the lowest threshold for decomposition (methane loss). Surprisingly, the relatively high isomerization barrier between tert- and sec-butyl ions is not reflected in the collisional activation spectra of these ions which are almost indistinguishable [212,220].

This result suggests that in this specific case either the isomerization after collision is fast compared with competing decompositions or, less likely, the CA technique samples exclusively ions near the threshold for decomposition.

The above generalizations still hold approximately for large unsaturated aliphatic hydrocarbon ions (> C_7) as well. For instance, collisional activation [221,222] and metastable ion spectra [223-225] as well as isotopic labelling [86,226-228] demonstrate that the open-shell isomeric molecular ions of the general formula $C_nH_{2n}^{+\cdot}$ (n = 7,8) and $C_nH_{2n-2}^{+\cdot}$ (n = 6,8) have a much lower tendency to isomerize than the corresponding closed-shell fragment ions $C_nH_{2n-1}^{+}$ (n = 3-7), $C_nH_{2n-3}^{+}$ (n = 6,7), $C_nH_{2n-5}^{+}$ and $C_nH_{2n-7}^{+}$ (n = 6), which show extensive or even complete equilibration to interconverting structures. This mixture of interconverting ions also includes cyclic structures, at least in the case of $C_6H_7^{+}$ and $C_6H_5^{+}$ ions [222].

Unfortunately the situation is more complex with small unsaturated hydrocarbon ions. Thus according to isotopic labelling data [229-232], metastable ion spectra [233,234], field ionization kinetics [235], photoelectron-photoion-coincidence measurements [210] and heat of formation data [68,236] open-shell $C_4H_8^{+\cdot}$ and $C_4H_6^{+\cdot}$ ions with sufficient energy to decompose have completely equilibrated to a mixture of interconverting structures whilst on the other hand MO calculations [195, 196], appearance potential measurements [68,202], kinetic energy release data [197-199] and collisional activation spectra [201] suggest that the closed-shell $C_3H_3^{+}$ ions exist as two distinct species, the propargyl and cyclopropenyl ion.

The influence of the molecular size on the isomerization of unsaturated or alicyclic hydrocarbon ions will be discussed more quantitatively for $C_nH_{2n}^{+\cdot}$ ions, the energetics of which have recently been studied in detail by Holmes et al. [99,100] and by Bowen and Williams [237].

Fig. V-15 shows the energy diagram for $C_3H_6^{+\cdot}$ and $C_4H_8^{+\cdot}$ fragmentations as reported by Holmes et al. [99,100]. In $C_3H_6^{+\cdot}$ ions the lowest threshold for decomposition of the least stable isomer is 1.3 eV, in $C_4H_8^{+\cdot}$ still 0.7 eV. These internal energies should be more than sufficient to cause interconversion between the various isomers although the diagrams in Fig. V-15 do not give any direct information about the highest point on the potential energy surface of interconverting $C_3H_6^{+\cdot}$ and $C_4H_8^{+\cdot}$ ions (i.e. the barrier for isomerization). With increasing molecular size, however, the difference between the heats of formation of the least stable isomer and the most favorable products decreases (0.5 eV for $C_5H_{10}^{+\cdot}$ ions, 0.04 eV for $C_6H_{12}^{+\cdot}$ ions [237]). Thus Bowen and Williams [237] concluded that in $C_5H_{10}^{+\cdot}$ ions one of the plausible

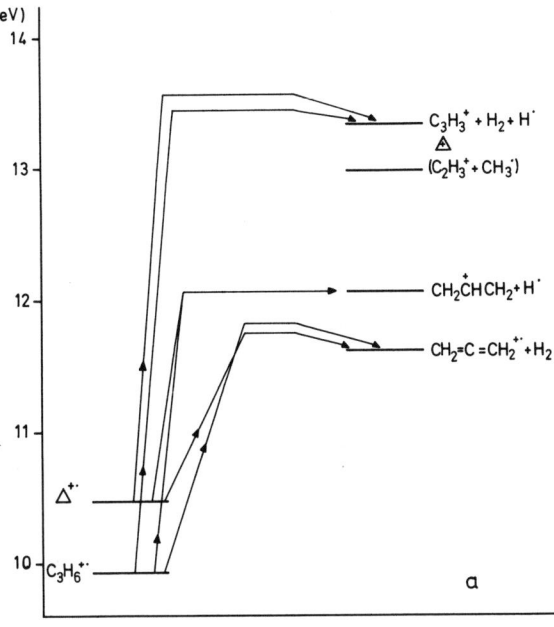
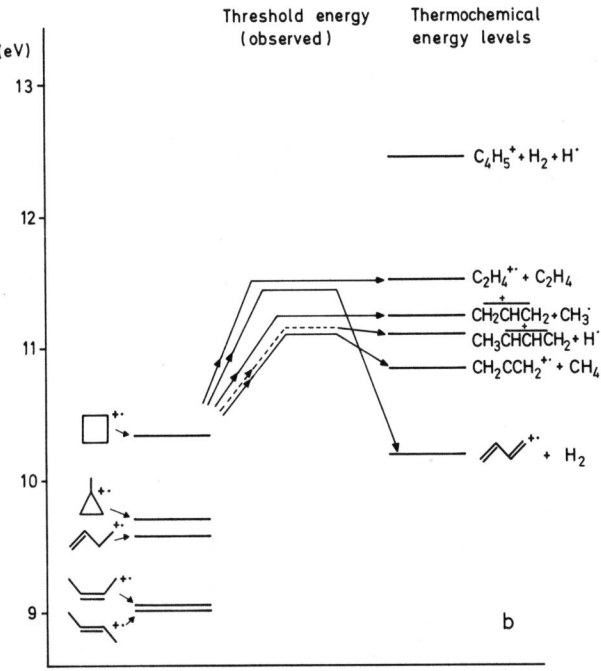

Fig. V-15. Energy diagram for fragmentations of $C_3H_6^{+\cdot}$ and $C_4H_8^{+\cdot}$ isomers [99,100]. (By courtesy of Heyden & Son Ltd.)

intermediates for the interconversion (a configuration with a secondary cationic site and a primary radical site) is still ~ 0.5 eV lower in energy than the most favorable product combination, whilst in $C_6H_{12}^{+\cdot}$ ions the corresponding intermediate is comparable in energy to the most favorable products. Thus interconversion and direct decomposition even of long-living $C_6H_{12}^{+\cdot}$ ions may compete with each other. Summarizing, in the homologous series of $C_nH_{2n}^{+\cdot}$ ions interconversion of the various isomers with sufficient energy to decompose and a lifetime of $\sim 10^{-6}$ s is observed up to n = 5*) while for n > 5 direct decomposition from the unrearranged precursor may compete with the isomerization more and more effectively as n increases [14,237].

4.1.2 The Influence of a Heteroatom on the Isomerization of a Hydrocarbon Chain

Heteroatom-containing ions are reviewed in Section V-4.2. Here only the influence of a heteroatom in directing the isomerization of a carbon skeleton will be discussed. Whilst the collisional activation spectra of isomeric pentyl ions initially differing in the branching of the hydrocarbon chain are identical (Fig. V-13), indicating complete equilibration of these isomers [212], substantial differences in the relative abundances are observed if such pentyl ions are linked to a heteroatom or a heteroatom-containing moiety [240]. This is illustrated in Fig. V-16 in which the CA spectra of n-$C_5H_{11}CO^+$ and iso-$C_5H_{11}CO^+$ (which are both closed-shell ions) are compared. Similar results were reported for n- and iso-$C_5H_{11}OCH_2^+$ as well as n- and iso-$C_5H_{11}NHCH_2^+$ ions [240]. It is obvious that the presence of a heteroatom reduces the tendency for an isomerization of the hydrocarbon chain. Energetic data suggest that the presence of a heteroatom leads to an increase in the isomerization threshold of the hydrocarbon chain [240]. The CA spectra only allow the exclusion of a *complete* equilibration e.g. between the n-, iso-, sec- and tert - $C_5H_{11}CO^+$ ions *prior* to decomposition, whilst chain branching may still occur *during* decomposition.

The stabilizing influence of a heteroatom can be used to distinguish between isomeric alkyl ions in the gas phase. Dymerski and McLafferty [220] demonstrated that gas phase derivation of n-, iso-, sec- and tert-butyl ions in a high pressure source e.g. by reaction with CH_3COCl leads to isomeric $C_6H_{11}O^+$ ions which give distinct CA spectra.

---

*) The non-decomposing $C_4H_8^{+\cdot}$ and $C_5H_{10}^{+\cdot}$ isomers may, however, still be differentiated using the collisional activation method [14,238,239].

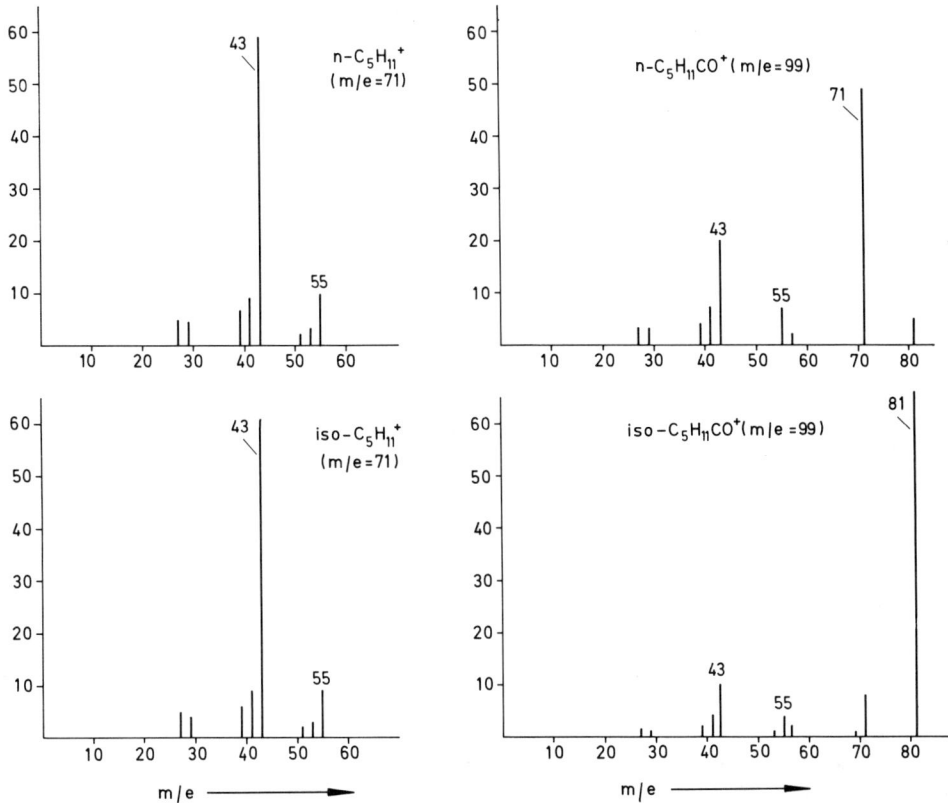

Fig. V-16. Collisional activation spectra of n-$C_5H_{11}CO^+$ and iso-$C_5H_{11}CO^+$ ions generated from the corresponding dipentyl ketones [240].

4.1.3 The Time Scale for Isomerization of Hydrocarbon Ions

Information about the time scale for the isomerization of hydrocarbon ions can be obtained using the field ionization technique. Qualitative information on the progress of an isomerization process as function of the ion lifetime may be obtained from time-dependent metastable ion spectra as discussed in Section V-3.4.3.5 whilst the quantitative determination of the relative rates of formation of the various fragments yields information about the time scale for such isomerization reactions. This will be demonstrated with linear octene ions as example [86]. If in these double bond isomers the radical site were localized after ionization one would expect prominent peaks due to allylic cleavage with charge retention or charge migration as shown in Scheme V-19 for 1-, 2-, 3- and 4-octene. The time dependent metastable ion spectra

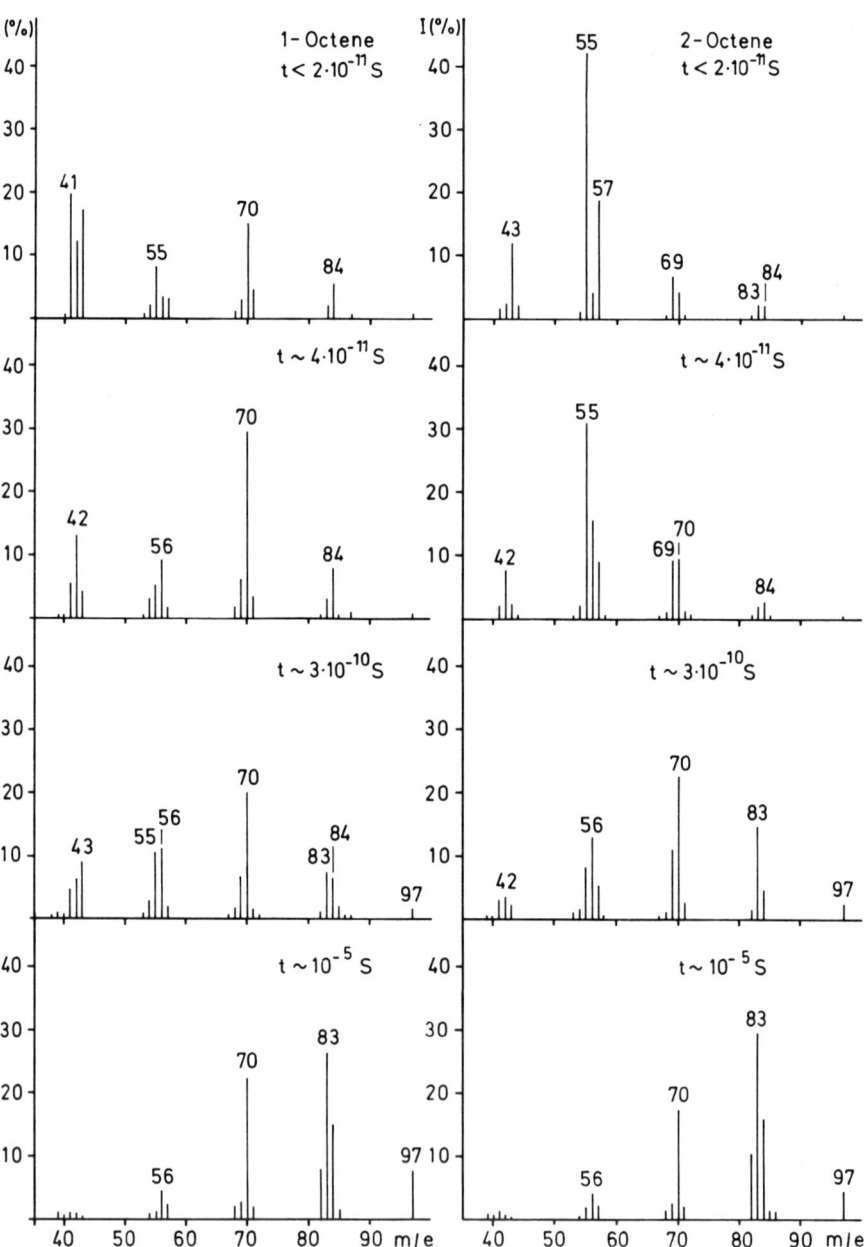

Fig. V-17. Time-dependent metastable ion spectra of 1-, 2-, 3- and 4- octene ions [86].

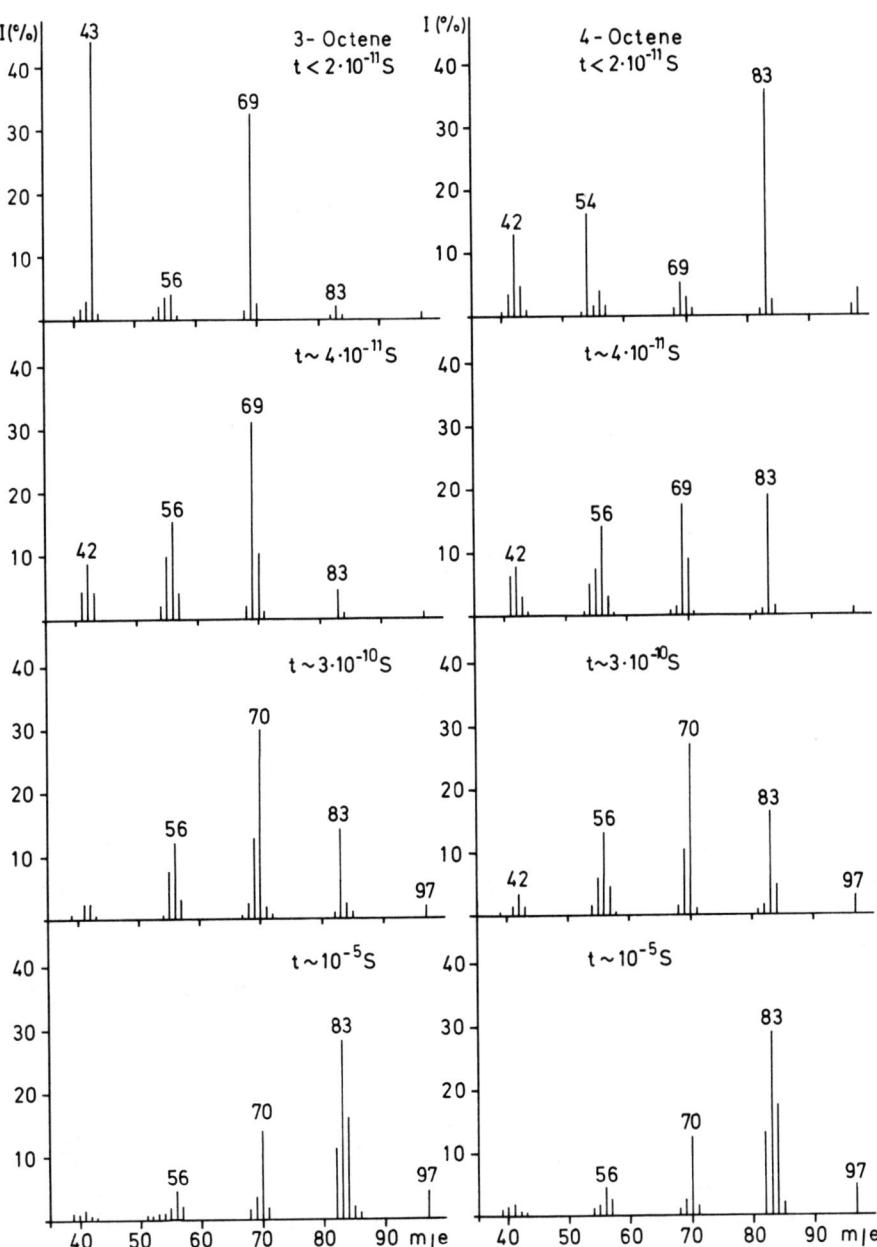

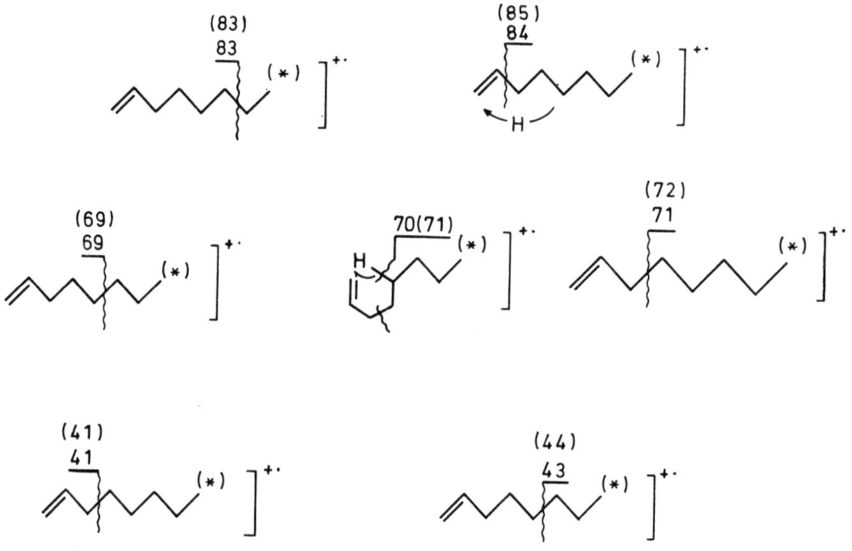

276   V   *The Structure of Gaseous Ions*

Scheme V-19

of these four double bond isomers are shown in Fig. V-17. Identical spectra are observed at ∼ 10^{-5} s after ionization demonstrating that complete isomerization to a mixture of interconverting 1-, 2-, 3- and 4- octene molecular ions has occurred*). With decreasing ion lifetime pronounced differences between the metastable ion spectra of the four isomers become apparent. At the shortest resolvable lifetime the spectra are dominated by fragments due to allylic cleavages, demonstrating that within 2×10^{-11} s after ionization the radical site is largely localized, i.e. the ions retain their original structure. In Fig. V-18 the relative rates of formation of four fragment ions are plotted as function of the precursor ion lifetime. The pronounced differences in the relative rates observed at the shortest ion lifetime (2×10^{-11} s) are again consistent with decomposition from the initial structure with localized radical site. With increasing ion lifetime the relative rates become more similar for the four isomers and, within the limits of reproducibility of the measurements, are the same after roughly 10^{-9} s after ionization, demonstrating that the isomerization process is complete at that time. Isotopic labelling [86] supports this conclusion

---

*
Additional collisional activation studies demonstrate that neither branched [221] nor cyclic [29] isomeric structures take part in this isomerization process to a significant degree.

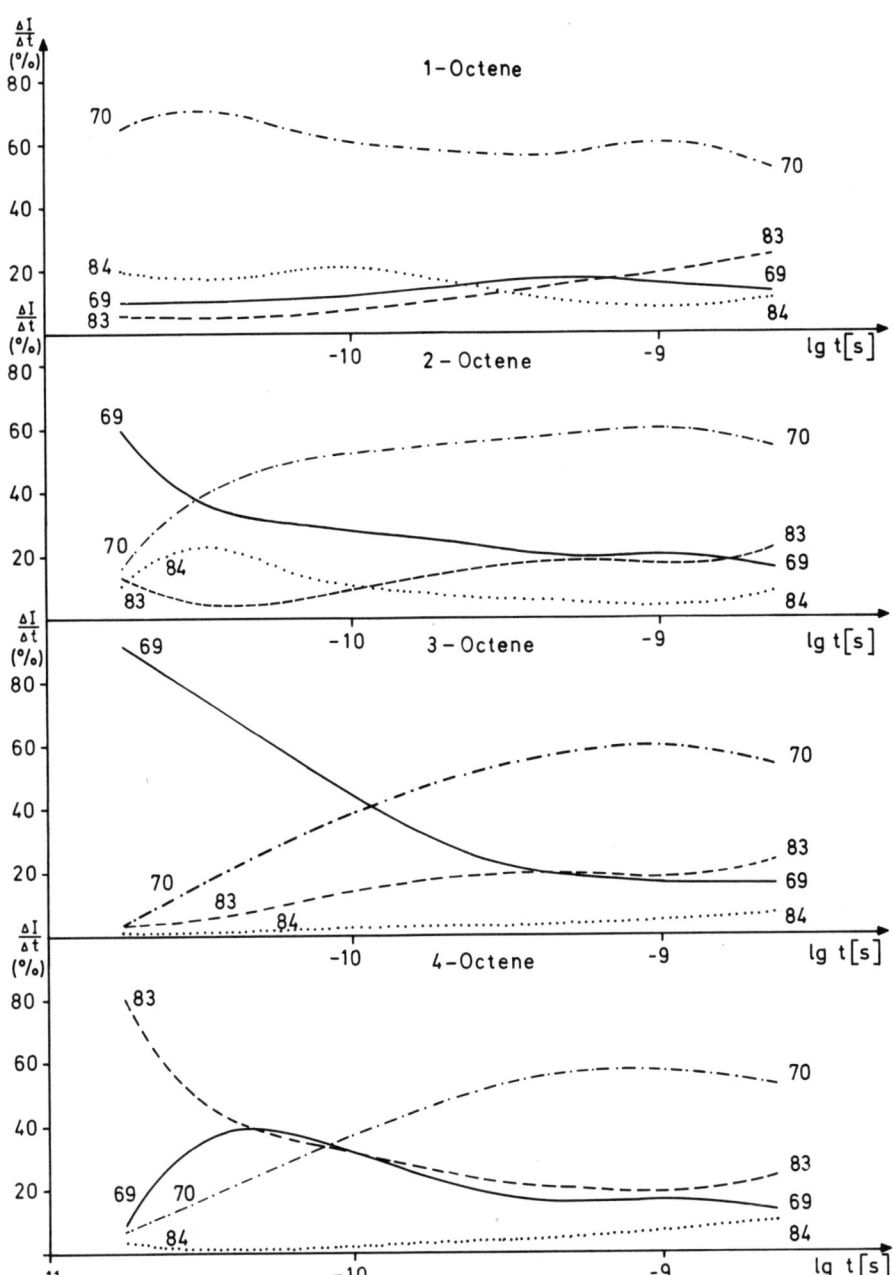

Fig. V-18. Relative rates for loss of C_2H_4, $C_2H_5\cdot$, C_3H_6 and $C_3H_7\cdot$ from 1-, 2-, 3- and 4- octene ions as a function of time [86].

and suggests that the interconversion between 1-, 2-, 3- and 4-octene ions does not result primarily from a skeletal isomerization, but, as expected, from radical site migration accompanied by hydrogen rearrangement[*].

On the other hand, $C_4H_8^{+\cdot}$ ions with sufficient energy to decompose also undergo complete skeletal isomerization as shown in the previous section. Morgan and Derrick [235] demonstrated that the molecular ions of 1-butene, 2-butene, 2-methylpropene and methylcyclopropane isomerize within 10^{-9} s to a mixture of interconverting structures whilst the isomerization process in cyclobutane is slower. Isotopic labelling suggests that interconversion between the branched and linear butene ions may occur via the methylcyclopropane ion as an intermediate [241].

Finally, ion lifetime measurements have demonstrated that the isomerization of alkylcycloalkanes with 3- to 5-membered rings to 1-octene is complete after 10^{-9} s (see Section V-3.4.3.5).

4.1.4 Aromatic Hydrocarbon Ions

Although some generalizations regarding the isomerization at least of large aliphatic hydrocarbon ions were possible the situation is more complex with aromatic hydrocarbon ions. Again the tendency for complete equilibration between isomeric (odd-electron) molecular ions is usually lower than that of even-electron fragments generated from these precursors. However, the even-electron fragments differ considerably in their behavior: whilst it was possible to detect several stable isomers in the case of $C_3H_3^+$ [201], $C_7H_7^+$ [123,124], $C_8H_9^+$ [242-244], $C_9H_{11}^+$ [245], $C_{10}H_{13}^+$ [246] and $C_{11}H_{11}^+$ [247] using the collisional activation method the large majority of non-decomposing isomeric $C_6H_5^+$ [206,207,222], $C_6H_7^+$ [222], $C_7H_9^+$ [248], $C_9H_7^+$ [249,250] and $C_{11}H_9^+$ ions [251] seem to interconvert freely[**]. According to the precursor, the mixture of interconverting structures also includes open chain structures in several instances[***].

A comparison of the isomerization behavior of n-, sec-, iso-, tert-butyl ions and the corresponding phenyl-substituted ions, $C_6H_5C_4H_8^+$ ions, generated by methyl loss from n-, iso-, sec- and tert-pentylbenzenes is of interest [246]. Whilst the CA spectra of the iso-

---

[*] The terminal hydrogens do not take part in this isomerization process.

[**] In spite of the almost identical CA spectra of these ions there may still be finite barriers for interconversion between the various isomers.

[***] Thus the differentiation between acyclic and aromatic hydrocarbon ions is to some extent arbitrary.

meric butyl ions are indistinguishable, those of the phenyl-substituted butyl ions give distinct spectra demonstrating that introduction of a phenyl ring reduces the tendency for isomerization of the hydrocarbon chain. Equilibration among the isomeric phenyl-substituted butyl ions is not even complete for those ions with sufficient energy to decompose and a lifetime of 10^{-5} s as shown by isotopic labelling [246]. Energetic data suggest that in this specific case the reduced tendency for isomerization results at least in part from the presence of decomposition channels with lower activation energy than in the unsubstituted butyl ions.

4.2 Heteroatom-containing Ions

4.2.1 Stability of Heteroatom-containing Ions

The stabilizing effect of heteroatoms on carbenium ion centers is well known. Thus molecular orbital calculations show that CH_2OH^+ and $CH_2NH_2^+$ ions, for instance, are stabilized with respect to CH_2H^+ ions by 201 and 276 kJ mol^{-1}, respectively [307]. Hence, in contrast to isomeric alkyl ions, $C_nH_{2n+1}^+$, which interconvert freely prior to decomposition, several non-interconverting isomeric oxonium, sulphonium and immonium ions, $C_nH_{2n+1}X^+$ (X = O, S, NH) exist, even at internal energies corresponding to the lowest threshold for decomposition (Table V-9a). If techniques which sample non-decomposing ions (CA, ICR) are employed, an even larger number of stable isomers can be detected, as summarized in Table V-9a. However, ions of the general formula $^+(CH_2)_nXR$ (46) and $CH_3(CH_2)_{n-1}X^+$ (47) (X = S, O, NH; R = H, CH_3, C_6H_5) have not been observed as distinct species [15,41,79,108,252-255] and thus probably do not correspond to a minimum on the potential energy surface. These ions might even not be formed initially. Thus the fragmentation of $C_6H_5OCH_2CH_2OCH_3^{+\cdot}$ ions does not lead to the ion $^+CH_2CH_2OCH_3$ (48) but directly to the ion $CH_3CH=OCH_3^+$ (49), probably by reciprocal hydrogen shift [255] (see Scheme IV-26) whilst decomposition of n-alkylthiols, $CH_3(CH_2)_nSH^{+\cdot}$, does not lead even initially to formation of $^+CH_2CH_2SH$ (50), but to protonated ethylene sulfide $\overline{CH_2CH_2}SH^+$ (51) in a displacement reaction [46] (see Scheme V-3).

A variety of stable isomers have not only been detected for closed-shell heteroatom-containing ions, but also for oxygen and nitrogen containing open-shell ions (radical cations), $C_nH_{2n}X^{+\cdot}$ (X = O, NH; n = 2,3) as summarized in Table V-9a.

4.2.2 Potential Energy Surfaces

In spite of our growing knowledge on gaseous ion structures little was known about the detailed potential energy surfaces of these ions until recently although this information is of crucial importance when discussing the isomerization of ions as shown in Section V-2.2. In that section four general types of potential energy surfaces were distinguished. In principle partial or even complete potential energy surfaces of organic ions can be obtained by molecular orbital calculations, at least for small ions, and interesting results have already been reported (see Section V-3.6). Moreover, potential energy surfaces of several closed-shell ions have recently been reported by Williams et al. [15,18,20-23,256] some of which have already been discussed in other sections of this book.

The potential energy surfaces are constructed using known or calculated heats of formation of the various isomeric ions and intermediates whilst the transition state energies for decomposition or isomerization are determined, if possible, from appearance potential measurements. Furthermore, the potential energy diagram rests on the assumption that symmetry allowed 1,2-methyl and 1,2-hydrogen shifts occur without activation energy where these processes are exothermic (or with no reverse activation energy if endothermic). Isotopic labelling, metastable abundance ratios and kinetic energy release values are used to support the proposed energy surfaces [15]. Some potential energy surfaces typical for heteroatom-containing ions will be discussed in more detail.

$C_2H_5O^+$ *Ions.* The $C_2H_5O^+$ ions are the most thoroughly studied heteroatom-containing ions. As discussed in Section V-3 it has been shown by various techniques that three stable isomers exist: $CH_3CH=OH^+$ (5), $CH_3O=CH_2^+$ (6) and $\overline{CH_2CH_2OH}^+$ (7). A probable mechanism for interconversion of these ions is shown in Scheme V-20. Using published or calcula-

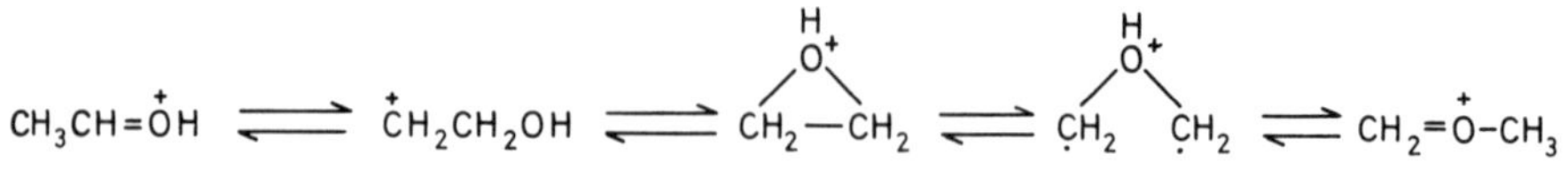

$$CH_3CH=\overset{+}{O}H \;\rightleftharpoons\; \overset{+}{C}H_2CH_2OH \;\rightleftharpoons\; \underset{(7)}{CH_2-CH_2}\overset{\overset{H}{\underset{|}{\overset{+}{O}}}}{} \;\rightleftharpoons\; \underset{}{\overset{\overset{H}{\underset{|}{\overset{+}{O}}}}{CH_2\;\;CH_2}} \;\rightleftharpoons\; CH_2=\overset{+}{O}-CH_3$$

(5)　　　　(8)　　　　(7)　　　　　　　　　　(6)

Scheme V-20

ted heats of formation of these species and their lowest thresholds for decomposition one arrives at a potential energy surface shown in Fig.

V-19*) [15,18]. It is obvious that the highest energy species on this surface is the diradicalic ion $\cdot CH_2O(H)CH_2^+\cdot$ (52) the heat of formation of which is well above the threshold for decomposition of both $CH_3CH=OH^+$ (5) and $CH_3O=CH_2^+$ (6), so that (5) and (6) do not interconvert prior to metastable decomposition**). However, if the potential

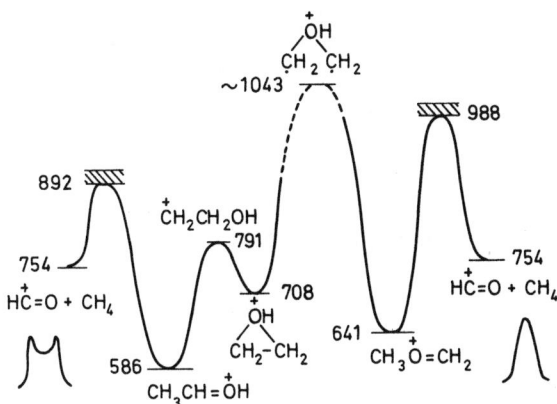

Fig. V-19. Potential energy surface for the unimolecular reactions of $C_2H_5O^+$ ions (energies in kJ mol^{-1}) [15]. (By courtesy of the American Chemical Society.)

energy diagram in Fig. V-19 is correct interconversion between the protonated acetaldehyde (5) and protonated ethylene oxide (7) should be observed prior to metastable decomposition. Such interconversion should lead to an equilibration of both carbons and all hydrogens except that bound to oxygen. Isotopic labelling demonstrated that this is indeed the case.

Thus the complete potential energy surface of $C_2H_5O^+$ ions includes the two contrasting cases where the isomerization barrier between two ions is either higher or lower than the decomposition thresholds of both ions (corresponding to Fig. V-2a and b). The potential energy diagram in Fig. V-19 illustrates why all techniques which sample non-decomposing ions (CA,ICR) led to the detection of three distinct $C_2H_5O^+$ ions while with those techniques which reflect the structure of the decomposing ions (MI,T) only two distinct $C_2H_5O^+$ ions have been observed.

---

*
Further details of this potential energy surface are reported in the original publication [18].
**
Isomerization will be also suppressed for ions of higher internal energy, as decomposition will be much faster than isomerization.

$C_2H_6N^+$ *Ions.* The potential energy surface of the $C_2H_6N^+$ ions $CH_3CH=NH_2^+$ (55), $CH_2CH_2NH_2^+$ (56), $\overline{CH_2CH_2NH_2}^+$ (57) and $CH_2=NHCH_3^+$ (58) can be constructed in analogy to that of the $C_2H_5O^+$ ions shown in Fig. V-19 [18, 19]. In correspondence to the oxonium ions (5) - (7), the $C_2H_6N^+$ ions (55), (57) and (58) are characterized by pronounced wells on the energy surface, while the β-aminoethyl cation (56) corresponds to a maximum on this surface, i.e. this ion collapses without activation energy to (55) or (57). In support of this conclusion the ions (55), (57) and (58) but not (56) have been identified as stable species in the gas phase using the collisional activation method [19,79]. However, further comparison of the potential energy surfaces of $C_2H_5O^+$ and $C_2H_6N^+$ ions reveals characteristic differences:

(1) In contrast to the oxonium ion (6) the analogous immonium ion (58) undergoes a slow isomerization to (55) prior to metastable decomposition.

(2) The energy requirements for this rate determining isomerization are similar to those for decomposition of the ion (55).

(3) While the cyclic and acyclic oxonium ions (5) and (7) interconvert prior to metastable decomposition, no such interconversion is predicted and observed for the corresponding cyclic and acyclic $C_2H_6N^+$ ions (55) and (57) prior to acetylene loss.

(4) The gas phase rearrangement of the β-hydroxyethyl cation (8) to protonated acetaldehyde (5) and protonated ethylene oxide (7) is markedly less exothermic than the corresponding rearrangement of the β-aminoethyl cation (56) to the protonated imine of acetaldehyde (55) or protonated aziridine (57) demonstrating that the positive charge is better stabilized by the NH_2 than by the OH function.

$C_3H_7O^+$ *Ions.* The potential energy surfaces for the ion pairs $CH_3CH_2O=CH_2^+$ (16) and $CH_3CH=OCH_3^+$ (17) as well as $(CH_3)_2C=OH^+$ (53) and $CH_3CH_2CH=OH^+$ (54) have been reported [20-21]. They demonstrate that slow rate-determining isomerizations are an important feature of metastable $C_3H_7O^+$ ions, i.e. these ions can be described by the schematic potential energy curve shown in Fig. V-2d. Thus (17) rearranges to (16) prior to metastable decomposition as discussed in Section V-3.4.3.4 (Fig. V-6). The potential energy surface of (53) and (54) is illustrated in Fig. V-20*). In this diagram it is assumed that protonated acetone (53) undergoes a slow isomerization to protonated propionaldehyde (54) prior to metastable ethylene and water loss. This conclusion is supported by

---

*
The barrier for a slow isomerization of (53) to (54), 803 kJ mol^{-1}, may include a small "kinetic barrier" since it has been concluded by Holmes and Lossing that at least a fraction of (53) may form $CH_2=OH^+$ and C_2H_4 at threshold [257]. Further details of the potential energy surface of (53) and (54) are reported in Ref. 22.

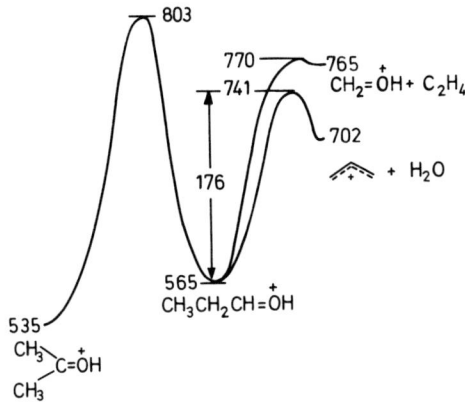

Fig. V-20. Partial potential surface for the dissociation of protonated acetone (energies in kJ mol^{-1}) [21]. (By courtesy of the Chemical Society, London.)

the following experimental evidence [21]:

(1) The thresholds for loss of H_2O and C_2H_4 from protonated acetone (54) are the same (and correspond to the energy required to isomerize (53) to (54)).

(2) This isomerization energy is greater than the thresholds for H_2O and C_2H_4 loss from protonated propionaldehyde.

(3) The kinetic energy released during metastable H_2O and C_2H_4 loss is higher for (53) than for (54) as a result of the higher excess energy of decomposing ions (53).

(4) Metastable loss of H_2O from (54) is dominant over C_2H_4 loss but H_2O loss competes less effectively if (53) is the precursor.

$C_4H_9O^+$ *Ions*. Potential energy surfaces have also been deduced for the homologous $C_4H_9O^+$ ions [23], which indicate that these isomers are again characterized by slow, rate-determining isomerizations. Thus in order to decompose the metastable ion $(CH_3)_2C=OCH_3^+$ (59) must undergo a rate determining isomerization to $CH_3CH_2CH=OCH_3^+$ (60) followed by a further rearrangement to $CH_3CH_2CH_2O=CH_2^+$ (61).

Summarizing, the construction of potential energy surfaces, even if only semiquantitative in some instances, permit the chemistry of organic ions in the gas phase, in particular their isomerization and fragmentation behavior, to be better understood.

4.3 Examples for Ion Structure Elucidation

A variety of ion structure studies in which the techniques of metastable ion characteristics (MI), kinetic energy release (T), collisional activation (CA) or ion cyclotron resonance spectroscopy (ICR) were employed is summarized in Table V-9a for hetero-atom-containing ions and in Table V-9b for hydrocarbon ions. The use of additional techniques such as isotopic labelling (L), heat of formation data (H), electron impact studies (EI), molecular orbital calculations (MO), charge stripping (CS), field ionization kinetics (FIK) or photo-electron-photoion-coincidence measurements (PPC) is indicated in the last columns. The structure of the non-decomposing and the decomposing ions, if well established, is indicated. A single arrow ($\rightarrow$) indicates a slow, rate-determining isomerization, two arrows ($\rightleftharpoons$) interconverting structures.

Table V-9a. Ion Structure Studies on Heteroatom-containing Ions

Ion	Ion Structures		References				Label, Others [a]
	Non-decomp. Ions	Decomposing Ions	MI	T	CA	ICR	
$C_2H_3N^{+\cdot}$	4 stable isomers ($CH_3CN^{+\cdot}$, $CH_3N{=}C^{+\cdot}$ identified)	interconverting structures (?)	258	258	258		34 (H)
$C_2H_4O^{+\cdot}$	$\overline{CH_2CH_2O}^{+\cdot}$ (24) $CH_3CH{=}O^{+\cdot}$ (25) $CH_2{=}CHOH^{+\cdot}$ (26)	$\overline{CH_2CH_2O}^{+\cdot}$ (24) $CH_3CH{=}O^{+\cdot}$ (25) $CH_2{=}CHOH^{+\cdot}$ (26)	259	97	1o2		26o (H)
$C_2H_4O_2^{+\cdot}$	$CH_3COOH^{+\cdot}$ (18) $CH_2{=}C(OH)_2^{+\cdot}$ (2o)	$CH_2COOH_2^{+\cdot}$ (?) $CH_3COOH^{+\cdot}$ (18) $CH_2{=}C(OH)_2^{+\cdot}$ (2o)	76, 96	96	17		17 (L)
$C_2H_4S^{+\cdot}$	$CH_3CH{=}S^{+\cdot}$ (62) $CH_2{=}CHSH^{+\cdot}$ (63)	(62 $\rightleftharpoons$ 63)			261	262	
$C_2H_4N_4^{+\cdot}$	interconverting tautomers of C-aminotriazole$^{+\cdot}$				263		263 (L)

Table V-9a. Continued

Ion	Non-decomp. Ions	Decomposing Ions	MI	T	CA	ICR	Label, Others [a]
$C_2H_5O^+$	$CH_3CH=OH^+$ (5) $\overline{CH_2CH_2}OH^+$ (7) $CH_3O=CH_2^+$ (6)	(5) ⇌ (7) ←∦− (6)	75, 264	75	46, 108, 265	43,45 160, 266, 267	46,268 269(L); 4o-44(H); 18,27o(PE)
$C_2H_5N^{+\cdot}$		$\overline{CH_2CH_2}NH^{+\cdot}$ $CH_3CH=NH^{+\cdot}$ $CH_2=CHNH_2^{+\cdot}$ $CH_2=NCH_3^{+\cdot}$	98				98(L)
$C_2H_5S^+$	$CH_3CH=SH^+$ $\overline{CH_2CH_2}SH^+$ (51) $CH_3S=CH_2^+$				46, 252		46,271, 272(L); 4o9(MO)
$C_2H_6N^+$	$CH_3CH=NH_2^+$ (55) $CH_3NH=CH_2^+$ (58) $\overline{CH_2CH_2}NH_2^+$ (57)	(58) → (57) → (55)	18 79			79	18,19(PE) 4o,434(H) 273(MO)
$C_3H_5O^+$	$CH_3CH_2CO^+$ (64) $CH_2=CHCH=OH^+$ (65) $CH_2=CHO=CH_2^+$ (66) $CH\equiv CCH_2OH_2^+$ (67) $CH_3\overline{CH=OCH_2}^+$ (68) $CH_3\overline{C=CHOH}^+$ (69) $CH_2=\overline{CCH_2OH}^+$ (7o) $\overline{CH=CHOCH_3}^+$ (71) $\overline{CH_2CH_2O=CH}^+$ (72)	$CH_3CH_2CO^+$ (64) (65) ⇌ (66) ⇌ (68) ⇌ (72)	274			4o5	274(L); 412(MO)
$C_3H_5O_2^+$		$\overline{OCH_2CH_2O=CH}^+$ $CH_2=CHC(OH)_2^+$	274				
$C_3H_6O^{+\cdot}$	$\overline{CH_2CH(CH_3)O}^{+\cdot}$ $\overline{CH_2CH_2CH_2O}^{+\cdot}$ $CH_2=CHOCH_3^{+\cdot}$ $C_2H_5CHO^{+\cdot}$ $CH_3CH=CHOH^{+\cdot}$ $CH_3COCH_3^{+\cdot}$ $CH_2=CH(OH)CH_3^{+\cdot}$	$\overline{CH_2CH(CH_3)O}^{+\cdot}$ $\overline{CH_2CH_2CH_2O}^{+\cdot}$ $C_2H_5CHO^{+\cdot}$ $CH_3COCH_3^{+\cdot}$ $CH_2=CH(OH)CH_3^{+\cdot}$	77 259, 275	11	1o 1o8	8,9 276, 418	34(H); 277(L)

286   V   The Structure of Gaseous Ions

Table V-9a.   Continued

Ion	Non-decomp. Ions	Decomposing Ions	MI	T	CA	ICR	Label, Others [a]
$C_3H_7O^+$	$(CH_3)_2C=OH^+$ (53) $C_2H_5CH=OH^+$ (54) $CH_2=OC_2H_5^+$ (16) $CH_3CH=OCH_3^+$ (17)	(53) → (54) (17) → (16)	77, 78, 255, 278		254, 255	406	20-22 (PE); 34,41,44 (H); 255,278 280 (L)
$C_3H_7O_2^+$		$CH_3O=CHOCH_3^+$ $CH_3CH_2C(OH)_2^+$	274				
$C_3H_7S^+$	$CH_3CH=SCH_3^+$ (73) $CH_2=SC_2H_5^+$ (74) $C_2H_5CH=SH^+$ (75) $(CH_3)_2C=SH^+$ (76) $CH_3\overline{CHCH_2SH}^+$ (77) $\overline{CH_2CH_2SCH_3}^+$ (78) $\overline{CH_2CH_2CH_2SH}^+$ (79)	(73) ⇌ (78)	255, 281		253 255	147 151 406	34 (H); 255,281(L)
$C_3H_8N^+$	$CH_2=NHCH_2CH_3^+$ (11) $(CH_3)_2C=NH_2^+$ (12) $CH_3CH_2CH=NH_2^+$ (13) $CH_3CH=NHCH_3^+$ (14) $(CH_3)_2N=CH_2^+$ (15)	$CH_2=NHCH_2CH_3^+$ (11) $(CH_3)_2C=NH_2^+$ (12) $CH_3CH_2CH=NH_2^+$ (13) (14) ⇌ (15)	79 81	81	79	161	34 (H); 81,161(L); 256 (PE)
$C_4H_4O^{+\cdot}$	pyrone$^{+\cdot}$-CO → furan$^{+\cdot}$		77, 282		283	284	285,286(L)
$C_4H_8Cl^+$	$\overline{CH_2CH_2CH_2CH_2Cl}^+$				287	407	287
$C_4H_7O_2^+$		$\overline{OCH_2CH_2O=C}-CH_3^+$ $CH_3CH=CHC(OH)_2^+$	274				
$C_4H_8O^{+\cdot}$			288		288		288 (L)
$C_4H_9O^+$	$C_3H_7CH=OH^+$ $C_2H_5C(CH_3)=OH^+$	$CH_3CH_2CH_2CH=OH^+$ (80) 289 $(CH_3)_2CHCH=OH^+$ (81) $C_2H_5C(CH_3)=OH^+$ (82) $C_3H_7O=CH_2^+$ (61) $CH_3CH=OC_2H_5^+$ (83) $C_2H_5CH=OCH_3^+$ (60) $(CH_3)_2C=OCH_3^+$ (59) (59) → (60) → (61) (81) ⇌ (82)			254		23 (PE); 289 (L); 41,44 (H)

Table V-9a. Continued

Ion	Non-decomp. Ions	Decomposing Ions	MI	T	CA	ICR	Label, Others
$C_5H_7O^+$					299		
$C_5H_7O_2^+$					290		
$C_5H_8O^{+\cdot}$	2 distinct ions; reketonization	2 distinct ions		31	31	31 291	31(L)
$C_5H_{10}O^{+\cdot}$				11			11(L)
$C_5H_{11}O^+$	$(C_2H_5)_2C=OH^+$ $C_3H_7C(CH_3)=OH^+$				254		41(H)
$C_6H_5O^+$					89		
$C_6H_5O_2^+$					292		
$C_6H_6O^{+\cdot}$	5 stable isomers, inter alia: phenol$^{+\cdot}$, cyclohexadienone$^{+\cdot}$			89, 157, 293, 416	156, 157, 294	155, 162	2o3,294 295,296 297(L)
$C_6H_8N_2^+$		$C_6H_5NHNH_2^+$	298				298(L)
$C_6H_9O^+$					299		
$C_6H_{10}O^{+\cdot}$	cyclohexanone$^{+\cdot}$ 1-cyclohexene-1-ol$^{+\cdot}$				3oo	3o1	3o2(L)
$C_7H_6F^+$		interconverting structures	3o3				
$C_7H_5NO_2^{+\cdot}$	several stable isomers, but o-nitrobenzaldehyde$^{+\cdot}$ and 2,1-benzisoxazoline-3-one$^{+\cdot}$ isomerize			31o	72		71(L)
$C_7H_6NO_2^+$		$m\text{-}NO_2C_6H_4CH_2^+$ $p\text{-}NO_2C_6H_4CH_2^+$	3o5				3o6(L)
$C_8H_5O_2^+$					3o8		

Table V-9a. Continued.

Ion	Non-decomp. Ions	Decomposing Ions	MI	T	CA	ICR	Label, Others
$C_8H_6O^{+\cdot}$			76, 282				309(L)
$C_8H_7NO_3^{+\cdot}$					72		304(L)
$C_8H_8O^{+\cdot}$		$C_6H_5COCH_3^{+\cdot}$ $C_6H_5C(OH)=CH_2^{+\cdot}$			89, 92		
$C_8H_8O_2^{+\cdot}$	o-$HOC_6H_4C(OH)=CH_2^{+\cdot}$ o-$HOC_6H_4COCH_3^{+\cdot}$	ketonization of enolic ion			417	417	417(L)
$C_8H_8O_3^{+\cdot}$			311				312(L)
$C_8H_9O^+$	$C_6H_5O=CHCH_3^+$ $C_6H_5C(OH)CH_3^+$		255		255		255(L)
$C_8H_9O_2^+$					308		
$C_8H_9S^+$	$C_6H_5S=CHCH_3^+$ (84) $C_6H_5\overline{SCH_2}CH_2$ (85)	(84)⇌(85)	255		255		255(L)
$C_8H_{10}O^{+\cdot}$		several distinct isomers	313				
$C_8H_{12}NO^+$					314		315(L)
$C_8H_{12}O^{+\cdot}$		several distinct isomers	313				
$C_8H_{14}O^{+\cdot}$	cyclooctanone$^{+\cdot}$ 2-ethylcyclohexanone$^{+\cdot}$				300		
$C_9H_9O_2^+$					308		
$C_9H_9O_3^+$					311		312(L)
$C_9H_{10}N^+$			408	408			
$C_9H_{11}O^+$					316		317(L)
$C_{10}H_{10}O_2^{+\cdot}$					318		

Table V-9a. Continued.

Ion	Non-decomp. Ions	Decomposing Ions	MI	T	CA	ICR	Label Others
$C_{12}H_9O^+$	$C_{12}H_9O^+$ from (86) and (87) interconvert				1o9		
$C_{12}H_{10}O^{+\cdot}$	$C_6H_5OC_6H_5^{+\cdot}$ (86) $C_6H_5-C_6H_4OH^{+\cdot}$ (87) $C_6H_5OCO_2C_6H_5^{+\cdot}$ $-CO_2 \to$ (86)	$C_6H_5OC_6H_5^{+\cdot}$ $C_6H_5C_6H_4OH^{+\cdot}$	8o	89	1o9, 11o		38(H)

^{a} PE = potential energy surface

Table V-9b. Ion Structure Studies on Hydrocarbon Ions

Ion Series				References			
Ions	MI	T	CA	ICR	Label	Other Techniques	
$C_nH_{2n+2}^{+\cdot}$							
$C_6H_{14}^{+\cdot}$	216				213-215, 319-323	34(H)	
$C_8H_{18}^{+\cdot}$	216		212				
$C_nH_{2n+1}^{+}$							
$C_3H_7^+$	216		212,324, 325	326	327,328	66,219,351(H); 188, 33o,419(MO)	
$C_4H_9^+$	2	197	212,22o		2,3,5, 82	15,66,219(H);188,329(MO)	
$C_5H_{11}^+$			212		217	7o,219(H);425(MO)	
$C_6H_{13}^+$	218		212		217	7o,219(H)	
$C_7H_{15}^+$			212		213,217	7o,219(H)	
$C_8H_{17}^+$	218		212		213,217		

Table V-9b. Continued.

Ion Series				References			
Ions	MI	T	CA	ICR	Label	Other Techniques	

Ions	MI	T	CA	ICR	Label	Other Techniques
$C_nH_{2n}^{+\cdot}$						
$C_3H_6^{+\cdot}$	99,234	99	208	331,332	230,333-336	34,237(H);208(CS)
$C_4H_8^{+\cdot}$	100,233, 237,238, 338	100	14,238		229,232, 337	34,237(H); 235,241,342,343 (FIK)
$C_5H_{10}^{+\cdot}$	237,239, 338		14,239 14,239	339 339	231 231	34,237(H) 34,237(H)
$C_6H_{12}^{+\cdot}$	233,237 239,338		239		83,84	34,237(H); 85,340(FIK)
$C_7H_{14}^{+\cdot}$	224,338		14		84,228	34,237(H); 341(FIK)
$C_8H_{16}^{+\cdot}$	86,338		14,29, 86,221		86	34,237(H); 29,86(FIK)
$C_nH_{2n-1}^{+}$						
$C_3H_5^{+}$		197	208,221			44,67,68,197(H); 208(CS),344,420, 421(MO)
$C_4H_7^{+}$	225		221		226	68(H);422,423(MO)
$C_5H_9^{+}$	225		221		225,227	69,345,424(H); 411(MO)
$C_6H_{11}^{+}$	225		221			
$C_7H_3^{+}$			221		86	
$C_nH_{2n-2}^{+\cdot}$						
$C_4H_6^{+\cdot}$						210(PPC);34,236(H)
$C_5H_8^{+\cdot}$	346	346			346	34,346(H)
$C_6H_{10}^{+\cdot}$	76	413	222			34(H);347(FIK)
$C_8H_{14}^{+\cdot}$			222			34(H)
$C_nH_{2n-3}^{+}$						
$C_3H_3^{+}$		197-199	201			195,196,414(MO); 68,202(H)
$C_4H_5^{+}$						410(MO)

Table V-9b. Continued.

Ion Series				References		
Ions	MI	T	CA	ICR	Label	Other Techniques
$C_5H_7^+$	225					69,424(H)
$C_6H_9^+$	225		222			
$C_7H_{11}^+$	313	348	222		348	429(MO)
$C_8H_{13}^+$	313					
$C_nH_{2n-4}^{+\cdot}$						
$C_5H_6^{+\cdot}$			208			353(H);208(CS);349
$C_6H_8^{+\cdot}$	350		350	171		34(H);352
$C_7H_{10}^{+\cdot}$	313,353	353			353	
$C_8H_{12}^{+\cdot}$	313					34(H);354
$C_nH_{2n-5}^+$						
$C_5H_5^+$			208		131	69(H);208(CS); 425-427(MO);428
$C_6H_7^+$		104,105	222			34(H),355(MO)
$C_7H_9^+$		94,104,105	248			34(H);429(MO)
$C_8H_{11}^+$		105				
$C_9H_{13}^+$		105				
$C_nH_{2n-6}^{+\cdot}$						
$C_6H_6^{+\cdot}$		89,356	206,207 357	158,358	357,359-363	34(H);73(EI);206(CS); 364-369;355,370(MO)
$C_7H_8^{+\cdot}$		94	108,371	127,134, 165-167	33,121, 204,205, 372-375	34(H);376(MO)
$C_8H_{10}^{+\cdot}$				163	373,378	

Table V-9b. Continued.

Ion Series	MI	T	CA	ICR	Label	Other Techniques
$C_nH_{2n-7}^+$						
$C_6H_5^+$		89	206,207, 222	430	33	73,379-381; 382(MO)
$C_7H_7^+$		94	123,124	125,126, 137,141	113	111-135; 67,136,137,141(H); 139-141,414
$C_8H_9^+$	383,384		242-244		242-244 373,378, 385-388	389,390(H); 391-394; 415,431-433(MO)
$C_9H_{11}^+$	395		245		395	
$C_{10}H_{13}^+$			246		246	
$C_nH_{2n-8}^{+\cdot}$						
$C_8H_8^{+\cdot}$	396		207	168,397	388,396, 398-400	401(FIK); 402(H)
$C_9H_{10}^{+\cdot}$				168		
$C_{10}H_{12}^{+\cdot}$			403	168	403	
$C_{11}H_{14}^{+\cdot}$			404		404	
$C_{12}H_{16}^{+\cdot}$			404		404	
Other Ions						
$C_9H_7^+$			249,250		435,438	436(H)
$C_{10}H_{10}^{+\cdot}$	437	437				
$C_{11}H_9^+$			257		438	436(H)
$C_{11}H_{11}^+$			247			
$C_{13}H_9^+$			108		439,440	

5 References

1. G.A. Olah, R.J. Spear and D.A. Forsyth, J. Am. Chem. Soc., **98**, 6284 (1976) and references herein.
2. B. Davis, D.H. Williams and A.N.H. Yeo, J. Chem. Soc., (B), 81 (1970).
3. A.N.H. Yeo and D.H. Williams, Chem. Commun., 737 (1970).
4. K. Levsen, Org. Mass Spectrom., **10**, 43 (1973).
5. H.W. Leung, C.W. Tsang and A.G. Harrison, Org. Mass Spectrom., **11**, 664 (1976).
6. J.L. Holmes, A.D. Osborne and G.M. Weese, Org. Mass Spectrom., **10**, 867 (1975).
7. F.W. McLafferty, Anal. Chem., **31**, 82 (1959).
8. J. Diekman, J.K. MacLeod, C. Djerassi and J.D. Balde-Schwieler, J. Am. Chem. Soc., **91**, 2069 (1969).
9. G. Eadon, J. Diekman and C. Djerassi, J. Am. Chem. Soc., **92**, 6205 (1970).
10. C.C. Van de Sande and F.W. McLafferty, J. Am. Chem. Soc., **97**, 4617 (1975).
11. J.H. Beynon, R.M. Caprioli and R.G. Cooks, Org. Mass Spectrom., **9**, 1 (1974).
12. T.W. Bentley in "Mass Spectrometry" Vol. 3 (R.A.W. Johnstone, Ed.), The Chemical Society, London, 1975, Chapter 2.
13. T.W. Bentley and R.A.W. Johnstone, Adv. Phys. Org. Chem., **8**, 151 (1970).
14. K. Levsen and J. Heimbrecht, Org. Mass Spectrom., **12**, 131 (1977).
15. D.H. Williams, Acc. Chem. Res., **10**, 280 (1977).
16. F. Borchers and K. Levsen, Org. Mass Spectrom., **10**, 584 (1975).
17. K. Levsen and H. Schwarz, J. Chem. Soc. Perkin Trans. II, 1231 (1976).
18. R.D. Bowen, D.H. Williams and G. Hvistendahl, J. Am. Chem. Soc., **99**, 7509 (1977).
19. C.C. Van de Sande, S.Z. Ahmad, F. Borchers and K. Levsen, in preparation.
20. G. Hvistendahl and D.H. Williams, J. Am. Chem. Soc., **97**, 3097 (1975).
21. G. Hvistendahl, R.D. Bowen and D.H. Williams, J. Chem. Soc. Chem. Commun., 294 (1976).
22. R.D. Bowen, J.R. Kalman and D.H. Williams, J. Am. Chem. Soc., **99**, 5481 (1977).
23. R.D. Bowen and D.H. Williams, J. Am. Chem. Soc., **99**, 6822 (1977).
24. M. Allan and J.P. Maier, Chem. Phys. Letters, **34**, 442 (1975).
25. M. Allan and J.P. Maier, Chem. Phys. Letters, **43**, 94 (1976).
26. M. Allan, E. Kloster-Jensen and J.P. Maier, Chem. Phys., **7**, 11 (1976).
27. M. Allan, J.P. Maier, O. Marthaler and E. Kloster-Jensen, Chem. Phys., in press.
28. M. Allan, J.P. Maier and O. Marthaler, Chem. Phys., **26**, 131 (1977).
29. F. Borchers, K. Levsen, H. Schwarz, C. Wesdemiotis and R. Wolfschütz, J. Am. Chem., **99**, 1716 (1977).
30. D.H. Williams and I. Howe, "Principles of Organic Mass Spectrometry", McGraw-Hill, London, 1972.
31. J.R. Hass, R.G. Cooks, J.F. Elder, M.M. Bursey and D.G.I. Kingston, Org. Mass Spectrom., **11**, 697 (1976).
32. K. Levsen and H. Schwarz, Angew. Chem. (Int. Ed.), **15**, 509 (1976).
33. H.M. Grubb and S. Meyerson in "Mass Spectrometry of Organic Ions", Academic Press, New York, 1963, Chapter 10.
34. H.M. Rosenstock, K. Draxl, B.W. Steiner and J.T. Herron, J. Phys. Chem. Ref. Data, Vol. 6, Suppl. 1, 1977 and references herein.

35. K. Hartmann, S. Lias, P.J. Ausloos and H.M. Rosenstock, National Bureau of Standards, NBSIR 76-1061 (1976).
36. J.F. Elder, J.H. Beynon and R.G. Cooks, Org. Mass Spectrom., 11, 415 (1976).
37. F. Benoit, Org. Mass Spectrom., 7, 1407 (1973).
38. P. Natalis and J.L. Franklin, J. Phys. Chem., 69, 2943 (1965).
39. K.B. Wiberg and W.J. Bartley, J. Am. Chem. Soc., 84, 3980 (1964).
40. B.H. Solka and M.E. Russell, J. Phys. Chem., 78, 1268 (1974).
41. F.P. Lossing, J. Am. Chem. Soc., 99, 7526 (1977).
42. K.M.A. Refaey and W.A. Chupka, J. Chem. Phys., 48, 5205 (1968).
43. J.L. Beauchamp and R.C. Dunbar, J. Am. Chem. Soc., 92, 1477 (1970).
44. R. Botter, J.M. Pechine and H.M. Rosenstock, Int. J. Mass Spectrom. Ion Phys., 25, 7 (1977).
45. R.H. Staley, R.R. Corderman, M.S. Foster and J.L. Beauchamp, J. Am. Chem. Soc., 96, 1260 (1974).
46. B. Van de Graaf, P.P. Dymerski and F.W. McLafferty, J. Chem. Soc. Chem. Commun., 978 (1975).
47. F.M. Benoit, A.G. Harrison and F.P. Lossing, Org. Mass Spectrom., 12, 78 (1977).
48. S. Tajima and T. Tsuchiya, Bull. Chem. Soc. Japan, 46, 3291 (1973).
49. G.R. Branton and C.N.K. Pua, Can. J. Chem., 51, 624 (1973).
50. C. Köppel, H. Schwarz and F. Bohlmann, Org. Mass Spectrom., 9, 324 (1974).
51. C. Köppel, H. Schwarz and F. Bohlmann, Org. Mass Spectrom., 9, 321 (1974).
52. C. Köppel, H. Schwarz and F. Bohlmann, Org. Mass Spectrom., 8, 25 (1974).
53. H. Schwarz and F. Bohlmann, Org. Mass Spectrom., 7, 395 (1973).
54. G. Conde-Caprace and J.E. Collin, Org. Mass Spectrom., 6, 415 (1972).
55. G. Conde-Caprace and J.E. Collin, Org. Mass Spectrom., 6, 341 (1972).
56. H. Schwarz and F. Bohlmann, Org. Mass Spectrom., 7, 1197 (1973).
57. C. Köppel, H. Schwarz and F. Bohlmann, Org. Mass Spectrom., 9, 567 (1974).
58. F. Bohlmann, C. Köppel, B. Müller, H. Schwarz and P. Weyerstahl, Tetrahedron, 30, 1011 (1974).
59. H. Schwarz and F. Bohlmann, Tetrahedron Lett., 3703 (1973).
60. H. Schwarz, K. Praefcke and J. Martens, Tetrahedron, 29, 2877 (1973).
61. H. Schwarz, C. Wesdemiotis, B. Hess and K. Levsen, Org. Mass Spectrom., 10, 595 (1975).
62. H. Schwarz, R.D. Petersen and C.C. Van de Sande, Org. Mass Spectrom., 12, 391 (1977).
63. R.F. Pottie, A.G. Harrison and F.P. Lossing, J. Am. Chem. Soc., 83, 3204 (1961).
64. R.F. Pottie and F.P. Lossing, J. Am. Chem. Soc., 83, 2634 (1961).
65. R. Taubert and F.P. Lossing, J. Am. Chem. Soc., 84, 1523 (1962).
66. F.P. Lossing and G.P. Semeluk, Can. J. Chem., 48, 955 (1970).
67. F.P. Lossing, Can. J. Chem., 49, 357 (1971).
68. F.P. Lossing, Can. J. Chem., 50, 3973 (1972).
69. F.P. Lossing and J.C. Traeger, J. Am. Chem. Soc., 97, 1579 (1975).
70. F.P. Lossing and A. Maccoll, Can. J. Chem., 54, 990 (1976).
71. F. Benoit and J.L. Holmes, Org. Mass Spectrom., 3, 993 (1970).

72. H. Schwarz, R. Sezi, K. Levsen, H. Heimbach and F. Borchers, Org. Mass Spectrom., 12, 569 (1977).
73. J. Momigny, L. Brakier and L. D'OR, Bull. Cl. Sci. Acad. Roy. Belg., 48, 1002 (1962).
74. H.M. Rosenstock, V.H. Dibeler and F.N. Harlee, J. Chem. Phys., 40, 591 (1964).
75. T.W. Shannon and F.W. McLafferty, J. Am. Chem. Soc., 88, 5021 (1966).
76. J.L. Occolowitz, J. Am. Chem. Soc., 91, 5202 (1969).
77. A.N.H. Yeo and D.H. Williams, J. Am. Chem. Soc., 93, 395 (1971).
78. C.W. Tsang and A.G. Harrison, Org. Mass Spectrom., 7, 1377 (1973).
79. K. Levsen and F.W. McLafferty, J. Am. Chem. Soc., 96, 139 (1974).
80. D.H. Williams, S.W. Tam and R.G. Cooks, J. Am. Chem. Soc., 90, 2150 (1968).
81. N.A. Uccella, I. Howe and D.H. Williams, J. Chem. Soc. (B), 1933 (1971).
82. R. Liardon and T. Gäumann, Helv. Chim. Acta, 54, 1968 (1971).
83. D.P. Stevenson, J. Am. Chem. Soc., 80, 1571 (1958).
84. S. Meyerson, T.D. Nevitt and P.N. Rylander, Adv. Mass Spectrom., 2, 313 (1963).
85. A.M. Falick and A.L. Burlingame, J. Am. Chem. Soc., 97, 1525 (1975).
86. F. Borchers, K. Levsen, H. Schwarz, C. Wesdemiotis and H.U. Winkler, J. Am. Chem. Soc., 99, 6359 (1977).
87. R.G. Cooks, J.H. Beynon, R.M. Caprioli and G.R. Lester, "Metastable Ions", Elsevier, Amsterdam, 1973.
88. D.M. Mintz and T. Baer, J. Chem. Phys., 65, 2407 (1976).
89. E.G. Jones, L.E. Bauman, J.H. Beynon, and R.G. Cooks, Org. Mass Spectrom., 7, 185 (1973).
90. M. Medved, R.G. Cooks and J.H. Beynon, Int. J. Mass Spectrom. Ion Phys., 19, 179 (1976).
91. B.H. Solka, J.H. Beynon and R.G. Cooks, J. Phys. Chem., 79, 859 (1975).
92. J.H. Beynon, R.M. Caprioli and T.W. Shannon, Org. Mass Spectrom., 5, 967 (1971).
93. E.G. Jones, J.H. Beynon and R.G. Cooks, J. Chem. Phys., 57, 2652 (1972).
94. R.G. Cooks, J.H. Beynon, M. Bertrand and M.K. Hoffman, Org. Mass Spectrom., 7, 1303 (1973).
95. J.L. Holmes, A.D. Osborne and G.M. Weese, Org. Mass Spectrom., 10, 867 (1975).
96. J.L. Holmes, Org. Mass Spectrom., 7, 341 (1973).
97. J.L. Holmes and J.K. Terlouw, Can. J. Chem., 53, 2076 (1975).
98. J.L. Holmes and J.K. Terlouw, Can. J. Chem., 54, 1007 (1976).
99. J.L. Holmes and J.K. Terlouw, Org. Mass Spectrom., 10, 787 (1975).
100. J.L. Holmes, G.M. Weese, A.S. Blair and J.K. Terlouw, Org. Mass Spectrom., 12, 424 (1977).
101. J.L. Holmes, P. Wolkoff and J.K. Terlouw, J. Chem. Soc. Chem. Commun., 492 (1977).
102. C.C. Van de Sande and F.W. McLafferty, J. Am. Chem. Soc., 97, 4613 (1975).
103. D.H. Williams and G. Hvistendahl, J. Am. Chem. Soc., 96, 6753 (1974).
104. D.H. Williams and G. Hvistendahl, J. Am. Chem. Soc., 96, 6755 (1974).
105. L.E. Brady, D.H. Williams, S.C. Traynor and K.J. Crowley, Org. Mass Spectrom., 10, 116 (1975).
106. H. Schwarz, F. Borchers and K. Levsen, Z. Naturforsch., 31b, 935 (1976).
107. F.W. McLafferty, P.F. Bente, III, R. Kornfeld, S.-C. Tsai and I. Howe, J. Am. Chem. Soc., 95, 2120 (1973).

108. F.W. McLafferty, R. Kornfeld, W.F. Haddon, K. Levsen, I. Sakai, P.F. Bente III, S.-C. Tsai and H.D.R. Schüddemage, J. Am. Chem. Soc., 95, 3886 (1973).

109. K. Levsen and F.W. McLafferty, Org. Mass Spectrom., 8, 353 (1974).

110. R. Robbiani, T. Kuster and J. Seibl, Angew. Chem. (Int. Ed.), 16, 115 (1977).

111. P.N. Rylander, S. Meyerson and H.M. Grubb, J. Am. Chem. Soc., 79, 842 (1957).

112. S. Meyerson and P.N. Rylander, J. Chem. Phys., 27, 901 (1957).

113. J.T. Bursey, M.M. Bursey and D.G.I. Kingston, Chem. Rev., 73, 191 (1973) and references herein.

114. K.L. Rinehart, A.C. Buchholz, G.E. Van Lear and H.L. Cantrill, J. Am. Chem. Soc., 90, 2983 (1968).

115. A. Siegel, J. Am. Chem. Soc., 96, 1251 (1974).

116. M.K. Hoffman and J.C. Wallace, J. Am. Chem. Soc., 95, 5064 (1973).

117. A.G. Harrison, P. Kebarle and F.P. Lossing, J. Am. Chem. Soc., 83, 777 (1961).

118. P. Brown, J. Am. Chem. Soc., 90, 2694 (1968).

119. S. Tajima, Y. Niwa, M. Nakajima and T. Tsuchiya, Bull. Chem. Soc. Japan, 44, 2340 (1971).

120. S. Meyerson, H. Hart and L.C. Leitch, J. Am. Chem. Soc., 90, 3419 (1968).

121. Y. Yamamoto, S. Takamuku and H. Sakurai, J. Am. Chem. Soc., 94, 661 (1972).

122. S. Takamuku, N. Sagi, K. Nagaoka and H. Sakurai, J. Am. Chem. Soc., 94, 6217 (1972).

123. J. Winkler and F.W. McLafferty, J. Am. Chem Soc., 95, 7533 (1973).

124. F.W. McLafferty and J. Winkler, J. Am. Chem. Soc., 96, 5182 (1974).

125. A. Venema and N.M.M. Nibbering, Tetrahedron Lett., 35, 3013 (1974).

126. R.C. Dunbar, J. Am. Chem. Soc., 97, 1382 (1975).

127. J. Shen, R.C. Dunbar and G.A. Olah, J. Am. Chem. Soc., 96, 6227 (1974).

128. M.K. Hoffman and M.D. Friesen, Org. Mass Spectrom., 9, 1081 (1974).

129. M.K. Hoffman and T.L. Amos, Tetrahedron Lett., 5235 (1972).

130. M.K. Hoffman and M.M. Bursey, Chem. Commun., 824 (1971).

131. R.A. Davidson and P.S. Skell, J. Am. Chem. Soc., 95, 6843 (1973).

132. F.E. Tibbetts, M.M. Bursey, W.F. Little, B.R. Willeford, S.A. Benezra, M.K. Hoffman and P.W. Jennings, Org. Mass Spectrom., 6, 475 (1972).

133. J.J. Solomon and F.H. Field, J. Am. Chem. Soc., 98, 1567 (1976).

134. M.K. Hoffman and M.M. Bursey, Tetrahedron Lett., 27, 2539 (1971).

135. A.N.H. Yeo and D.H. Williams, Chem. Commun., 886 (1970).

136. J.C. Traeger and R.G. McLoughlin, J. Am. Chem. Soc., 99, 7351 (1977).

137. J.A. Jackson, S.G. Lias and P. Ausloos, J. Am. Chem. Soc., 99, 7515 (1977).

138. C. Köppel, H. Schwarz and F. Bohlmann, Org. Mass Spectrom., 9, 332 (1974).

139. M.S.J. Dewar and D. Landman, J. Am. Chem. Soc., 99, 4633 (1977).

140. C. Cone, M.J. Dewar and D. Landman, J. Am. Chem. Soc., 99, 372 (1977).

141. J.L.M. Abboud, W.J. Hehre and R.W. Taft, J. Am. Chem. Soc., 98, 6072 (1976).

142. J.H. Futrell and T.O. Tiernan, Science, 162, 415 (1968).

143. F.H. Field, Accts. Chem. Res., 1, 42 (1968).

144. D.A. Durden, P. Kebarle and A. Good, J. Chem. Phys., 50, 805 (1969).

145. A.A. Herod and A.G. Harrison, Int. J. Mass Spectrom. Ion Phys., 4, 415 (1970).

146. L.R. Anders, J.L. Beauchamp, R.C. Dunbar, and J.D. Baldeschwieler, J. Chem. Phys.,

<u>45</u>, 1062 (1966).

147. J.D. Baldeschwieler, Science, <u>159</u>, 263 (1968).
148. J.L. Beauchamp, Ann. Rev. Phys. Chem., <u>22</u>, 527 (1971).
149. J.H. Futrell, in "Dynamic Mass Spectrometry", Vol.2, (D. Price, Ed.), Heyden, London (1971), p. 97.
150. G.C. Goode, K.R. Jennings, and C.J. Drewery, in "Mass Spectrometry", (A. Maccoll, Ed.), International Review of Science, Phys. Chem. Series I, Butterworths, London, 1972, p. 183
151. J.D. Baldeschwieler and S.S. Woodgate, Acc. Chem. Res., <u>4</u>, 114 (1971).
152. H. Hartmann, K.H. Lebert, and K.P. Wanczek, Fortschr. Chem. Forsch., <u>43</u>, 57 (1973).
153. T.A. Lehman and M.M. Bursey, "Ion Cyclotron Resonance Spectrometry", Wiley-Interscience, New York, 1976.
154. J.M.S. Henis, in "Ion-Molecule Reactions", (J.L. Franklin, Ed.), Plenum, New York, 1972, p. 395.
155. K.B. Tomer and C. Djerassi, Tetrahedron, <u>29</u>, 3491 (1973).
156. F. Borchers, K. Levsen, C.B. Theissling and N.M.M. Nibbering, Org. Mass Spectrom., <u>12</u>, 746 (1977).
157. A. Maquestiau, Y.V. Haverbeke, R. Flammang, C.D. Meyer, C.G. Das and G.S. Reddy, Org. Mass Spectrom., <u>12</u>, 631 (1977).
158. M.L. Gross, D.H. Russell, R.J. Aerni and S.A. Bronczyk, J. Am. Chem. Soc., <u>99</u>, 3603 (1977).
159. M.L. Gross and F.W. McLafferty, J. Am. Chem. Soc., <u>93</u>, 1267 (1971).
160. A.S. Blair and A.G. Harrison, Can. J. Chem., <u>51</u>, 703 (1973).
161. K.B. Tomer, Org. Mass Spectrom., <u>9</u>, 686 (1974).
162. N.M.M. Nibbering, Tetrahedron, <u>29</u>, 385 (1973).
163. R.T. McIver, Org. Mass Spectrom., <u>10</u>, 396 (1975).
164. R.C. Dunbar in "Ion-Molecule Reactions", (P. Ausloos, Ed.), Plenum Press, New York, N.Y., 1975, and references herein.
165. R.C. Dunbar, J. Am. Chem. Soc., <u>95</u>, 472 (1973).
166. R.C. Dunbar and E.W. Fu, J. Am. Chem. Soc., <u>95</u>, 2716 (1973).
167. R.C. Dunbar and R. Klein, J. Am. Chem. Soc., <u>99</u>, 3744 (1977).
168. E.W. Fu and R.C. Dunbar, J. Am. Chem. Soc., in press.
169. R.C. Dunbar, J. Am. Chem. Soc., <u>95</u>, 6191 (1973).
170. E.W. Fu, P.P. Dymerski and R.C. Dunbar, J. Am. Chem. Soc., <u>98</u>, 337 (1976).
171. R.C. Dunbar and H.H. Teng, J. Am. Chem. Soc., in press.
172. R.C. Dunbar, E.W. Fu and G.A. Olah, J. Am. Chem. Soc., <u>99</u>, 7502 (1977).
173. P.P. Dymerski, E.W. Fu and R.C. Dunbar, J. Am. Chem. Soc., <u>96</u>, 4109 (1974).
174. R.C. Dunbar, J. Am. Chem. Soc., <u>98</u>, 4671 (1976).
175. R.C. Dunbar, Anal. Chem., <u>48</u>, 723 (1976).
176. R.C. Dunbar and R. Klein, J. Am. Chem. Soc., <u>98</u>, 7994 (1976).
177. M. Riggin, R. Orth and R.C. Dunbar, J. Chem. Phys., <u>65</u>, 3365 (1976).
178. R. Orth, R.C. Dunbar and M. Riggin, Chem. Phys., <u>19</u>, 279 (1977).
179. R.G. Orth and R.C. Dunbar, J. Chem. Phys., <u>66</u>, 1616 (1977).
180. J.M. Kramer and R.C. Dunbar, J. Chem. Phys., <u>60</u>, 5122 (1974).
181. M. Riggin and R.C. Dunbar, Chem. Phys. Lett., <u>31</u>, 539 (1975).

182. R.C. Dunbar, Chem. Phys. Lett., 32, 508 (1975).

183. B.N. McMaster in "Mass Spectrometry", Vol. 3 and 4, R.A.W. Johnstone (Ed.), The Chemical Society, London 1975 and 1977, Chapter 1.

184. R.C. Bingham, M.J.S. Dewar and D.H. Lo, J. Am. Chem. Soc., 97, 1285 (1975).

185. J.A. Pople, J. Am. Chem. Soc., 97, 5306 (1975).

186. W.J. Hehre, J. Am. Chem. Soc., 97, 5308 (1975).

187. M.J.S. Dewar, J. Am. Chem. Soc., 97, 6591 (1975).

188. P.K. Bischof and M.J.S. Dewar, J. Am. Chem. Soc., 97, 2278 (1975).

189. R. Sustmann, J.E. Williams, M.J.S. Dewar, L.C. Allen and P.V.R. Schleyer, J. Am. Chem. Soc., 91, 5350 (1969).

190. H. Kollmar and H.O. Smith, Theor. Chim. Acta, 20, 65 (1971).

191. W.A. Lathan, W.J. Hehre and J.A. Pople, J. Am. Chem. Soc., 93, 808 (1971).

192. P.C. Hariharan, W.A. Lathan and J.A. Pople, Chem. Phys. Lett., 14, 385 (1972).

193. B. Zurawski, R. Ahlrichs and W. Kutzelnigg, Chem. Phys. Lett., 21, 309 (1973).

194. J. Weber, M. Yoshimine and A.D. McLean, J. Chem. Phys., 64, 4159 (1976).

195. L. Radom, P.C. Hariharan, J.A. Pople and P.V.R. Schleyer, J. Am. Chem. Soc., 98, 10 (1976).

196. R. Kebabcioglu and V. Dyczmons, Z. Naturforsch., 30a, 1680 (1975).

197. J.L. Holmes, A.D. Osborne and G.M. Weese, Org. Mass Spectrom., 10, 867 (1975).

198. P. Goldberg, J.A. Hopkinson, A. Mathias and A.E. Williams, Org. Mass Spectrom., 3, 1009 (1970).

199. D.K. Sen-Sharma, K.R. Jennings and J.H. Beynon, Org. Mass Spectrom., 11, 319 (1976).

200. B.P. Tsai, A.S. Werner and T. Baer, J. Chem. Phys., 63, 4384 (1975).

201. C. Köppel and F.W. McLafferty, private communication.

202. A.C. Parr, A.J. Jason and R. Stockbauer, Int. J. Mass Spectrom. Ion Phys., 26, 23 (1978).

203. I. Howe and D.H. William, Chem. Commun., 1195 (1971).

204. I. Howe and F.W. McLafferty, J. Am. Chem. Soc., 92, 3797 (1970).

205. K. Levsen, F.W. McLafferty and D.M. Jerina, J. Am. Chem. Soc., 95, 6332 (1973).

206. R.G. Cooks, J.H. Beynon and J.F. Litton, Org. Mass Spectrom., 10, 503 (1975).

207. F. Borchers and K. Levsen, Org. Mass Spectrom., 10, 584 (1975).

208. A. Maquestiau, Y. Van Haverbeke, R. Flammang, C. De Meyer and A. Menu, Org. Mass Spectrom., 12, 706 (1977).

209. H.S.W. Massey and H.B. Gilbody, "Electronic and Ionic Phenomena", 2nd Edn, Vol. IV, Oxford University Press, London, 1974.

210. A.S. Werner and T. Baer, J. Chem. Phys., 62, 2900 (1975).

211. S. Meyerson, T.D. Nevitt and P.N. Rylander, Adv. Mass Spectrom., 1, 313 (1959).

212. K. Levsen, Org. Mass Spectrom., 10, 43 (1975).

213. K. Levsen, H. Heimbach, G.J. Shaw and G.W.A. Milne, Org. Mass Spectrom., 12, 663 (1977).

214. R. Liardon and T. Gäumann, Helv. Chim. Acta, 52, 528 and 1042 (1969).

215. C. Corolleur, S. Corolleur and F.G. Gault, Bull. Soc. Chim. France, 158 (1970).

216. F.W. McLafferty and T.A. Bryce, Chem. Commun., 1215 (1967).

217. A. Fiaux, B. Wirz and T. Gäumann, Helv. Chim. Acta, 57, 525 and 708 (1974).

218. W.G. Cole and D.H. Williams, Chem. Commun., 784 (1969).

219. R.D. Bowen and D.H. Williams, J. Chem. Soc. Perkin Trans. II, 1479 (1976).
220. P.P. Dymerski and F.W. McLafferty, J. Am. Chem. Soc., 98, 6070 (1976).
221. K. Levsen, Org. Mass Spectrom., 10, 55 (1975).
222. K. Levsen and E. Hilt, Liebigs. Ann. Chem., 257 (1976).
223. A.J. Dale, W.D. Weringa and D.H. Williams, Org. Mass Spectrom., 6, 501 (1972).
224. A.M. Falick, P. Tecon and T. Gäumann, Org. Mass Spectrom., 11, 409 (1976).
225. M.A. Shaw, R. Westwood and D.H. Williams, J. Chem. Soc. (B), 1773 (1970).
226. J.L. Holmes, D.C.M. Tong and R.T.B. Rye, Org. Mass Spectrom., 6, 897 (1972).
227. K.B. Tomer, J. Turk and R.H. Shapiro, Org. Mass Spectrom., 6, 235 (1972).
228. A.M. Falick and T. Gäumann, Helv. Chim. Acta, 59, 987 (1976).
229. W.A. Bryce and P. Kebarle, Can. J. Res., 34, 1249 (1956).
230. W.H. McFadden, J. Phys. Chem., 67, 1074 (1963).
231. B.J. Millard and D.F. Shaw, J. Chem. Soc. (B), 664 (1966).
232. G.G. Meisels, J.Y. Park and B.G. Giessner, J. Am. Chem. Soc., 91, 1555 (1969).
233. G.A. Smith and D.H. Williams, J. Chem. Soc. (B), 1529 (1970).
234. M.L. Gross and P.H. Lin, Org. Mass Spectrom., 7, 795 (1970).
235. R.P. Morgan and P.J. Derrick, Org. Mass Spectrom., 10, 563 (1975).
236. M.L. Vestal in "Fundamental Processes in Radiation Chemistry" (P. Ausloos, Ed.), Interscience, New York, 1968, p. 59.
237. R.D. Bowen and D.H. Williams, Org. Mass Spectrom., 12, 453 (1977).
238. T. Nishishita, F.M. Bockhoff and F.W. McLafferty, Org. Mass Spectrom., 12, 16 (1977).
239. T. Nishishita and F.W. McLafferty, Org. Mass Spectrom., 12, 75 (1977).
240. K. Levsen, Tetrahedron, 31, 2431 (1975).
241. R.P. Morgan, P.J. Derrick and A.G. Harrison, J. Am. Chem. Soc., 99, 4189 (1977).
242. N.M.M. Nibbering, T. Nishishita, C.C. Van de Sande and F.W. McLafferty, J. Am. Chem. Soc., 96, 5668 (1974).
243. C. Köppel and F.W. McLafferty, J. Chem. Soc. Chem. Commun., 20, 810 (1976).
244. C. Köppel, C.C. Van de Sande, N.M.M. Nibbering, T. Nishishita and F.W. McLafferty, J. Am. Chem. Soc., 99, 2883 (1977).
245. C. Köppel and F.W. McLafferty, to be published.
246. J. Heimbrecht, K. Levsen and H. Schwarz, Z. Naturforsch., 31b, 1299 (1976).
247. M. Bobrich, H. Schwarz, K. Levsen and P. Schmitz, Org. Mass Spectrom., 12, 549 (1977).
248. H. Schwarz, F. Borchers and K. Levsen, Z. Naturforsch., 31b, 935 (1976).
249. H. Schwarz, C. Wesdemiotis, K. Levsen and F. Borchers, in "Advances in Mass Spectrometry, Vol. 7, Heyden, London (1977).
250. C. Wesdemiotis, H. Schwarz, K. Levsen and F. Borchers, Liebigs Ann. Chem., 1889 (1976).
251. H. Schwarz, C. Wesdemiotis, H. Heimbach and K. Levsen, Org. Mass Spectrom., 12, 213 (1977).
252. B. Van de Graaf and F.W. McLafferty, J. Am. Chem. Soc., 99, 6806 (1977).
253. B. Van de Graaf and F.W. McLafferty, J. Am. Chem. Soc., 99, 6810 (1977).
254. F.W. McLafferty and I. Sakai, Org. Mass Spectrom., 7, 971 (1973).
255. K. Levsen, H. Heimbach, C.C. Van de Sande and J. Monstrey, Tetrahedron, 33, 1785 (1977).

256. D.H. Williams and R.D. Bowen, J. Am. Chem. Soc., 99, 3192 (1977).
257. D.H. Williams, J.L. Holmes and F.P. Lossing, private communication.
258. J. van Thuijl, J.J. van Houte, A. Maquestiau, R. Flammang and C. De Meyer, Org. Mass Spectrom., 12, 196 (1977).
259. J.G. Pritchard, Org. Mass Spectrom., 8, 103 (1974).
260. J.L. Holmes, J.K. Terlouw and F.P. Lossing, J. Phys. Chem., 80, 2860 (1976).
261. J.L. Holmes, P. Wolkoff and J.K. Terlouw, J. Chem. Soc., Chem. Commun., 492 (1977).
262. K.B. Tomer and C. Djerassi, J. Am. Chem. Soc., 95, 5335 (1973).
263. A. Maquestiau, Y. Van Haverbeke and R. Flammang, Tetrahedron Lett., 41, 3747 (1976).
264. F.W. McLafferty and W.T. Pike, J. Am. Chem. Soc., 89, 5951 (1967).
265. F.W. McLafferty and H.D.R. Schüddemage, J. Am. Chem. Soc., 91, 1866 (1969).
266. D. Van Raalte and A.G. Harrison, Can. J. Chem., 41, 3118 (1963).
267. M.T. Bowers and P.R. Kemper, J. Am. Chem. Soc., 93, 5352 (1971).
268. A.G. Harrison and B.G. Keyes, J. Am. Chem. Soc., 90, 5046 (1968).
269. B.G. Keyes and A.G. Harrison, Org. Mass Spectrom., 9, 221 (1974).
270. R.D. Bowen and D.H. Williams, J. Chem. Soc. Chem. Commun., 378 (1977).
271. D. Amos, R.G. Gillis, J.L. Occolowitz and J.F. Pisani, Org. Mass Spectrom., 2, 209 (1969).
272. B.G. Keyes and A.G. Harrison, J. Am. Chem. Soc., 90, 5671 (1968).
273. F. Jordan, J. Phys. Chem., 80, 76 (1976).
274. T.J. Mead and D.H. Williams, J. Chem. Soc., (B) 1654 (1971).
275. F.W. McLafferty and W.T. Pike, J. Am. Chem. Soc., 89, 5953 (1967).
276. G. Eadon, J. Diekman and C. Djerassi, J. Am. Chem. Soc., 91, 3986 (1969).
277. F.W. McLafferty, D.J. McAdoo, J.S. Smith and R. Kornfeld, J. Am. Chem. Soc., 93, 3720 (1971).
278. C.W. Tsang and A.G. Harrison, Org. Mass Spectrom., 3, 647 (1970).
279. C.W. Tsang and A.G. Harrison, Org. Mass Spectrom., 5, 877 (1971).
280. A.S. Siegel, Org. Mass Spectrom., 3, 1417 (1970).
281. W.J. Broer and W.D. Weringa, Org. Mass Spectrom., 12, 326 (1977).
282. W.T. Pike and F.W. McLafferty, J. Am. Chem. Soc., 89, 5954 (1967).
283. H. Heimbach and K. Levsen, unpublished results.
284. M.L. Gross, D.H. Russel, R. Phongbetchara and P.H. Lin, Adv. Mass Spectrom., 7, (1977).
285. P. Brown and M.M. Green, J. Org. Chem., 32, 1681 (1967).
286. W.H. Pirkle and M. Dines, J. Am. Chem. Soc., 90, 2318 (1968).
287. C.C. Van de Sande and F.W. McLafferty, J. Am. Chem. Soc., 97, 2298 (1975).
288. D.J. McAdoo, F.W. McLafferty and T.E. Parks, J. Am. Chem. Soc., 94, 1601 (1972).
289. T.J. Mead and D.H. Williams, J. Chem. Soc. Perkin Trans. II, 876 (1972).
290. H. Bornowski, v. Feistkorn, H. Schwarz, K. Levsen and P. Schmits, Z. Naturforsch., 32b, 664 (1977).
291. J.R. Hass, M.M. Bursey, D.G.I. Kingston and H.P. Tannenbaum, J. Am. Chem. Soc., 94, 5095 (1972).
292. J.H. Beynon, R.A. Saunders and A.E. Williams, Ind. Chim. Belge, 29, 311 (1964).

293. M.K. Hoffman, M.D. Friesen and G. Richmond, Org. Mass Spectrom., **12**, 150 (1977).
294. F. Borchers, K. Levsen and H.D. Beckey, Int. J. Mass Spectrom. Ion Phys., **21**, 125 (1976).
295. F.W. McLafferty and L.J. Schiff, Org. Mass Spectrom., **2**, 757 (1969).
296. F.M. Benoit and A.G. Harrison, Org. Mass Spectrom., **11**, 599 (1976).
297. J.K. MacLeod and C. Djerassi, J. Am. Chem. Soc., **88**, 1840 (1966).
298. P. Wolkoff and S. Hammerum, Org. Mass Spectrom., **11**, 375 (1976).
299. C.C. Van de Sande, C. De Meyer and A. Maquestiau, Bull. Soc. Chem. Belge, **85**, 79 (1976).
300. H. Schwarz, Org. Mass Spectrom., **12**, 470 (1977).
301. J.K. MacLeod in "Advances in Mass Spectrometry", Vol. 7, Heyden, London, 1977.
302. H.E. Audier, M. Fetizon and J.C. Tabet, Org. Mass Spectrom., **10**, 347 (1975).
303. K.R. Jennings and A. Whiting, Org. Mass Spectrom., **6**, 917 (1972).
304. M.M. Bursey, Tetrahedron Lett., 981 (1968).
305. R. Westwood, D.H. Williams and A.N.H. Yeo, Org. Mass Spectrom., **3**, 1485 (1970).
306. R.H. Shapiro and J.W. Serum, Org. Mass Spectrom., **2**, 533 (1969).
307. P.A. Kollmann, W.F. Trager, S. Rothenberg and J.E. Williams, J. Am. Chem. Soc., **95**, 458 (1973).
308. H. Schwarz, R. Sezi, U. Rapp, H. Kaufmann and S. Meier, Org. Mass Spectrom., **12**, 39 (1977).
309. J.L. Occolowitz and G.L. White, Aust. J. Chem., **21**, 997 (1968).
310. P.C. Vijfhuizen, W. Heerma and N.M.M. Nibbering, Org. Mass Spectrom., **11**, 787 (1976).
311. K. Levsen and H. Schwarz, Org. Mass Spectrom., **10**, 752 (1975).
312. H. Schwarz, C. Wesdemiotis and F. Bohlmann, Org. Mass Spectrom., **9**, 1226 (1974).
313. A.J. Dale, W.D. Weringa and D.H. Williams, Org. Mass Spectrom., **6**, 501 (1972).
314. K. Levsen, G.E. Berendsen, N.M.M. Nibbering and H. Schwarz, Org. Mass Spectrom., **12**, 125 (1977).
315. B. Richter and H. Schwarz, Org. Mass Spectrom., **10**, 522 (1975).
316. N.M.M. Nibbering, C.C. Van de Sande, T. Nishishita and F.W. McLafferty, Org. Mass Spectrom., **9**, 1059 (1974).
317. M.A. Th. Kerhoff and N.M.M. Nibbering, Org. Mass Spectrom., **7**, 37 (1973).
318. F. Borchers, H. Heimbach, K. Levsen, H. Morhenn and H. Schwarz, Org. Mass Spectrom., **12**, 573 (1977).
319. M.V. Gur'ev and M.V. Tikhomirov, Zh. Fiz. Khim., **32**, 2731 (1958).
320. M.V. Gur'ev, M.V. Tikhomirov and N.N. Tunitskii, Dokl. Akad. Nauk SSSR, **123**, 120 (1958).
321. J.H. Beynon, R.A. Saunders, A. Topham and A.E. Williams, J. Phys. Chem., **65**, 114 (1961).
322. S. Meyerson, J. Chem. Phys., **42**, 2181 (1965).
323. N. Dinh-Nguyen, R. Ryhage, S. Ställberg-Stenhagen and E. Stenhagen, Ark. Kemi, **18**, 393 (1961).
324. P.P. Dymerski, R.M. Prinstein, P.F. Bente and F.W. McLafferty, J. Am. Chem. Soc., **98**, 6834 (1976).
325. W.F. Haddon and F.W. McLafferty, J. Am. Chem. Soc., **90**, 4745 (1968).
326. D.J. McAdoo, F.W. McLafferty and P.F. Bente, J. Am. Chem. Soc., **94**, 2027 (1972).
327. Ch. Ottinger, J. Chem. Phys., **47**, 1452 (1967).

328. M. Vestal and J.H. Futrell, J. Chem. Phys., 52, 978 (1970).

329. N. Bodor, M.J.S. Dewar and D.H. Lo, J. Am. Chem. Soc., 94, 5303 (1972).

330. P.C. Hariharan, L. Radom, J.A. Pople and P.v.R. Schleyer, J. Am. Chem. Soc., 96, 599 (1974) and references herein.

331. M.L. Gross and F.W. McLafferty, J. Am. Chem. Soc., 93, 1267 (1971).

332. M.L. Gross, J. Am. Chem. Soc., 94, 3744 (1972).

333. H.H. Voge, C.D. Wagner, and D.P. Stevenson, J. Catalysis, 2, 59 (1963).

334. S.R. Smith, R. Schor, and W.P. Norris, J. Phys. Chem., 69, 1615 (1965).

335. A.Y.-K. Lau, B.H. Solka, and A.G. Harrison, Org. Mass Spectrom., 9, 555 (1974).

336. A.G. Harrison and P.P. Dymerski, Org. Mass Spectrom., in press.

337. M.S.-H. Lin and A.G. Harrison, Can. J. Chem., 52, 1813 (1974).

338. K. Levsen, R. Weber and F. Borchers, to be published.

339. M.L. Gross, P.H. Lin and S.J. Franklin, Anal. Chem., 44, 974 (1972).

340. A.M. Falick and T. Gäumann, 24th Annual Conference on Mass Spectrometry, San Diego, California, 1976.

341. P. Tecon, D. Stahl and T. Gäumann, Int. J. Mass Spectrom. Ion Phys., in press.

342. P.J. Derrick and A.L. Burlingame, J. Am. Chem. Soc., 96, 4909 (1974).

343. P.J. Derrick, A.M. Falick and A.L. Burlingame, in "Advances in Mass Spectrometry", Vol. 6 (A.R. West, Ed), Applied Science Publisher, Barking, 1974, p. 877.

344. L. Radom, P.C. Hariharan, J.A. Pople and P.v.R. Schleyer, J. Am. Chem. Soc., 95, 6531 (1973).

345. S. Pignataro, S. Cassuto and F.P. Lossing, J. Am. Chem. Soc., 89, 3693 (1967).

346. J.L. Holmes, Org. Mass Spectrom., 8, 247 (1974).

347. P.J. Derrick, A.M. Falick and A.L. Burlingame, J. Am. Chem. Soc., 94, 6794 (1972).

348. J.L. Holmes, D. McGillivray and N.S. Isaacs, Org. Mass Spectrom., 9, 510 (1974).

349. J.L. Occolowitz and G.L. White, Aust. J. Chem., 21, 997 (1968).

350. T.E. Smith, S.R. Smith and F.W. McLafferty, Org. Mass Spectrom., in press.

351. J.L. Holmes and G. Weese, Org. Mass Spectrom., 9, 618 (1974).

352. A.G. Harrison, P. Haynes, S. McLean and F. Meyer, J. Am. Chem. Soc., 87, 5099 (1965).

353. J.L. Holmes and D.M. Gillivray, Org. Mass Spectrom., 5, 1349 (1971).

354. E.F.H. Brittain, C.H.J. Wells and H.M. Paisley, J. Chem. Soc. (B), 503 (1969).

355. W.C. Ermler, R.S. Mulliken and E. Clementi, J. Am. Chem. Soc., 98, 388 (1976).

356. T. Keough, T. Ast, J.H. Beynon and R.G. Cooks, Org. Mass Spectrom., 7, 245 (1973).

357. C. Köppel, H. Schwarz, F. Borchers and K. Levsen, Int. J. Mass Spectrom. Ion Phys., 21, 15 (1976).

358. M.L. Gross and R.J. Aerni, J. Am. Chem. Soc., 95, 7875 (1973).

359. K.R. Jennings, Z. Naturforsch., 22a, 454 (1967).

360. I. Horman, A.N.H. Yeo and D.H. Williams, J. Am. Chem. Soc., 92, 2131 (1970).

361. W.O. Perry, J.H. Beynon, W.E. Baitinger, J.W. Amy, R.M. Caprioli, R.N. Renaud, L.C. Leitch and S. Meyerson, J. Am. Chem. Soc., 92, 7236 (1970).

362. J.H. Beynon, R.M. Caprioli, W.O. Perry and W.E. Baitinger, J. Am. Chem. Soc., 94, 6828 (1972).

363. R.J. Dickinson and D.H. Williams, J. Chem. Soc. (B), 249 (1971).

364. B. Andlauer and Ch. Ottinger, Z. Naturforsch., 27a, 293 (1972).
365. H.M. Rosenstock, K.E. McCulloh and F.P. Lossing, Adv. Mass Spectrom., 7, (1977) and references herein.
366. H.M. Rosenstock, J.T. Larkins and J.A. Walker, Int. J. Mass Spectrom. Ion Phys., 11, 309 (1973).
367. H.M. Rosenstock, K.E. McCulloh and F.P. Lossing, Int. J. Mass Spectrom. Ion Phys., 25, 327 (1977).
368. J.H.D. Eland, Int. J. Mass Spectrom. Ion Phys., 13, 457 (1974).
369. J.H.D. Eland, R. Frey, H. Schulte and B. Brehm, Int. J. Mass Spectrom. Ion Phys., 21, 209 (1976).
370. G.A. Gallup, D. Steinheider and M.L. Gross, Int. J. Mass Spectrom. Ion Phys., 22, 185 (1976).
371. K. Levsen and H.D. Beckey, Org. Mass Spectrom., 9, 570 (1974).
372. R. Renaud and L.C. Leitch, Can. J. Chem., 34, 98 (1956).
373. F. Meyer and A.G. Harrison, J. Am. Chem. Soc., 86, 4757 (1964).
374. J.H. Beynon, J.E. Corn, W.E. Baitinger, R.M. Caprioli and R.A. Benkeser, Org. Mass Spectrom., 3, 1371 (1970).
375. M.A. Baldwin, F.W. McLafferty and D.M. Jerina, J. Am. Chem. Soc., 97, 6169 (1975).
376. M.J.S. Dewar and D. Landmann, J. Am. Chem. Soc., 99, 2446 (1977).
377. R.C. Dunbar, J. Shen and G.A. Olah, J. Am. Chem. Soc., 94, 6862 (1972).
378. A. Venema, N.M.M. Nibbering and Th. de Boer, Tetrahedron Lett., 2141 (1971).
379. I.P. Fisher, T.F. Palmer and F.P. Lossing, J. Am. Chem. Soc., 86, 2741 (1964).
380. R.A.W. Johnstone and F.A. Mellon, J. Chem. Soc. Faraday Trans. II, 1209 (1972).
381. Y.L. Sergeev, M.E. Akopyan, F.I. Vilesov and V.I. Kleimenov, Opt. Spectrosc. (USSR), 29, 63 (1970).
382. J.D. Dill, P.v.R. Schleyer, J.S. Binkley, R. Seeger, J.A. Pople and E. Haselbach, J. Am. Chem. Soc., 98, 5428 (1976).
383. A. Venema, N.M.M. Nibbering, K.H. Maurer and U. Rapp, Int. J. Mass Spectrom. Ion Phys., 17, 89 (1975).
384. N. Uccella, Org. Mass Spectrom., 10, 494 (1975).
385. S. Meyerson and P.N. Rylander, J. Phys. Chem., 62, 2 (1958).
386. S. Meyerson and P.N. Rylander, J. Am. Chem. Soc., 79, 1058 (1957).
387. N.M.M. Nibbering and Th.J. de Boer, Org. Mass Spectrom., 2, 157 (1969).
388. A. Venema, N.M.M. Nibbering and Th.J. de Boer, Org. Mass Spectrom., 3, 1589 (1970).
389. A.G. Harrison, P. Kebarle and F.P. Lossing, J. Am. Chem. Soc., 83, 777 (1961).
390. F. Meyer, P. Haynes, S. McLean and A.G. Harrison, Can. J. Chem., 43, 211 (1965).
391. H.Fr. Grützmacher, Org. Mass Spectrom., 3, 131 (1970).
392. R. Shapiro and T.F. Jenkins, Org. Mass Spectrom., 2, 771 (1969).
393. W.J. Richter and W. Vetter, Org. Mass Spectrom., 2, 781 (1969).
394. J.B. Thomson and C. Djerassi, J. Org. Chem., 32, 3905 (1967).
395. N.A. Uccella and D.H. Williams, J. Am. Chem. Soc., 94, 8778 (1972).
396. H.F. Grützmacher and M. Puschmann, Chem. Ber., 104, 2079 (1971).
397. C.L. Wilkins and M.L. Gross, J. Am. Chem. Soc., 93, 895 (1971).
398. E.I. Quinn and F.L. Mohler, J. Res. Nat. Bur. Std. A62, 39 (1959).
399. M.I. Gorfinkel, N.S. Bugreeva, I.S. Isaev, Izv. Sibirsk. Otd. Akad. Nauk SSSR,

88 (1974).
400. R. Stolze and H. Budzikiewicz, Org. Mass Spectrom., 13, 25 (1978).
401. K. Levsen, F. Borchers, R. Stolze and H. Budzikiewicz, to be published.
402. J.L. Franklin and S.R. Carrol, J. Am. Chem. Soc., 91, 5940 (1969).
403. H. Schwarz, C. Wesdemiotis and B. Hess, Org. Mass Spectrom., 10, 595 (1975).
404. K. Levsen, H. Heimbach, M. Bobrich, J. Respondek and H. Schwarz, Z. Naturforsch., 32b, 880 (1977).
405. P.P. Dymerski and F.W. McLafferty, in preparation.
406. J.K. Kim, M.C. Findlay, W.H. Henderson and M.C. Caserio, J. Am. Soc., 95, 2184 (1973).
407. R.D. Wieting, R.H. Staley and J.L. Beauchamp, J. Am. Chem. Soc., 96, 7552 (1974).
408. G. Cum, G. Sindona and N.A. Uccella, Org. Mass Spectrom., 12, 8 (1977).
409. J.W. Gordon, G.H. Schmid and I.G. Czismadia, J. Chem. Soc. Perkin Trans. II, 1722 (1975).
410. W.J. Hehre and A.J.P. Devaquet, J. Am. Chem. Soc., 98, 4370 (1976).
411. G.A. Olah, R.J. Spear, P.C. Hilberty and W.J. Hehre, J. Am. Chem. Soc., 98, 7470 (1976).
412. J.D. Pulfer and M.A. Whitehead, Can. J. Chem., 51, 2220 (1973).
413. D.H. Russell, M.L. Gross and N.M.M. Nibbering, J.L. Holmes, Euchem Conference on "The Chemistry of Ion Beams", Nordwijk, Holland, 1977.
414. L. Radom and H.F. Schaefer, J. Am. Chem. Soc., 99, 7522 (1977).
415. M.J.S. Dewar and D. Landman, J. Am. Chem. Soc., 99, 7439 (1977).
416. D.H. Russell, M.L. Gross and N.M.M. Nibbering, J. Am. Chem. Soc., in press.
417. T.A. Molenaar-Langeveld, N.M.M. Nibbering, R.P. Morgan and J.H. Beynon, Org. Mass Spectrom., 13, 172 (1978).
418. R.D. Smith, D.A. Herold, T.A. Elwood and J.H. Futrell, J. Am. Chem. Soc., 99, 6042 (1977).
419. L. Radom, J.A. Pople and P.v.R. Schleyer, J. Am. Chem. Soc., 94, 5935 (1972).
420. P. Merlet, S.D. Peyerimhoff, R.J. Buenker and S. Shin, J. Am. Chem. Soc., 96, 959 (1974).
421. D.H. Aue, W.R. Davidson and M.T. Bowers, J. Am. Chem. Soc., 98, 6700 (1976).
422. W.J. Hehre and P.C. Hiberty, J. Am. Chem. Soc., 94, 5917 (1972).
423. W.J. Hehre and P.C. Hiberty, J. Am. Chem. Soc., 96, 302 (1974).
424. F.P. Lossing and J.C. Traeger, Int. J. Mass Spectrom. Ion Phys., 19, 9 (1976).
425. M.J.S. Dewar and R.C. Haddon, J. Am. Chem. Soc., 95, 5836 (1973).
426. W.J. Hehre and P.v.R. Schleyer, J. Am. Chem. Soc., 95, 5837 (1973).
427. M.S.J. Dewar and R.C. Haddon, J. Am. Chem. Soc., 96, 255 (1974).
428. T. Keough, J.H. Beynon and R.G. Cooks, J. Am. Chem. Soc., 95, 1695 (1973).
429. M.J. Dewar, R.C. Haddon, A. Komornicki and H. Rzepa, J. Am. Chem. Soc., 99, 377 (1977).
430. M. Speranza, M.D. Sefcik, J.M.S. Henis and P.P. Gaspar, J. Am. Chem. Soc., 99, 5583 (1977).
431. W.J. Hehre, J. Am. Chem. Soc., 94, 5919 (1972).
432. W.J. Hehre, J. Am. Chem. Soc., 96, 5207 (1974).
433. J.B. Collins, J.D. Dill, E.D. Jemmis, Y. Apeloig, P.v.R. Schleyer, R. Seeger and J.A. Pople, J. Am. Chem. Soc., 98, 5419 (1972).

434. A.G. Loudon and K.S. Webb, Org. Mass Spectrom., 12, 283 (1977).
435. R.T. Aplin and S. Safe, Chem. Commun., 14o (1967).
436. H. Schwarz and F. Bohlmann, Org. Mass Spectrom., 7, 395 (1973).
437. M.L. Gross, E. Chin, D. Pokorny and F.L. DeRoos, Org. Mass Spectrom. 12, 55 (1977).
438. H. Schwarz and F. Bohlmann, Org. Mass Spectrom., 7, 23 (1973).
439. J.H. Bowie and T.K. Bradshaw, Aust. J. Chem., 23, 1431 (197o).
44o. J.H. Bowie, P.F. Donaghue, H.J. Rodda and B.K. Simons, Tetrahedron, 24, 3965 (1968).

Subject Index

Ab initio MO calculations 260-262
Absolute Rate Theory 26, 27
Activated complex 27
  see also "transition state"
  loose, tight 33, 91
  and Phase Space Theory 37-38
Activation Energy
  definition 27
  direct cleavage versus rearrangement 92
  and isotope effects 115
  and QET calculations 37
Amplification factor 131
Angular momentum
  conservation of 38
  role of, in collision-induced dissociations 138
  role of, in kinetic energy release distributions 78
Anchimeric assistance 177
Appearance potentials
  dependence on ion lifetimes 111
  determination of 108-113
  determination of heats of formation from 220
  elucidation of reaction mechanisms using 182
  extrapolation methods 108
  influence of the kinetic shift on 108
  influence of the competitive shift on 112
  influence of the thermal shift on 113
  of metastable ions 110
  and QET calculations 37
  unusual, of metastable ions 46
Atom randomization
  see "atom scrambling"
Atom scrambling
  activation energy for 101
  carbon scrambling 102
  definition 98
  degree of 163
  energy dependence of 99-100
  hydrogen scrambling 99-101
  lifetime dependence of 100-101
  mechanisms for 187
  MO calculations 198
  and non-specific hydrogen rearrangements 192
  suppression of 19o
Autoionization 7, 25
  and energy deposition functions 14
  in PEPICO studies 8
Avoided crossing 43

Barber-Elliott technique 54, 56
Bethe-Born theory 17

Breakdown graph
  calculation of 40
  and collision-induced dissociation 141
  comparison of calculated and experimental graphs using
    charge exchange 67
    electron impact 64
    photoelectron- photoion coincidence 69
    photoionization 66
  definition 40
  determination of appearance potentials from 112
  experimental determination 64-71
  time dependence 112
Breakdown scheme 39
Bond density 196
Bond order 196
Bond strength 152, 196

Certifugal barrier 28, 79
Charge distribution, MO calculations of 161
Charge exchange
  determination of breakdown graphs 67
  determination of rate constants 71
Charge localization
  and QET calculations 62
  rationalization of reaction mechanisms 160-162
Charge stripping
  fundamental studies 143
  ion structure determination 264, 265
Charge transfer reactions 141-143
  charge exchange 142
  charge inversion 143
  charge stripping 143, 264
  determination of appearance potentials 143
  determination of electronic states 143
Chemical activation 43
Classical approximation of the QET 29
Collisional activation (CA)
  see also "collision-induced dissociation"
  elucidation of reaction mechanisms 199
  examples for ion structure determination 245-248, 266-273
  experimental methods 244
  fundamental aspects 138-141
  ion structure determination by 243-248
Collision complex 38, 44, 76-78
Collision-induced dissociation 138
  see also "collisional activation"
  dependence on translational energy 140

Collision-induced dissociation, continued
  determination of electronic states 141
  elucidation of reaction mechanisms 178, 180, 186, 200, 201, 225
  energy distribution after collision 139
  and QET 138
  pressure dependence 141
  transferred excitation energy 140
Collision processes 138-143
  charge transfer reactions 141-143
  collision-induced dissociation 138-141
Comparison of mass spectra, for ion structure elucidation 226
Competing reactions
  direct cleavage versus rearrangement 62
Competitive shift 37, 104, 112
Composite metastable peaks 167-169
Concerted reactions 171, 176, 194
Configuration
  of the molecular ion 33
  of the transition state 33
Configuration interaction (CI) 260-262
Consecutive reactions, experimental tests of 166
Continuum, of rate constants 52-57
Correlation energy 260
Critical slope method 10
Cross section
  for collision 38
  for ionization 4
Crossed beam experiments 44
Cycloidal mass spectrometer 50

DADI spectra
  see "direct analysis of daughter ions"
Decomposing ions, definition 212
Decomposition mechanism, and ion structure 225
Deflection technique 75
Degree of freedom effect 105-108
  of collision-induced fragments 108
  on the kinetic energy release 238
Degradation reactions, use of, for ion structure determination 225-248
Density of states 28, 29-32
  criterion of minimum local density of states 35
  influence of the activation energy on 91
  influence of the transition state geometry on 91, 213
  and ionization potential 3
  and thermal energy distribution 19
Diatomic ions, potential curves 25
Direct analysis of daughter ions 166
Displacement reaction 224
Dissociation energy 156
Doubly charged ions 142
  kinetic energy release 142

Doubly charged ions, continued
  H/D randomization in 142

Electron impact, cross section 4
Electron monochromator 11
Electron spectroscopy 10
  see also "photoelectron spectroscopy"
Energy compensation method 10
Energy deposition function 12-17
  determination of,
    by electron impact 14
    by photoelectron spectroscopy 15-17
    by photoionization 13
Energy diagram, for fragmentation of ions 270, 271
Energy distribution, internal 12-19, 25, 89, 90
  see also "energy deposition function"
  between fragments 102-108
    factors influencing the 102-108
    predictions of the QET 102
Energy distribution, thermal 17-19
Energy distribution difference (EDD) technique 10, 110
Energy fluctuation, during decomposition 66, 104
Energy loss function 17
Energy partitioning 74, 102, 134-138
  and degree of freedom effect 105-108
  empirical correlations 136
  equipartioning 103, 104
  experimental determination of 135-136
  and QET calculations 136
  theoretical determination of 137
  and transition state geometry 137
Energy partitioning quotient 135
  as transition state probe 169
Energy randomization 43-45, 77
Energy relaxation 42
  rate for 44
  vibrational 42
  vibronic 42
Enumeration of states 29-32
  assuming anharmonic oscillators 30
  assuming harmonic oscillators 29-31
  classical approximation 29, 58, 62, 64
  direct state counting 30, 32
  including internal rotors 30, 31
  Vestal's approximation 31, 59-61, 64-70
Equilibrium constants, of ion molecule reactions 255
Equilibrium geometry of ions 258-262
Even-electron rule 153-154
Extrapolated voltage difference 10

Field ionization
  determination of time-dependent metastable ion characteristics 235
  elucidation of reaction mechanisms 186-196
  see also "field ionization kinetics"
  principle of ion lifetime measurements 56

Field ionization, continued
  time dependence of isomerization reactions 273-278
Field ionization kinetics 186-196
  differentiation between hydrogen randomization and non-specific hydrogen rearrangements 192
  elucidation of atom randomization mechanisms 187
  study of isomerization reactions 235, 273-278
  study of McLafferty rearrangements 195
  suppression of atom randomization 190
Fluorescence spectra, of ions 43, 215
  and isolated electronic states 47
Fragment abundance,
  dependence on
    the activation energy 92
    the chain length 180
    the geometry of the activated complex 91
    the internal energy 92, 177, 192
    the ion lifetime 96
Fragmentation pathways, qualitative prediction from MO calculations 196
Franck-Condon envelope 2
Franck-Condon factors 1-3
  and energy deposition 12
  and ionization cross section 5, 6
  and photoelectron spectroscopy 12
  and radiationless transitions 43
Franck-Condon principle 1, 2
Frequencies, vibrational and rotational,
  of activated complexes 36
  guidelines for the choice of 36
  of ions 35
Frequency factor 29, 62
Functional group interaction, detection of 177-186
  using
    abundances of metastable ions 179
    appearance potentials 182
    energy dependence measurements 177
    relative fragment abundances 180
    symmetry arguments 183

Geometry of the transition state 33
  see also "transition state", "activated complex"
  loose, tight 33
  dependence of the rate constant on 91

Inductive effect, on ionization potentials 19
Interconverting structures 214, 215, 266-272, 279
Internuclear distance 1, 2, 5
Ion cyclotron resonance (ICR)
  determination of appearance potentials 110

Ion cyclotron resonance, continued
  determination of vibrational frequencies of ions 35
  instrumental details 249
  study of ion-molecule reactions for structure determination 249-258
Ionization cross section 4-8
  see also "ionization efficiency curves"
Ionization efficiency curves 4-8
  for atoms 4
  and determination of
    breakdown graphs 64-66
    energy deposition functions 14
    ionization potentials 9
  for electron impact 4, 6
  first, second derivative of 13, 14, 64-66
  influence of autoionization 8
  for molecules 5
  for photoionization 4, 5
  step function 4
  temperature dependence 113
Ionization potential 1
  adiabatic, vertical 3
  definition 1
  determination of 8-12
  first, second 1
  influence of hot bands on the determination of 6
  substituent effects on the,
    aliphatic compounds 19
    aromatic compounds 20
Ionization potential, determination of 8-12
  electron impact 9
  extrapolation methods 10
  optical spectroscopy 9
  photoionization 9
Ionization process 1-12
Ion lifetimes
  dependence of competing direct cleavages and rearrangements on 96-98
  dependence of fragment abundances on 96
  and isolated electronic states 45
  and QET 52-57
Ion lifetime determination 48-57
  determination of rate constants 71-74
  of energy selected ions using
    charge exchange 71
    photoelectron-photoion coincidence 72-74
  long lifetimes 50
  principle 48
  pulse technique 50
  short ion lifetimes 53-57
  use of
    cycloidal mass spectrometers 50
    field ionization 55, 56
    molecular beams 53-55

Ion lifetime determination, continued
    tandem mass spectrometers 51, 52
    time-of-flight instruments 50
    variation of repeller voltage 50
Ion-molecule reactions, structure elucidation by 249-258
    equilibrium constants 255
    examples for structure determination 252
    instrumental aspects 249
    photodissociation technique 256
    principle 249
    proton transfer reactions 254
Ion pair formation 7, 25
Ion trapping
    and appearance potential measurements 111, 112
    in an ICR cell 256
    in a space charge 111, 253
Isolated electronic states 45-47, 167
Isomerization of ions
    aliphatic hydrocarbon ions 266-272
    aromatic hydrocarbon ions 278
    cycloalkanes 236
    experimental results 266-283
    fundamental considerations 212-215
    heteroatom-containing ions 279
    influence of a heteroatom on the 272
    and isolated electronic states 46, 47
    parameters determining the 212-215
    and radiationless transitions 42
    rate-determining isomerization 232
    time scale for 273
Isotope effects 113-118
    definitions
        intermolecular 115
        intramolecular 113
        normal 115
        primary 113, 117
        reverse 115
        secondary 114, 117
    determination of 163
    determination of ion structures 264
    energy dependence 116
    in the kinetic energy release 117
    lifetime dependence 117
    as mechanistic probe 172-176
    in metastable ions 117
    QET-calculations 116-117
Isotopic labelling 163, 183

Hammett equation 119
Hammond's postulate 152
Hartree-Fock self consistent field (HF-SCF) procedure 260
Heat of formation
    determination of ion structures 219-225
    examples for structure elucidation 221-224
    experimental determination of 219
    and isolated electronic states 46
    MO calculations of 258-263
    of molecular ions 1

Hot bands 6
    influence on ionization potential measurements 9
    in photoelectron spectra 12
    vibrational 7
Hot ions 134
K versus E curves 89
    competing direct cleavages and rearrangement reactions 92
    dependence on the
        activation energy 92
        transition state geometry 91
KERD, see kinetic energy release distribution
Kinetic energy release 129-133
    amplification factor 131
    determination of 130
    of doubly charged ions 142
    energy dependence 132, 238
    examples for structure elucidation 239-242
    as ion structure probe 237
    and molecular orbital symmetry considerations 170-172, 242
    as potential surface probe 170
    sources for 134
    temperature dependence 132
    as transition state probe 166-169
    vibrational fine structure 131
Kinetic energy release distribution 74-79, 132-133
    comparison of theoretical and experimental results 76-79
    from collision complexes 76-78
    experimental determination of 75
    role of angular momentum 77-79
    test of the QET 74-79
    theoretical 75
    types of 132
Kinetic shift 37, 108-112
    calculations of the 109
    experimental determination 110
    influence on AP measurements 108, 221
    parameters determining the 109

Langevin collision model 38
Lifetimes, see "ion lifetimes"
Linked scan 166

Mass-analyzed ion kinetic energy spectra 166
Mass spectra
    comparison of calculated and experimental mass spectra 58-63
    energy dependence 60
    2E spectra 142
    E/2 spectra 143
    -E spectra 143
    QET calculations 39-41
    temperature dependence 61
McLafferty rearrangement 33, 194
Mesomeric effect, on ionization potentials 20

Metastable ion characteristics 228-237
  energy dependence 229
  examples 230
  influence of rate-determining isomerizations on 232
  principle 228
  time dependence 235
Metastable ions 128-138
  appearance potentials of 110
  composite peaks 167-169
  determination of the
    energy partitioning quotient 135-137
    intercharge distance 142
    kinetic energy release 131
    kinetic energy release distribution 75, 132
  elucidation of reaction mechanisms 165-172, 179
  experimental methods 166
  and isolated electronic states 45
  lifetimes of 128
  missing 66
  origin of 128
  peak shapes 129
  prediction of the QET 28
  sources for kinetic energy release 134
  structure elucidation of ions 228-242
  temperature dependence 17, 90
Metastable ions, abundance ratio of
  see also "metastable ion characteristics"
  dependence on the
    activation energy 96
    electron energy 105
    internal energy 90
Metastable window 92, 107
Microcanonical ensemble 26
Microscopic reversibility principle 37
MIKE, see mass-analyzed ion kinetic energy spectra
Molecular orbital calculations
  determination of
    transition state geometries 172
    vibrational frequencies of ions 35
  examples for ion structure calculations 261-264
  and ion structure 258-263
  methodical aspects 260
  and reaction mechanisms 196-199

Neighboring group interaction
  see "functional group interaction"
Non-classical structures, of ions 258-263
Non-decomposing ions, definition 212
Non-fixed energy 74
  in collisionally activated ions 140
  in metastable ions 134
Non-specific hydrogen rearrangements 192
Norrish type II reaction 47

Oscillator
  anharmonic 30
  anharmonicity 43
  effective number of 30, 62
  harmonic 29

PEPICO, see "photoelectron-photoion-coincidence"
Phase Space Theory 37, 38
  calculation of kinetic energy release distributions 75
  comparison with experimental rate constants 72
Photodissociation ICR 256-258
  detection of isolated electronic states 47
  ion structure determination 256-258
  time-resolved photodissociation studies 256
Photoelectron-photoion-coincidence
  detection of isolated electronic states 46
  determination of
    breakdown graphs 46
    energy partitioning quotients 137
    kinetic energy release distributions 75, 76, 132
    rate constants 72-74
    time-dependent breakdown graphs 112
  principle 46
Photoelectron spectrometer 11
Photoelectron spectroscopy 10
  determination of
    energy deposition functions 15
    vibrational frequencies of ions 35
    ionization potential determination 10
    threshold method 11
    zero kinetic energy method 11
Photoionization, cross section 4
Polarization function 260
Potential energy diagram
  see "potential energy surface"
Potential energy surface
  of diatomic ions 25, 27
  experimental determination 170, 233, 269, 280-283
  and kinetic energy release 170
  MO calculations 259-263
  of polyatomic ions, fundamental considerations 212-215
Predissociation 7, 8
  of metastable ions 128
  rotational 131
Principle of detailed balance 37
Product stability 152-156
Proton affinities
  determination of heats of formation from 219
  rationalization of reaction mechanisms 159
Proton transfer reactions 254
Pulsed ion sources 50, 111, 112

QET, see "Quasi-Equilibrium Theory"
Quasi-Equilibrium-Theory 25-80
   application to organic mass spectrometry 89
   basic assumptions 26
   calculation of rate constants 32-40
   enumeration of states 29-32
   experimental tests 41-74
   kinetic energy release distribution, prediction of the 75
   minimum rate constant 28
   modified QET 37
   parameters for calculation of rate constants 32-37
   rate expression 27-28, 79
Quadrupole mass filter 51, 73

Radiationless transition, mechanism for 42
Radiative transitions, in ions 43, 215
Radical site localization 160-166
Rate constants
   calculation of 32-37
      comparison with experiments 72-73
      necessary parameters 33-37
   determination of 71-74
      by charge exchange 71
      by photoelectron-photoion coincidence 72-74
   internal energy dependence 89
   and isolated electronic states 46-47
   structure elucidation of ions 265
Rate of reaction, determination of 56
Reaction coordinate 27, 35
Reaction mechanisms 152-202
   methods for the elucidation of,
      abundant metastable ions 179
      appearance potentials 182
      energy dependence 177
      field ionization kinetics 186
      fragmentation pathways 165
      isotope effects 172-176
      isotopic labelling 163
      kinetic energy release and energy partitioning 166-169
      kinetic energy release and MO symmetry considerations 170-172
      molecular orbital calculations 196
      steric blocking 164
      structure of ions 199
      structure of neutrals 202
      symmetry arguments 183
Recombination energy 67
Resonance photoionization 11
Retarding potential difference (RPD) 10
Reverse activation energy 74
   determination of 136, 156
   and energy partitioning 135
   influence on heats of formation determinations 221
   and isotope effects 118
   and kinetic energy release 134

Rice-Ramsperger-Kassel-Marcus (RRKM) Theory 28
   see also "Quasi-Equilibrium-Theory"
Rotational frequencies 35

Scrambling, see "atom scrambling"
Semilog plot 10, 110
Stable ions, definition 212
Steradiancy analyzer 11
Steric blocking 164
Stevenson's rule 156-159
Structure of ions 209-283
   definitions 210
   initial, final 211
   internal energy dependence 211, 212, 215-218
   lifetime dependence 215-218, 235-237, 273-277
   as mechanistic probe 199
   methods for structure elucidation 215-265
Structure of ions, examples
$C_2H_3^+$ 261
$C_2H_4O^{+\cdot}$ 240, 247
$C_2H_4O_2^{+\cdot}$ 234
$C_2H_5O^+$ 222, 228, 239, 246, 252, 280
$C_2H_6N^+$ 229, 282
$C_3H_3^+$ 222, 262
$C_3H_5^+$ 265
$C_3H_6^{+\cdot}$ 265, 270
$C_3H_6O^{+\cdot}$ 209
$C_3H_7O^+$ 232, 282
$C_3H_8N^+$ 231, 239, 246, 254
$C_4H_6^{+\cdot}$ 265
$C_4H_8^{+\cdot}$ 270
$C_4H_9^+$ 232, 269
$C_4H_9O^+$ 283
$C_5H_5^+$ 265
$C_5H_6^{+\cdot}$ 265
$C_5H_{11}^+$ 267
$C_6H_6^{+\cdot}$ 227
$C_6H_6O^{+\cdot}$ 251, 254
$C_6H_7^+$ 242
$C_7H_5NO_2^+$ 226
$C_7H_7^+$ 248, 254, 256, 262
$C_7H_8^{+\cdot}$ 256, 264
$C_8H_8^{+\cdot}$ 256
$C_8H_8O^{+\cdot}$ 240
$C_8H_{10}^{+\cdot}$ 255

Structure of ions, examples, continued
$C_8H_{16}^{+\cdot}$ 235, 273
$C_8H_{18}^{+\cdot}$ 266
$C_{12}H_{10}O^{+\cdot}$ 222, 231, 240, 247
$RCOOH_2^+$ 224
Structure of ions, methods for the elucidation of 215-265
    collisional activation 243-248
    comparison of mass spectra 226
    decomposition mechanisms 225
    equilibrium constants of ion-molecule reactions 255
    heats of formation 219-225, 258-263
    ion-molecule reactions 249-258
    kinetic energy release 237-242
    mechanism of formation 219
    metastable ion characteristics 228-237
    molecular orbital calculations 258-263
    photodissociation ICR 256
    proton transfer reactions 255
Structure of neutral fragments 202
Substituent constants
    Brown's 20, 118
    inductive 19
    Hammett's 118
    polar 19
Substituent effects
    on appearance potentials 118
    on competing fragmentations 125
    on fragment abundances, influence of
        the activation energy 119
        competing fragmentations 121
        the energy distribution 121
        $k(E)$ 122
        secondary decompositions 123
    on hydrogen scrambling 124
    on ionization potentials 19-21
    on the kinetic energy release 127
    MO calculations of 199
    on MO energies 20
Superexcited state 7, 25
Symmetry conservation, of molecular orbitals 170-172
Symmetry considerations
    and ion structures 226
    and reaction mechanisms 183
Symmetry factor 28, 36

Tandem mass spectrometer 51, 52, 68
Thermal shift 113
Time scale, mass spectrometric 41
Time-of-flight mass spectrometer 50, 70, 73, 75
Translational spectroscopy 132
Threshold behavior, for ionization
    for atoms 4
    and energy deposition function 12
    for molecules 5
Threshold laws 4

Threshold laws, continued
    for electron impact 4
    for photoionization 4
Threshold photoelectron spectroscopy 11
Transition probabilities, electronic 2
Tunneling
    and field ionization 56
    and metastable ions 129
    and the Quasi-Equilibrium-Theory 28

Unimolecular decompositions
    laws 40
    competing 40, 62
    consecutive 40, 62

Vibrational overlap integral 2, 43

Zero kinetic energy photoelectron spectroscopy 11
Zero point energy 115